Centre for Innovation in Mathematics Teaching
University of Exeter

FOUNDATION
MATHEMATICS

Writers Tim Cross
Jon Middle
William Thallon

Editor David Burghes

Assistant Andrea Perry
Editors Nigel Price
Dick Russell
Alan Sherlock

Heinemann Educational

ACKNOWLEDGEMENTS

This text has been written by a group of authors working as part of the Mathematics component of the Wessex Project. It has been a delight to work not only with the writers and editors of this text but with the complete team. It says much for their enthusiasm and dedication that we have both a revised AEB syllabus for A Level Mathematics, including the opportunity for two complete A Levels in mathematical subjects and a comprehensive, complete set of texts and resources. In particular, I would like to thank Bob Rainbow (Wessex Project director) for allowing me the opportunity to work with the Wessex project, John Commerford (AEB Mathematics Officer) for his cooperation and patience throughout the development, and Nigel Price for his efficient coordination of the project.

We are also very grateful for the help and assistance of the editorial staff at Heinemann, Philip Ellaway and Ruth Burdett, and to the many organisations which have helped us.

Major funding was provided by the Leverhulme Trust, and we have had generous donations from British Steel, through Colin Green, Tony Nicholson and Brian Taylor. We readily acknowledge their help.

Finally we are all indebted to the support staff, Bettina Brunati, Lisa Cole, Liz Holland, Margaret Roddick, Ann Tylisczuk and Sally Williams for turning our draft manuscript into an attractive, well-presented text. Their dedication and good humour despite impossible time schedules has been an inspiration to me.

David Burghes

(Project Director)

FOUNDATION MATHEMATICS

This is one of four Core Texts particularly developed for AEB's Mathematics Core and Options syllabus for A and AS level awards.

This text covers the **Foundation Mathematics Core** and is based on the Inter Board Common Core for A Level Mathematics. Students taking an A Level in Mathematics will also need to study the Calculus Option in order to cover all the topics in the Inter Board Common Core.

The development of these texts has been coordinated at the *Centre for Innovation in Mathematics Teaching* as part of the

Enterprising Mathematics for AS and A-level Project

in association with the Wessex Project and AEB. It has been partly funded by a grant from the *Leverhulme Trust* and a donation from *British Steel*. The project has been directed by David Burghes and coordinated by Nigel Price

Enquiries regarding this project and further details of the work of the Centre should be addressed to

Margaret Roddick
CIMT
School of Education
University of Exeter
Heavitree Road
EXETER EX1 2LU

CONTENTS

PREFACE

Mathematics is an important technique for the solution of problems in almost any area of society today. It has been of crucial importance in the advancement of space travel, telecommunications, computing, power generation and environmental issues. The use of mathematical models to predict whale populations has done much to conserve current stocks whilst the solution of the equations of motion for ballistics has helped design rockets of awesome power which can pinpoint targets. Mathematics in the hands of people and governments can be used for the benefit or destruction of mankind. Whatever view you take, however, its power cannot be ignored.

This particular text aims to develop mathematical topics and concepts, illustrating how the techniques can be used in practical ways. Some mathematical themes are important for their own sake and this is also recognised. Nevertheless, the main theme behind this book is to provide a comprehensive text which covers the main topics in the Inter Board Current Core, showing where appropriate their uses and applications.

The material in this text covers 40% of an A Level (80% of an AS Level award) and to complete an A Level in Mathematics a second Core must be studied together with two Options, one of which must be **Calculus**.

Students taking a single AS Level award in Mathematics must study this Core Text and any one Option from the list of twelve available. Resources for all these Options are published separately. Full details are available from Heinemann Educational.

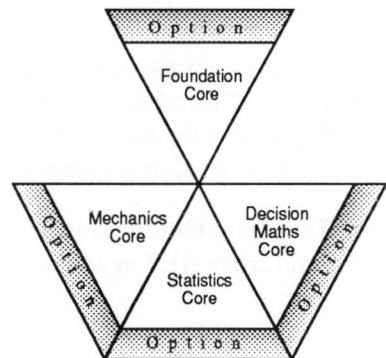

Full details of AEB syllabuses in Mathematics are available from :

> Associated Examining Board
> Stag Hill House
> GUILDFORD
> Surrey
> GU2 5XJ

This text has been produced for students and includes examples, activities and exercises. It should be noted that the activities are **not** optional but are an important part of the learning philosophy in which you are expected to take a very active part. The text integrates

- **Exposition** in which the concept is explained;
- **Examples** which show how the techniques are used;
- **Activities** which either introduce new concepts or reinforce techniques;
- **Discussion Points** which are essentially 'stop and think' points, where discussion with other students and teachers will be helpful;
- **Exercises** at the end of most sections in order to provide further practice;
- **Miscellaneous Exercises** at the end of each chapter which provide opportunities for reinforcement of the main points of the chapter.

Discussion points are written in a special typeface as illustrated here.

Note that answers to all the exercises are given at the back of the book. You are expected to have a calculator available throughout your study of this text and occasionally to have access to a computer.

Some of the sections, exercises and questions are marked with an asterix (*). This means that they are either **not** central to the development of the topics in this text and can be omitted without causing problems, or they are regarded as particularly challenging.

The **Teacher's Guide** gives further details and advice on optional material for the AEB syllabus.

1 THE NATURE OF MATHEMATICS

Objectives

After studying this chapter you should

* appreciate the different aspects of mathematics;
* be able to understand how simple mathematical models are constructed to solve problems;
* understand the power and limitations of mathematical analysis.

1.0   Introduction

You are starting off on a course of mathematical study leading to an A-level or AS award in Mathematics. Although the **starting point** of your course is clearly defined from the work that you have done in your GCSE Mathematics course, the **end point** is not so clearly defined, since it depends on your interests and ability.

Those who are more interested in Physics and Technology will probably take further mathematical study with **Mechanics**, whilst those interested in Biology or Geography need a good grounding in techniques in **Statistics**, and those specialising in Business Studies will need further work in **Decision Mathematics**.

All these areas are mathematical and the extent of the topic can be illustrated by the fact that it is possible to take not just one A-level, but two or even three A-levels in mathematical subjects. This text is based on the current 'common core', which all A-levels in mathematics have to cover. Whilst the syllabus for the common core is quite prescriptive, the aim is to show why and how to apply the various topics under study. Of course, not all mathematical topics are immediately useful and some have historically been developed for their own sake with their applications coming later.

The first task, though, will be to see some of the ways in which mathematics is developed and used; in particular its uses

* to explain

* to predict

* to make decisions

will be illustrated in the next section.

1.1 Case studies

Bode's Law

In 1772, the German astronomer, *Johann Bode*, investigated the pattern formed by the distances of planets from the sun.

At the time, only six planets were know, and the pattern he devised is shown below. The distances are measured on a scale that equates 10 units to the Sun - Earth distance.

The fit between actual distances and Bode's pattern is remarkably good.

Planet	Actual distance	Bode's Pattern
Mercury	4	0 + 4 = 4
Venus	7	3 + 4 = 7
Earth	10	6 + 4 = 10
Mars	15	12 + 4 = 16
Asteroids -	-	-
Jupiter	52	48 + 4 = 52
Saturn	96	96 + 4 = 100

What do you think is the missing entry?

There are also planets further out than Saturn.

Find the next two numbers in Bode's pattern.

In fact, the data continues as shown here

Planet	Actual distance
Uranus	192
Neptune	301
Pluto	395

Can you give an explanation of how Bode's Law can be adapted for this extra data?

Wind Chill

When the temperature drops near zero, it is usual for weather forecasters to give both the expected air temperature, and the **wind chill** temperature - this is the temperature actually felt by someone, which depends on the wind speed and air temperature. So, for example, the wind chill temperature for an actual temperature of

$0°$ C and wind speed of 10 mph is given by $-5.5°$ C.

For $v > 5$ mph, the wind chill temperature is given by

$$T = 33 + \left(0.45 + 0.29\sqrt{v} - 0.02v\right)(t - 33)$$

where $t°$ C is the air temperature and v mph the wind speed. This formula was devised by American scientists during the Second World War, and is based on experimental evidence.

sub into formulae.

Example

Find the wind chill temperature when

(a) $t = 2°$C, $v = 20$ mph;

(b) $t = 10°$C, $v = 5$ mph;

(c) $t = 0°$C, $v = 40$ mph.

Solutions

(a)   When $t = 2$, $v = 20$,

$$T = 33 + \left(0.45 + 0.29\sqrt{20} - 0.02 \times 20\right)(-31)$$

$$= -8.8°\,\text{C}.$$

(b)   When $t = -10$, $v = 5$,

$$T = 33 + \left(0.45 + 0.29\sqrt{5} - 0.02 \times 5\right)(-43)$$

$$= -9.9°\,\text{C}.$$

(c)   When $t = 0°$C, $v = 40$ mph,

$$T = 33 + \left(0.45 + 0.29\sqrt{40} - 0.02 \times 40\right)(-33)$$

$$= -16.0°\,\text{C}.$$

What is the significance of a wind speed of 5 mph?

Heptathlon

The Heptathlon is a competition for female athletes who take part in **seven** separate events (usually spread over a two day period). For each event, there is a point scoring system, based on the idea that a good competitor will score 1000 points in each event. For example, the points scoring system for the 800 m running event is

$$P = 0.11193 \, (254 - m)^{1.88}$$

where m is the time taken in seconds for the athlete to run 800 m.

What points are scored for a time of 254 seconds?

Example

What points are scored for a time of 124.2 seconds, and what time would give a point score of 1000?

Solution

For $m = 124.2$,

$$P = 0.11193 \, (254 - 124.2)^{1.88}$$

$$\Rightarrow \quad P = 1051.$$

(Scores are always rounded down to the nearest whole number.)

Now, to score 1000 points requires a time of m seconds where

$$1000 = 0.11193 \, (254 - m)^{1.88}$$

$$\Rightarrow \quad (254 - m)^{1.88} = 8934.15$$

$$\Rightarrow \quad 254 - m = (8934.15)^{\frac{1}{1.88}}$$

$$\Rightarrow \quad m = 254 - 126.364$$

giving $\qquad m = 127.64.$

All track events use a points scoring system of the form

$$P = a(b - m)^c$$

with suitable constants a, b and c.

Suggest an appropriate formula for the points system in the track events in the Heptathlon.

Simple pendulum

The great Italian scientist, *Galileo*, was the first to make important discoveries about the behaviour of swinging weights. These discoveries led to the development of pendulum clocks.

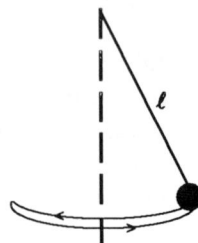

Activity 1   Period of pendulum swing

Attach a weight at one end of a light string, the other end being fixed. Let the pendulum swing freely in a vertical plane and for various lengths of pendulum, ℓ, in metres, find the corresponding times in seconds of one complete oscillation (known as the **period**) - it is more accurate to time, say, five oscillations and then divide the total time by 5. On a graph, plot the period, T, against the square root of the pendulum length, $\ell^{\frac{1}{2}}$. What do you notice?

In fact, the two quantities are related by the formula

$$\boxed{T = 2.006\ \ell^{\frac{1}{2}}}$$

Example

What pendulum length gives a periodic time of 1 second?

Solution

If $T = 1$ then

$$1 = 2.006\ \ell^{\frac{1}{2}}$$

$$\Rightarrow\quad \ell^{\frac{1}{2}} = \frac{1}{2.006} = 0.4985$$

$$\Rightarrow\quad \ell \approx 0.706\ \text{m}.$$

Activity 2

Construct a simple pendulum with $\ell = 0.706\ \text{m}$, and check its periodic time.

Perfect numbers

These are numbers whose divisors (excluding the number itself) add up to the number. Excluding the number 1, the first perfect number is 6, since

$$6 = 3 \times 2 = 1 \times 6$$

and $3 + 2 + 1 = 6.$

Activity 3

Test the numbers 7, 8, ... , 30 in order to find the next perfect number. (You might find it useful to write a short computer program to test whether any number is perfect.)

You have probably realised by now that perfect numbers are pretty thin on the ground!

Example

Are the following numbers perfect :

(a) 220   (b) 284   (c) 496?

Solution

(a) $220 = 220 \times 1$

$= 110 \times 2$

$= 55 \times 4$

$= 44 \times 5$

$= 22 \times 10$

$= 20 \times 11$

and $1+2+4+5+10+11+20+22+44+55+110 = 284$.

Hence 220 is not a perfect number.

(b) $284 = 284 \times 1$

$= 142 \times 2$

$= 71 \times 4$

and $1+2+4+71+142 = 220$.

Hence 284 is not a perfect number (but note its connection with 220).

(c) $496 = 496 \times 1$

$= 284 \times 2$

$= 124 \times 4$

$= 62 \times 8$

$= 31 \times 16$

and $1+2+4+8+16+31+62+124+248 = 496$.

Hence 496 is a perfect number.

In fact 496 is the third perfect number, and 8128 is the fourth. Although there are still many unknown results concerning perfect numbers, it has been shown that

(a)   all **even** perfect numbers will be of the form

$$2^{n-1}\left(2^n - 1\right)$$

when n is a prime number. This number is in fact perfect when $2^n - 1$ is prime;

(b)   all even perfect numbers end in 6 or 8;

(c)   the sum of the inverses of all divisors of a perfect number add up to 1

e.g. for 6, $\frac{1}{6} + \frac{1}{3} + \frac{1}{2} = 1$.

Activity 4   Perfect numbers

Given that the fifth and sixth perfect number are 33 550 336 and 8 589 869 056 respectively, copy and complete the table below.

n	$2^n - 1$	prime	$2^{n-1}\left(2^n - 1\right)$	perfect
2	3	✓	6	✓
3				
5				
7				
11				
13				

You probably noticed in the example above that 220 and 284 are connected through their divisors. They are called **amicable pairs** (they are the smallest numbers that exhibit this property) and are regarded as tokens of great love. In the Bible, for example, Jacob gave Esau 220 goats to express his love (Genesis 32, verse 14).

Activity 5   Amicable pairs

Write a short program to generate amicable pairs, and use it to find the next lowest pair.

Day of the week

The algorithm below gives a method for determining the day of the week for any date this century. The date used as an example is 3 March, 1947.

1.	Write $y = $ year.	$y = 1947$
2.	Evaluate $\left[\dfrac{y-1}{4}\right]$ ignoring the remainder.	$\left[\dfrac{y-1}{4}\right] = \left[\dfrac{1946}{4}\right] = 486$
3.	Find $D = $ day of year. (Jan. 1st $= 1, \dots,$ Feb. 1st $= 32$, etc.)	$D = 31 + 28 + 3 = 62$
4.	Calculate $s = y + \left[\dfrac{y-1}{4}\right] + D$.	$s = 1947 + 486 + 62$ $= 2495$
5.	Divide by 7 and note remainder, R.	$\dfrac{s}{7} = \dfrac{2495}{7} = 356,$ with remainder $R = 3$
6.	The remainder is the key to the day : $R = 1 \Rightarrow$ Friday $R = 2 \Rightarrow$ Saturday, etc.	Hence 3 March 1947 was in fact a Sunday.

Activity 5

(a) Use the algorithm to find the day of the week on which you were born.

(b) Analysis how and why this algorithm works.

Bar code design

Nearly all grocery products now include an identifying Bar Code on their wrapper (supermarkets now use them both for sales checkout and stock control). There are two types of EAN (European Article Numbers) - 13 digit and 8 digit. The shortened 8 digit code will be considered here. A possible example is shown opposite. The number has three parts.

$$0\ 0 \qquad 0\ 1\ 1\ 5\ 2 \qquad 5$$

↑	↑	↑
retailer's code	product code	check digit

The check digit is chosen so that

$$3\times\left(1^{st}+3^{rd}+5^{th}+7^{th}\ \text{numbers}\right)+\left(2^{nd}+4^{th}+6^{th}+8^{th}\ \text{numbers}\right)$$

is exactly divisible by 10. For the numbers above

$$3\times(0+0+1+2)+(0+1+5+5)$$

$$=\ 3\times3+11=9+11=20$$

which is divisible by 10.

If the check digit is in error, the optical bar code reader will reject the code.

Example

Find the check digit for the EAN codes :

(a) $5021421x$ (b) $0042655x$.

Solution

(a) Denoting the check digit by x, the number

$$3\times(5+2+4+1)+(0+1+2+x)=3\times12+3+x=39+x$$

must be divisible by 10, so x must be 1.

(b) Similarly

$$3\times(0+4+6+5)+(0+2+5+x)=3\times15+7+x=52+x$$

must be divisible by 10, so x must be 8.

If the optical bar code reader makes one mistake in reading a number, will it always be detected?

Another 8 digit EAN is shown opposite. It has left and right hand guard bars and centre bars. In between there are 8 bars of varying thickness. Each number is represented by a unique set of 2 bars and 2 spaces. As can be seen in the magnified version of 5, each number code is made up of 7 **modules**.

The digit 5 is written as 0110001 to indicate whether a module is white (0) or black (1).

All left hand numbers start with 0 and end with 1, and use a total of 3 or 5 black modules. Right hand numbers are the complement of the corresponding left hand code e.g. right hand 5 is 1001110.

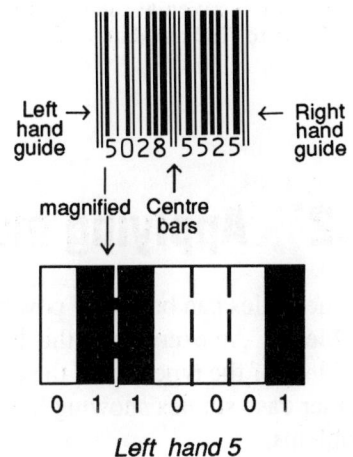

Left hand 5

Activity 6

Design all possible codes for left hand numbers, and use examples found on products to identify the code for each number.

The seven case studies in this section have demonstrated a variety of uses of mathematics. They have ranged from the practical design problem in Bar Codes and Bode's pattern for planetary distances to the development of perfect numbers (which as yet have no obvious applications). Mathematics embraces all these concepts, although it is the practical application side that will be emphasised where possible throughout this text. This aspect will be considered in greater depth in the next section.

Exercise 1A

1. Use the wind chill temperature formula to find its value where

   (a) $t = 0°C$, $v = 20$ mph

   (b) $t = 5°C$, $v = 20$ mph

   (c) $t = -5°C$, $v = 20$ mph.

   Plot a graph of wind chill temperature against air temperature, t, for $v = 20$ mph. Use your graph to estimate the wind chill temperature when $t = 10°C$ and $v = 20$ mph.

2. The points scoring system for the high jump event in the Heptathlon is given by

   $$p = a(m - b)^c$$

   where $a = 1.84523$, $b = 75.0$, $c = 1.348$ and m is the height jumped in centimetres. Find the points scored for a jump of 183 cm, and determine the height required to score 1000 points.

3. An algorithm for determining the date of Easter Sunday is given at the top of the next column. Use it to find the date of Easter next year.

Step	Number	Divide by	Answer	Remainder (if needed)
1	$x = $ year	100	$b = $	$c = $
2	$5(b + c)$	19	-	$a = $
3	$3(b + 25)$	4	$r = $	$s = $
4	$8(b + 11)$	25	$t = $	-
5	$19a + r - t$	30	-	$h = $
6	$a + 11h$	319	$g = $	-
7	$60(5 - s) + c$	4	$j = $	$k = $
8	$2j - k - h + g$	7	-	$m = $
9	$h - g + m + 110$	30	$n = $	$q = $
10	$q + 5 - n$	32	-	$p = $

*4. Write a computer program to determine the date of Easter Sunday for the next 100 years. Illustrate the data using a histogram.

5. Find the check digits for these EAN codes :

   (a) 0034548*   (b) 5023122*.

6. Determine whether these EAN codes have the correct check digit :

   (a) 00306678   (b) 06799205.

1.2   Applying mathematics

Mathematics can be a very powerful tool in solving practical problems. An example of this is given below with an optimisation problem of the type met in the commercial world, as well as two further case studies showing how mathematics is used to solve problems.

Metal cans

The most popular size of metal can contains a volume of about 440 ml. As they are produced in millions each week, any savings that can be made in their manufacture will prove significant. Part of the cost of making steel cans is based on the amount of material used, so it might be sensible to design a can which minimises the amount of metal used to enclose the required volume.

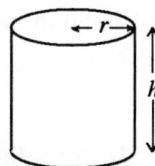

To analyse this problem, you must find an expression for the total surface area of a can. Suppose the cylindrical can has radius r and height h, then total surface area,

$$S = \text{curved surface area} + \text{top area} + \text{base area}.$$

What are the dimensions of the rectangle used for the curved surface area?

Assuming that no metal is wasted, an expression for the total surface area is given by

$$S = 2\pi rh + \pi r^2 + \pi r^2$$

$$\Rightarrow \quad S = 2\pi rh + 2\pi r^2. \tag{1}$$

The formula for S shows that it is a function of two variables, r and h. But in reality it is a function of only one variable since r and h are constrained by having to enclose a specified volume.

You should be familiar with the formula for the volume of a cylindrical can:

$$V = \text{area of cross section} \times \text{height}$$

or, in this case

$$440 = \pi r^2 h. \tag{2}$$

This equation can be used to find an expression for h which is substituted into (1) to eliminate h.

From (2)

$$h = \frac{440}{\pi r^2} \tag{3}$$

and substituting into (1) gives

$$S = 2\pi r \left(\frac{440}{\pi r^2} \right) + 2\pi r^2$$

giving

$$\boxed{S = \frac{880}{r} + 2\pi r^2} \tag{4}$$

What happens to S if r is very small or large? Does it make sense?

The problem is to find the value of r which minimises the total surface area S.

Activity 7   Minimising packaging costs

Draw a graph of S as given by equation (4), for x values between 5 and 10. (If you use a graphic calculator, you will need to enter the equation in the form

$$y = \frac{880}{x} + 2\pi x^2$$

where y replaces S and x replaces r.)

Use your graph to obtain an estimate of the base radius which would make the surface area of the can a minimum. (If you know how to 'magnify' a portion of the graph, you may be able to make a better estimate by concentrating on the part of the graph close to the minimum.)

Also find the corresponding optimum height from equation (3).

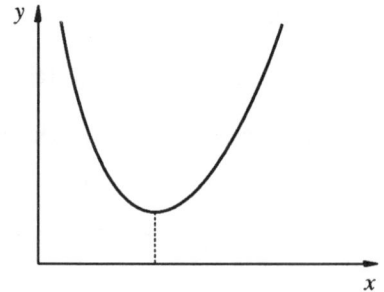

You have made a number of assumptions in finding this optimum value of r. To complete the analysis, you must find out whether this solution, based on minimising the total surface area, is in fact used in practice.

Activity 8   Validating the result

Find the dimensions of the usual size of a can containing 440 ml. Do they agree with your theoretical result? If not, suggest what assumptions should be modified.

In practice, the aesthetic look and feel of an object might be more important than minimising the total surface area of the packaging.

Can you suggest objects where the packaging is clearly not based on minimising the material used?

Returning to the metal can problem, there is one particular case where aesthetic appeal would not be important, and that is in the design of cans for trade use (e.g. hotels, caterers). You will see later that for these cans, your model does indeed provide the basis for deciding the optimum dimensions.

Reading age formula

Educationalists need to be able to assess the minimum reading age of certain books so that they can be appropriately catalogued, particularly for use with young children.

You are probably aware that, for example, it is much easier and quicker to read one of the tabloids (e.g. 'The Sun') than one of the quality 'heavies' (e.g. ' The Guardian').

What factors influence the reading age of a book, newspaper or pamphlet?

There have been many attempts at designing a formula for finding the reading age of a text. One example is known as the **FOG Index**. This is given by

$$R = \frac{2}{5}\left(\frac{A}{n} + \frac{100L}{A}\right) \qquad (5)$$

where the variables are defined for a sample passage of the text by

$A =$ number of words

$n =$ number of sentences

$L =$ number of words containing three or more syllables (excluding '-ing' and '-ed' endings).

Activity 9

Find four or five books of varying reading difficulty. First estimate the minimum reading ages for each of these, then use the FOG formula to compare the two sets of data.

Of course, the whole concept of a designated reading age for a particular book is perhaps rather dubious. Nevertheless, the problem is a real one, and teachers and publishers do need to know the appropriate order for their reading books.

The two case studies above illustrate the idea of a **mathematical model**; that is a mathematical description of the problem. For the metal can problem the mathematical model is described by the equations (1) and (3), leading to the mathematical problem of finding the value of r which minimises

$$S = \frac{880}{r} + 2\pi r^2.$$

In the second problem, the reading age formula, the mathematical model is essentially given by equation (5).

The translation of the problem from the real world to the mathematical world can be summarised in the diagram opposite. The model is formed from the real problem by making various assumptions, whilst the solution to the mathematical problem must be interpreted back in terms of the real problem. This will be illustrated in the next example.

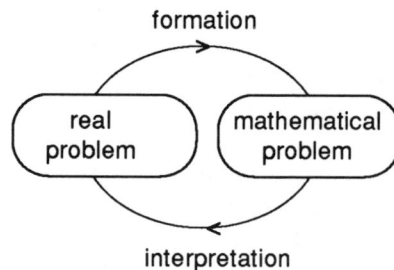

formation

real problem mathematical problem

interpretation

Handicapping weight lifters

In weight lifting, there are nine official body weight classes. For some competitions, it is important to be able to compare lifts made by competitors in different classes. This means that some form of handicapping must be used.

There are a number of models that have been used to provide a form of handicapping. For example if

$$L = \text{actual lift (in kg)}$$

$$W = \text{competitor's weight (in kg)}$$

$$L' = \text{handicapped lift,}$$

then two possible solutions are

(a) $L' = L - W$

(b) $L' = L / (W - 35)^{\frac{1}{3}}$.

The first method was used for some time in a television programme (TV Superstars) in which competitors of different weights competed against each other in a number of sports events. The second method, called the O'Carroll formula, is used in more serious competitions in order to find an overall winner.

Example

The best lifts (for the 'snatch' lift) for eight competitors are given on the following page, together with their weight. Use the two models to find an overall winner.

Competitor	Weight (in kg)	Lift (Snatch) (in kg)
1	52	105.1
2	56	117.7
3	60	125.2
4	67.5	135.2
5	75	145.2
6	82.5	162.7
7	90	170.3
8	110	175.3

Solution

The handicapped lifts are shown below.

Competitor	W	L	$L-W$	$L/(W-35)^{\frac{1}{3}}$
1	52	105.1	53.1	40.9
2	56	117.7	61.7	42.7
3	60	125.2	65.2	42.8
4	67.5	135.2	67.7	42.4
5	75	145.2	70.2	42.4
6	82.5	162.7	80.2	44.9 ←
7	90	170.3	80.3 ←	44.8
8	110	175.3	65.3	41.6

For the first method, the winner is competitor number 7, but the second method makes competitor number 6 the winner.

Whilst mathematics is a precise science, applications to real problems require both an understanding of the problem and an appreciation that, whilst mathematics can provide answers and give precise explanations based on particular assumptions and models, it cannot always solve the real problem. Mathematics can help to design multi-stage rockets that work, but it can't necessarily help to solve the problem of world peace. Often mathematical analysis can help in making the best decisions, and, for example the success of mathematical modelling is shown by the fact that man has stepped on the moon. You should, though, be aware that most problems in real life are more complicated than a single equation or formula!

Exercise 1B

1. Repeat the analysis for finding the minimum surface area of a metal can where the volume enclosed is 1000 ml. Determine the values of r and h which minimises the surface area.

2. A mathematical model for the reading age of a text is given by

$$R = 25 - \frac{N}{10}$$

where N is the average number of one syllable words in a passage of 150 words. Use this model to find the reading age of a number of books. Compare the results with those found in the case study outlined in Section 1.2.

3. Use the handicapping model

$$L' = L / W^{\frac{2}{3}}$$

to find the winner of the competition described in the case study in the section before this exercise.

4. Horseshoes are made by blacksmiths taking straight strips of iron and bending them into the usual horseshoe shape. To find what length of strip of iron is required, the blacksmith measures the width, W inches, of the shoe and uses a formula of the form

$$L = aW + b$$

to find the required strip length, L inches. Use the following data to find estimates for a and b:

Width W (inches)	Length L (inches)
5	12
5.75	13.50

2 USING GRAPHS

Objectives

After studying this chapter you should

- be able to illustrate simple functions with a graph;
- understand what is meant by mapping, domains and ranges;
- be able to identify if a function is odd or even or neither.

2.0 Introduction

Graphs can be used to quickly get an idea of how one quantity varies as another quantity changes. This can be very useful when trying to solve a wide range of problems. An illustration is given in the first activity which deals with the currency problem already met in Chapter 1.

Activity 1 Conversion rates

Conversion rates between different currencies are often displayed in bank windows. You may either assume the rates given below, or find out the current rates from a local bank, to draw 3 graphs showing the number of francs, dollars and Deutschmarks you can buy for any number of pounds sterling up to £50.

£1 = 10.5 Fr   £1 = $1.65   £1 = 2.90 DM

Use your graphs to do these conversions.

(a) £25 into francs, (b) $35 into pounds, (c) 80 DM into francs.

Banks usually charge a fixed commission of about £2 every time they change currency for you. On the same three pairs of axes you drew earlier, draw conversion graphs which take this commission into account. What are the values of the conversions given above now?

Another problem which can be readily illustrated graphically is that of temperature conversion. You are probably familiar with the rule for conversion from degrees Celsius (°C) to degrees Fahrenheit (°F). You multiply by 9, divide by 5 and add on 32.

This can be written as a mathematical formula

$$F = \frac{9}{5} \times C + 32.$$

Instead of using algebra you can draw a graph of F against C and use it to convert from degrees Celsius to degrees Fahrenheit.

Activity 2   Temperature conversions

Since $32°\,F = 0°\,C$ and $212°\,F = 100°\,C$, plot these two points on a graph and draw a straight line to join them. Use your graph to convert

(a) $30°\,C$ to $°\,F$ (b) $10°\,C$ to $°\,F$ (c) $100°\,C$ to $°\,F$.

The lowest possible temperature is $-273°\,C$. Can you use your graph to find the corresponding temperature in Fahrenheit?

2.1   Mappings, domains and ranges

Very often when trying to solve a problem you may produce a rule which links one quantity with another. For instance, the speed of a car may be linked with its braking distance, or two currencies may be linked by their the rate of exchange. Once a rule or formula has been produced, it is tempting to draw a graph illustrating it, to help solve the problem. However, not all of the graph may be relevant.

Activity 3   What happens at x = zero?

Use a graph plotting program or calculator to draw the graphs of

$$y = \frac{1}{x} \text{ and } y = \sqrt{x}.$$

What happens to the first graph when $x = 0$? Why do you think this happens?

Why do you think there is no graph when $x < 0$ for the second equation?

Another example is given by the relationship between pressure and volume. For a fixed amount of any gas, kept at a fixed temperature, pressure and volume are linked by the formula

$$p = \frac{k}{v},$$

where p is the pressure, v is the volume and k is a constant.

The graph for this formula is shown opposite. It shows that the pressure of the gas gets higher as the volume gets smaller in the right half of the graph. Then the pressure 'jumps', so that it is negative as soon as the volume is negative. This does not make sense. A gas cannot have a negative volume, so the left 'branch' does not exist. To show that $p = \frac{k}{v}$ can only be used for positive values of v, the formula can be written as

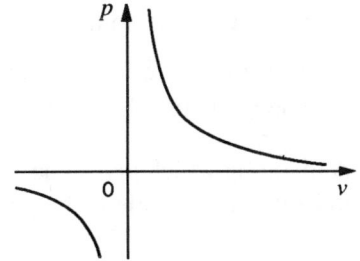

$$\boxed{p = \frac{k}{v}, \; v > 0}$$

The figure opposite shows the graph of this rule, which now makes sense. As the volume increases, the pressure gets closer to zero. As the volume gets closer to zero, so the pressure gets higher - but the volume never actually equals zero.

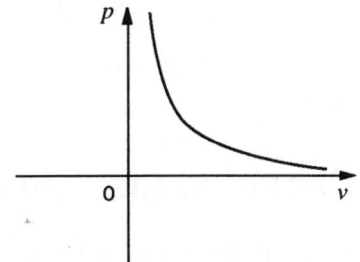

Mapping diagrams

The equation $p = \frac{k}{v}$ can be thought of as a way of linking a value of the volume, v, with a value of the pressure. For instance, if $k = 10$, then $p = \frac{10}{v}$. So if $v = 1$, $p = \frac{10}{1} = 10$.

Thus 1 is 'sent' or 'mapped' to 10 by the equation.

If $v = 2$, $p = \frac{10}{2} = 5$. So 2 is 'mapped' to 5.

The figure opposite shows how several possible values of v are mapped to corresponding values of p. The formula $p = \frac{k}{v}$ is a

mapping - it links the values in one set of numbers (here the set of volumes) with another set of numbers (the set of possible pressures). The set of all possible volumes, greater than zero, is called the **domain** of the mapping. The set of all possible pressures (again, any number greater than zero) is the **range** of the mapping.

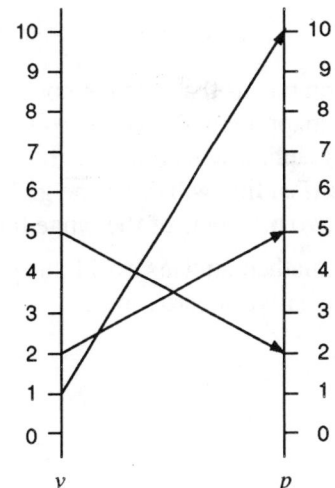

Another example of a possible mapping is $F = \pm\sqrt{T}$, $T \geq 0$. The domain is any number greater than or equal to zero. This mapping is rather different to $p = \dfrac{k}{v}$, however, as it gives two 'answers' for every value in the domain. For instance, $\pm\sqrt{4}$ is $+2$ or -2, since $(+2)^2 = 4$ and $(-2)^2 = 4$. The figure opposite shows the graph of F. The range of F is any number, positive or negative.

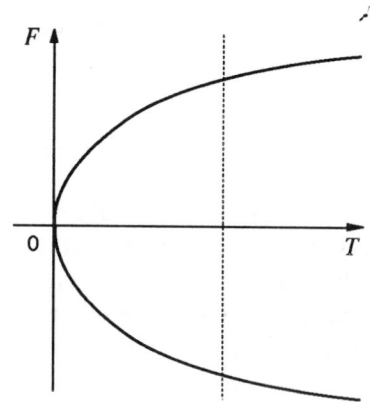

A mapping which gives one, and only one value, for every number in its domain is called a **function**. So the mapping

$$p = \frac{k}{v},\ v > 0$$

is a function, but the mapping

$$F = \pm\sqrt{T},\ T \geq 0$$

is not, as it gives more than one answer for some members of its domain.

Changing the rule to get a function

Notice that the domain can be vital when deciding if a mapping is a function or not. For example, if the domain of $p = \dfrac{k}{v}$ is any value of v, positive or negative, including zero, the mapping ceases to be a function. This is because when $v = 0$, p cannot be calculated.

However, it is possible to adapt the rule for F to make it into a function. Suppose

$$F = +\sqrt{T},\ T \geq 0$$

which means that F is the positive square root of T. The graph of this mapping is shown opposite. If a vertical line is drawn across the graph, it will cross the graph only once. (In the previous figure a vertical line will cross the graph only twice - showing that there are two members of the range for each member of the domain). In fact mathematicians avoid this problem by agreeing that $\sqrt{x}$ means the positive square root.

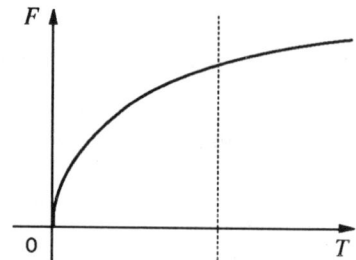

Number sets

Sets of numbers which are often used as domains or ranges have names, so that they can be described accurately.

Any number on a number line, including fractions and decimals and any number in between them, is called a **real** number. The set of 'real' numbers is denoted by $\mathbb{R}$. Unless otherwise stated, you can usually assume that a domain is the real number set, or a part of it. Other commonly used sets of numbers are the **integers**, which are the positive and negative 'whole' numbers

$$..., -2, -1, 0, 1, 2, ...$$

These are denoted by $\mathbb{Z}$. Other domains and ranges are usually described using set notation, like the range in this next example.

Example

The table below shows the current postal rates in the U.K. for first class letters.

Weight up to	60 g	100 g	150 g	200 g
Postage	24p	35p	43p	51p

This is a mapping from part of the real numbers (between 0 and 200), to a 'discrete' range. This means that the range contains only certain values. In this case, the range is the set of numbers $\{24, 35, 43, 51\}$. The graph of this mapping is shown opposite.

A vertical line drawn across the graph will cut it only once, so this mapping is a function, even though it does not appear to have a 'formula' of the usual sort. It is possible to write down a kind of formula, however ;

$$\text{Postage, } P = \begin{cases} 24 \text{ for } 0 < W < 60 \\ 35 \text{ for } 60 \le W < 100 \\ \cdots\cdots\cdots \text{etc.} \end{cases}$$

The value of P changes in steps as the weight increases, so it is necessary to have different rules for different parts of the domain.

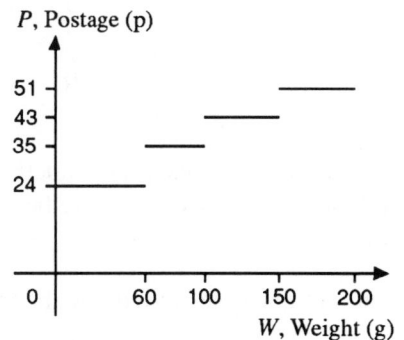

Activity 4

Use a graph plotting program or a graphic calculator to make sketches illustrating the following rules. By looking at the sketches, decide on a domain which will make the rule a function.

21

(a) $y = \dfrac{1}{x-5}$ (b) $y = \dfrac{3-x}{1-x}$ (c) $y = +\sqrt{x-2}$

(d) $y = 1 - x^2$ (e) $y = \dfrac{1}{x^2}$ (f) $y = \dfrac{2}{1-x^3}$.

Function notation

Returning to the mapping of volume to pressure met earlier in this section, another way of writing the mapping or function is illustrated by

$$p\!:\!v \mapsto \frac{k}{v},\ v \in \mathbb{R},\ v > 0.$$

This is read as p is the function which maps v to $\dfrac{k}{v}$ where v is any real number greater than zero. The more usual way of writing a function is

$$p(v) = \frac{k}{v},\ v \in \mathbb{R},\ v > 0$$

which illustrates that p is a function of v. In what follows the second method will generally be used, but you should be aware of the alternative.

Another example is given by the function

$$F\!:\!t \mapsto +\sqrt{t},\ t \in \mathbb{R},\ t \geq 0$$

or in the usual notation

$$F(t) = +\sqrt{t},\ t \in \mathbb{R},\ t \geq 0.$$

Hence, when $t = 4$, $F(4) = +\sqrt{4} = 2$, and in general if $t = a$,

$$F(a) = +\sqrt{a};$$

when $t = b^2$,

$$F\!\left(b^2\right) = +\sqrt{b^2} = b.$$

Exercise 2A

1. f is a function defined by the equation
   $f(x) = x^2 + 2$.

   Find the value of

   (a) $f(2)$ (b) $f(-1)$ (c) $f(0)$

   (d) $f(a^2)$ (e) $f(1-a)$

   (where a is a constant real number).

2. g is defined by $g(x) = \dfrac{1}{x}$

   Find the value of the following if possible

   (a) $g(1)$ (b) $g(-1)$ (c) $g(0)$

   (d) $g(a^2)$ (e) $g(1-a)$

   (where a is a constant real number).

3. For each of the following rules, use a graphic calculator or computer to make a sketch, for values of x between -10 and $+10$. Use your sketches to work out a domain for each mapping which will make each one a function.

   (a) $f : x \mapsto \dfrac{1}{x^3}$ (b) $g : x \mapsto x^5$

   (c) $h : x \mapsto \dfrac{1}{x+2}$ (d) $m : x \mapsto +\sqrt{1-x}$.

4. The wind chill temperature, $T°C$, depends on the actual temperature, $t°C$, and the wind speed, v mph; an appropriate formula for T is given by

   $$T = 33 + \left(0.45 + 0.29\sqrt{v} - 0.02v\right)(t - 33)$$

   for $t > -273°$, $v \geq 5$. Sketch a graph of T against t for varying wind speeds; for example $v = 10$, 15 and 20 mph.

2.2   Some important graphs

Many of the functions which arise from problems can be 'built up' from simpler functions. In this section you will see how some of these simpler functions behave by looking at their graphs. This will help you to sketch more complicated functions later on.

Activity 5   Some well known curves

(a) Use a computer or calculator to make sketches of the following curves on the same pair of axes. Make the sketches for values of x between -2 and $+2$, and the y values between -20 and $+20$.

$$y = x, \; y = x^2, \; y = x^3, \; y = x^4, \; y = x^5, \; y = x^6.$$

Which points do all the curves pass through? What happens to the curves between $x = 0$ and $x = 1$ as the power of x increases? What happens as the power increases for values of x greater than 1? Try to predict what the curves $y = x^7$ and $y = x^8$ will look like and check your answer on the computer or calculator.

(b) Sketch these curves on separate axes with x-axis from -2 to 2 and y from -10 to 10.

$$y = x^2, \; y = x^4, \; y = \frac{1}{x^2}, \; y = \frac{1}{x^4}$$

Describe any symmetry these graphs have.

(c) Now sketch these graphs on four pairs of axes like the ones you have just used.

$$y = x, \ y = x^3, \ y = \frac{1}{x}, \ y = \frac{1}{x^3}$$

Describe any symmetry these graphs have. (Use the term 'rotational symmetry' in the description).

Activity 6   Odd and even powers

The first four graphs in part (b) of Activity 5 were all for even powers of x, whilst the graphs in part (c) were all for odd powers of x. Use the knowledge you have gained from these graphs to describe the symmetry of the following graphs. Then check your answers by using a computer or calculator to see the graphs.

(a) $y = x^{10}$,   (b) $y = x^{11}$,   (c) $y = \dfrac{1}{x^5}$,

(d) $y = x^2 - 3$,   (e) $y = x^3 + 1$,   (f) $y = x - 2$.

Were you surprised by any of the graphs? If so, try to find out why you were wrong.

Activity 7   Fractional powers

It is possible to find the value of a fractional power. Chapter 9 covers this in more detail. Using your graph plotting device, sketch these curves on the same axes, with x values from -1 to $+2$ and y values from -1 to 8.

$$y = x^{\frac{1}{2}}, \quad y = x^{\frac{1}{3}}, \quad y = x^1, \quad y = x^{\frac{3}{2}},$$
$$y = x^2, \quad y = x^{\frac{5}{2}}, \quad y = x^3.$$

Which points do all the curves have in common?

What happens as the power of x increases when x is between 0 and 1, and when x is more than 1?

Do all the curves exist for x values less than zero?

Shapes of graphs

The graph for any power of x will pass through the points $(0, 0)$ and $(1, 1)$. When x is between 0 and 1, the higher the power of x, the lower the result becomes. For example

Power of $\frac{1}{2}$	$\left(\frac{1}{2}\right)^0$	$\left(\frac{1}{2}\right)^1$	$\left(\frac{1}{2}\right)^2$	$\left(\frac{1}{2}\right)^3$	$\left(\frac{1}{2}\right)^4$
Result	1	$\frac{1}{2}$	$\frac{1}{4}$	$\frac{1}{8}$	$\frac{1}{16}$

This means that the graphs of powers of x get 'flatter' as the power increases when x is between 0 and 1. This is shown in the figure opposite.

When x is larger than 1, the higher the power of x the larger the result:

Power of 2	2^0	2^1	2^2	2^3	2^4
Result	1	2	4	8	16

This means that the curve of a higher power of x will be higher than the curve of a lower power when x is greater than 1. So the curve of a higher power 'overtakes' the curve of a lower power when $x = 1$. This is illustrated opposite.

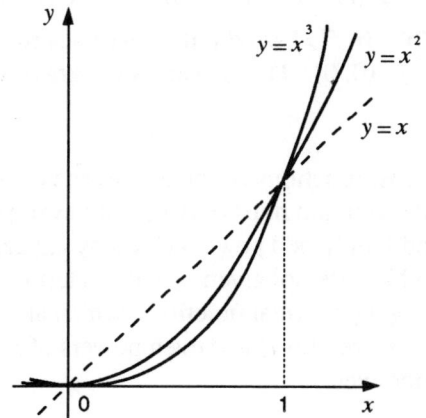

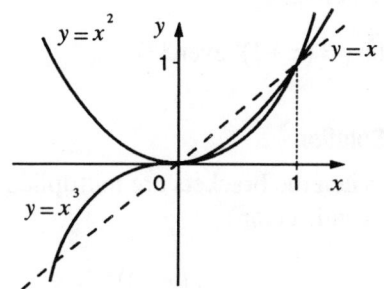

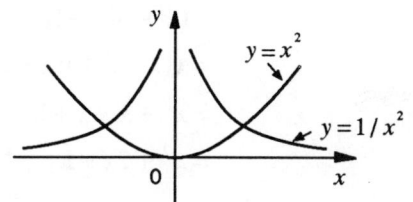

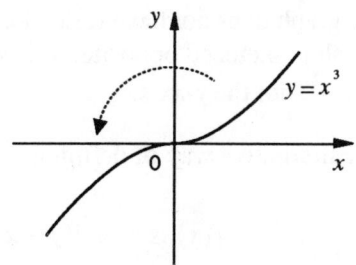

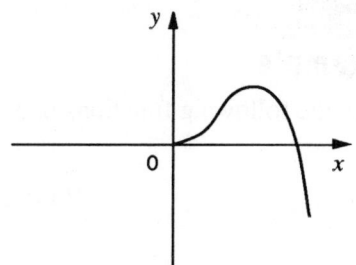

Odd and even functions

The curves for even powers of x are symmetrical about the y-axis. Any function whose curve has the y-axis as a line of symmetry is therefore called an **even** function. The graph of two even functions are shown opposite.

Curves for odd powers of x have two fold rotational symmetry about $(0, 0)$. That is, the right hand side of the curve can be rotated through $180°$ about $(0, 0)$ so that it fits onto the left hand side. This is illustrated for the graph of $y = x^3$. Functions having graphs with this kind of rotational symmetry are called **odd** functions.

Obviously, if you know already that a function is even or odd, you can easily sketch the whole of its graph if you know one half of it.

Example

One part of the graph of $y = f(x)$ is shown opposite. Complete the curve assuming that

(a) $f(x)$ is even,

(b) $f(x)$ is odd.

25

Solution

(a) If $f(x)$ is even, the curve must be symmetrical about the y-axis, so the curve will be the one shown opposite.

(b) If $f(x)$ is odd, the curve has twofold rotational symmetry about $(0, 0)$. This produces the graph below.

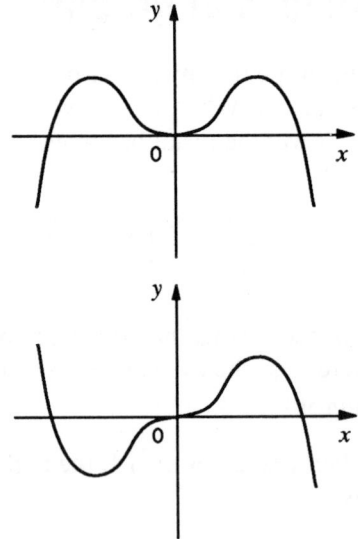

Most functions are neither even nor odd. For a function to be even, its equation must only contain even powers of x. For a graph to be odd, only odd powers of x may appear. (There are some functions which are not given usually in terms of powers of x, like the trigonometrical functions, which are even or odd however). A mixture of odd and even powers of x means the graph is neither odd nor even.

Example

Is $y = (x+1)^2$ even?

Solution

When the brackets are multiplied out the reason the function is **not** even is clear :

$$y = (x+1)^2$$
$$= (x+1)(x+1)$$
$$= x(x+1) + 1(x+1)$$
$$= x^2 + x + x + 1$$
$$= x^2 + 2x + 1.$$

This equation contains an even power (x^2) and an odd power (x), so the graph does not have reflection symmetry about the y-axis. The graph is sketched opposite. It is symmetrical, but about the line $x = -1$, not the y-axis.

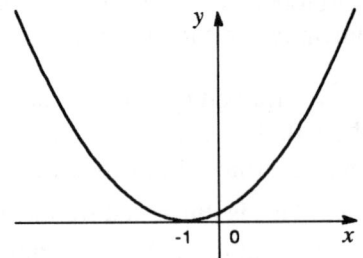

An alternative way of defining odd and even functions is to say that

$$f(x) \text{ is even if } f(-x) = f(x)$$
$$f(x) \text{ is odd if } f(-x) = -f(x).$$

Example

Are the following functions odd, even or neither?

(a) $f(x) = x^2 + 1$ (b) $f(x) = x^3$ (c) $f(x) = \frac{1}{x+1}$, $(x \neq -1)$.

Solution

(a) $f(-x) = (-x)^2 + 1 = x^2 + 1 = f(x)$ – hence even.

(b) $f(-x) = (-x)^3 = -x^3 = -f(x)$ – hence odd.

(c) $f(-x) = \dfrac{1}{-x+1} \neq \pm f(x)$ – hence neither even or odd.

Exercise 2B

1. Decide whether the graphs illustrate an even function, an odd function, or neither:

(a)

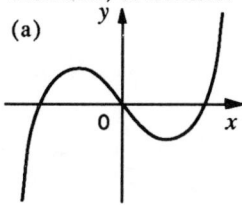

(b)

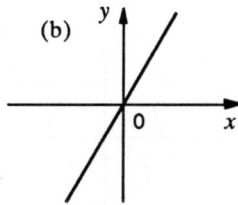

(c)

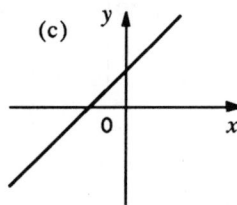

(d)

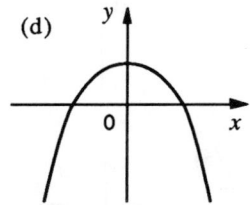

2. Copy and complete the following curves, assuming that the function each represents is i) even, and ii) odd.

(a)

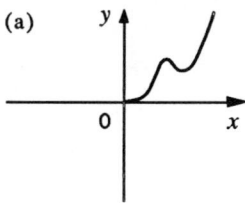

(b)

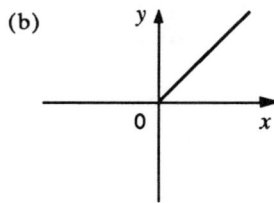

(c)

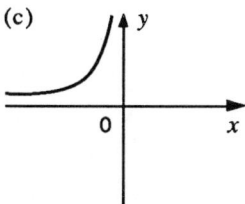

(d)

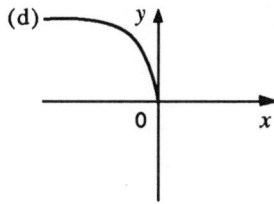

3. By looking only at the equation decide whether these functions are even, odd or neither.

(a) $y = 3x$ (b) $y = 3x + 1$

(c) $y = (x+2)^2$ (d) $y = x^2 + 2$

(e) $y = \left(x^2 + 2\right)^2$

(You will need to multiply out any bracket. Also note that a constant number can be thought of as an even power of x. For instance $2x^0 = 2 \times 1 = 2$ as $x^0 = 1$.)

2.3   Miscellaneous Exercises

1. Many shoes give both their U.K. size and European equivalent. For example,

   English adult size 12 = European size 47

   English adult size 5 = European size 38

   Use this information to construct a conversion graph between the two sizes. What does English size 0 correspond to in terms of European size?

   (English adult size 0 is in fact equivalent to junior size 13).

2. Give a reason why the domains for each of these functions are unsuitable, and give a domain that is acceptable. Also state the range.

   (a) $f(x) = \dfrac{1}{\sqrt{x}}, \ x \in \mathbb{R}$

   (b) $f(x) = \dfrac{1}{x-3}, \ x > 0$

   (c) $f(x) = \sqrt{6-x}, \ x > 0$

   (d) $f(x) = \dfrac{1}{(x-2)(x+3)}, \ x \in \mathbb{R}$.

3. Which of the graphs below represent functions?

    (a)

    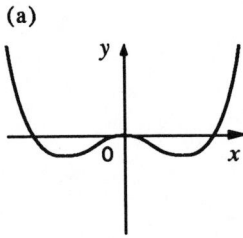

    (b)

    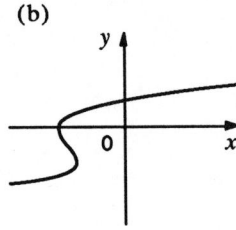

    (c)

    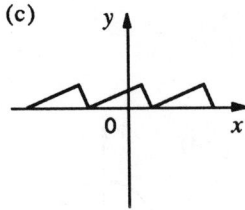

    (d)

    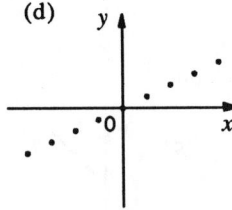

    (e)

    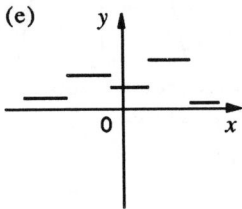

4. Let $f: x \mapsto \dfrac{1}{x^3 - 8}$, $x \in \mathbb{R}$, $x \neq 2$. Find:

    (a) $f(0)$ (b) $f(1)$ (c) $f(-1)$ (d) $f(-2)$.

5. State whether each of the following graphs represents an even or an odd function, or neither.

    (a)

    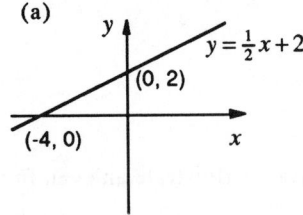

    $y = \frac{1}{2}x + 2$

    (0, 2)

    (-4, 0)

    (b)

    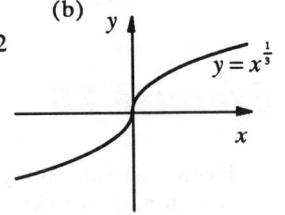

    $y = x^{\frac{1}{3}}$

    (c)

    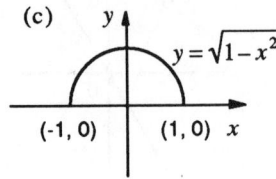

    $y = \sqrt{1 - x^2}$

    (-1, 0)   (1, 0) x

3 FUNCTIONS

Objectives

After studying this chapter you should

* understand what is meant by a composite function;

* understand the difference between $f(g(x))$ and $g(f(x))$;

* know what is meant by the inverse of a function;

* be able to sketch the graph of a function's inverse.

3.0   Introduction

You have now met a number of functions. Some have represented practical relationships whilst others have been simply mathematical functions. In this chapter, you will extend these ideas by looking at how two functions can be used to define another function, and considering how to find inverse functions, and what they represent. Whilst many of the functions to be studied will be quite complex mathematically, it is the practical application of mathematics which motivates the need for those extensions.

3.1   Composite functions

You have already met some quite complicated functions in the first two chapters. In this section you will see how **composite** functions can be built up, and why they are an important concept in mathematics. The idea of a composite function is introduced with a practical currency exchange rate example.

Example

A bank in the U.K. offers the exchange rate

$$£1 = \$1.7$$

plus an administration payment of £2 for each transaction. A similar shop in U.S.A. offers the exchange rate

$$\$1 = 1.6 \text{ marks}$$

plus an administration payment of $3 for each transaction. How many marks will you actually receive if you first exchange £10 into dollars in the U.K., and into marks in the U.S.A?

Solution

It is easy enough to solve this problem in two numerical stages. Firstly the £10 is changed to dollars. Taking away the £2 transaction payment, leaves you with $1.7 \times 8 = \$13.6$.

Secondly the $13.6 are changed into marks, remembering to take off 3 marks transaction payment. So you have 10.6×1.6 marks = 16.96 marks.

A more general approach is to form two functions to represent the two transactions. If t = amount in £, x = amount in $ and y = amount in marks, then for the first transaction

$$x = 1.7(t - 2)$$
$$= 1.7t - 3.4$$

and, since x is a function of t, write

$$x(t) = 1.7t - 3.4. \tag{1}$$

Similarly, for the second transaction

$$y = 1.6(x - 3)$$
$$= 1.6x - 4.8$$

and write

$$y(x) = 1.6x - 4.8 \tag{2}$$

to show that y is a function of x. Since x in turn is a function of t, you can write y as a function of t by substituting (1) in (2) to give

$$y(x) = y(x(t))$$
$$= 1.6x(t) - 4.8$$
$$= 1.6(1.7t - 3.4) - 4.8$$
$$= 2.72t - 5.44 - 4.8$$
$$= 2.72t - 10.24$$

i.e. $\qquad f(t) = 2.72t - 10.24 \tag{3}$

where $\qquad f(t) = y(x(t)).$

Does equation (3) give the correct solution when $t = 10$?

The composite function $y(x(t))$ is written for short as yx. Note that, in this context, yx does not mean the function y times the function x. This would be written as $y(x)x$.

Example

The functions f and g are defined by

$$f(t) = 4t - 3$$

$$g(t) = 2t - 1$$

Find the composite functions fg and gf.

Solution

$$fg = f\big(g(t)\big) = f(2t - 1)$$

$$= 4(2t - 1) - 3$$

$$= 8t - 7.$$

whereas

$$gf = g\big(f(t)\big) = g(4t - 3)$$

$$= 2(4t - 3) - 1$$

$$= 8t - 7.$$

In this case the two composite functions fg and gf are identical. This is not generally true as you will see in the activity below.

Activity 1

The function f and g are defined by

$$f(x) = x^2 + 1$$

$$g(x) = x - 1$$

Is $fg = gf$ for any value of x?

As you should have seen composite functions are **not** usually commutative. That is, in general $gf \neq fg$. For instance, if

$$f(x) = x + 3,\ x \in \mathbb{R}$$

and $\qquad g(x) = 2x,\ x \in \mathbb{R}$

then $f(g(x)) = f(2x)$

$$= 2x + 3 \text{ for } x \in \mathbb{R}.$$

Similarly $g(f(x)) = g(x+3)$

$$= 2(x+3)$$
$$= 2x + 6 \text{ for } x \in \mathbb{R}.$$

The two composite functions are clearly different for all values of x. Also note that, because the range of the function which is applied first is the domain for the second function, it is essential that the range of the first is suitable as a domain for the second.

Example

If $h(x) = \dfrac{1}{x}$, $x \in \mathbb{R}$, $x \neq 0$ and $m(x) = x - 2$, $x \in \mathbb{R}$, find suitable domains for the composite function mh and hm.

Solution

Although $m(h(x)) = \dfrac{1}{x} - 2$, $x \in \mathbb{R}$, $x \neq 0$ is a function, there is a problem with the composite function $h(m(x))$.

$$h(m(x)) = h(x - 2)$$

$$= \frac{1}{x - 2}.$$

If the domain is chosen to be $x \in \mathbb{R}$, $x \neq 0$ which is the domain of the first function m, then x can be equal to 2.

However, $hm(2) = \dfrac{1}{2 - 2} = \dfrac{1}{0}$, and this does not exist.

So $x \in \mathbb{R}$, $x \neq 0$ is not suitable as a domain for hm.

If $x \in \mathbb{R}$, $x \neq 2$ is chosen as the domain, hm will be defined for all the values given in the domain.

Exercise 3A

1. Work out composite functions *fg* and *gf* for each of the following pairs of functions. In each case, state a suitable domain for the composite function. (You may find a sketch of the composite function graph made with a computer or calculator helpful when checking your answer.)

   (a) $f(x)=x-1$ and $g(x)=x^3$

   (b) $f(x)=+\sqrt{x}$ and $g(x)=x-2$

   (c) $f(x)=\dfrac{1}{x}$ and $g(x)=x+1$

   (d) $f(x)=x^2-1$ and $g(x)=\dfrac{1}{x}$

   (e) $f(x)=x+3$ and $g(x)=x-3$

   (f) $f(x)=6-x$ and $g(x)=6-x$.

2. Find the composite function *fg* and *gf* if
   $$f(x)=1+\frac{1}{x} \text{ and } g(x)=x^2.$$

3. Find the composite function *hfg* if $f(x)=x-3$,
   $$g(x)=x^2 \text{ and } h(x)=\frac{1}{x}.$$

3.2 Inverse functions

In Chapter 2 you met the function that transforms degrees Celsius to degrees Fahrenheit, namely

$$F=\frac{9}{5}C+32. \qquad (1)$$

Suppose you wanted to find the formula that gives degrees Celsius in terms of degrees Fahrenheit, then taking 32 from both sides

$$F-32=\frac{9}{5}C$$

and multiplying by $\dfrac{5}{9}$ gives

$$\boxed{C=\frac{5}{9}(F-32)} \qquad (2)$$

This is an example of an **inverse** function.

Example

If $y=$ number of dollars and $x=$ equivalent number of pounds and $y=1.7x$, express x in terms of y.

Solution

You must make x the subject of the formula when $y=1.7x$.

This gives $x=\dfrac{y}{1.7}$

(in fact, exchange rates from dollars to pounds and pounds to dollars are not in practice equivalent, and there is usually a transaction charge)

Activity 2   Kepler's third law

In 1619 the astronomer *Kepler* announced his third law of planetary motion (dedicated to James II of England) which stated that the periodic time of a planet, T, is related its average radius of orbit, say R, by the formula

$$T = kR^{\frac{3}{2}}.$$

Find the inverse function which expresses R as a function of T.

A special notation is introduced for inverse functions. For example, the temperature conversion formula, with x now denoting degrees Celsius and y degrees Fahrenheit,

$$y = f(x) = \frac{9}{5}x + 32 \tag{3}$$

can be rearranged to give, $y - 32 = \frac{9}{5}x$ and $x = \frac{5}{9}(y - 32)$.

The inverse function is denoted by f^{-1}, so

$$f^{-1}(y) = \frac{5}{9}(y - 32).$$

Since y could be any variable, we can rewrite the inverse function as a function of the variable x as

$$f^{-1}(x) = \frac{5}{9}(x - 32). \tag{4}$$

Note that the meaning of the variable x is different in (3) and (4). In (3), it represents the temperature in degrees Celsius, so that, for example, $20°C$ will transform to $\frac{9}{5} \times 20 + 32 = 36 + 32 = 68°F$. Whilst in (4), x represents degrees Fahrenheit, so that $77°F$ transforms to $\frac{5}{9} \times (77 - 32) = \frac{5}{9} \times 45 = 25°C$.

Example

Find the inverse of $f(x) = \dfrac{1}{1-x} + 2$, $x \in \mathbb{R}$, $x \neq 1$, and state the domain of the inverse function.

Solution

Let $f(x) = y$, so that

$$y = \frac{1}{1-x} + 2$$

$$\Rightarrow \quad y - 2 = \frac{1}{1-x}$$

$$\Rightarrow \quad (1-x)(y-2) = 1$$

$$\Rightarrow \quad 1 - x = \frac{1}{y-2}$$

$$\Rightarrow \quad x = 1 - \frac{1}{y-2}.$$

This formula gives the inverse function as

$$f^{-1}(y) = 1 - \frac{1}{y-2}.$$

Replacing y by x, this becomes

$$f^{-1}(x) = 1 - \frac{1}{(x-2)}.$$

As $f^{-1}(x)$ is the inverse of $f(x)$, its **domain** will be the **range** of $f(x)$. This is because its task is to map members of the original range back onto the corresponding members of the domain.

The figure opposite shows the graph of $f(x)$. The range is all the real numbers, except 2. There is no value of x for which $f(x) = 2$, as shown by the dotted line.

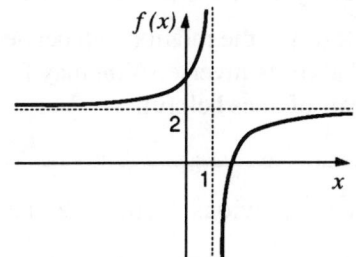

So the domain of $f^{-1}(x)$ is the set of real numbers, except 2. This means the full definition of $f^{-1}(x)$ is

$$f^{-1}(x) = 1 - \frac{1}{x-2}, \ x \in \mathbb{R}, \ x \neq 2.$$

The graph of $f^{-1}(x)$ is shown opposite.

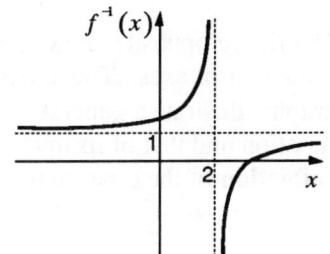

(Note also that the range of $f^{-1}(x)$ is the domain of $f(x)$).

Exercise 3B

Find the inverses of these functions.
State the domain of each inverse.

1. $f(x) = x + 2, x \in \mathbb{R}$

2. $f(x) = 4x - 1, x \in \mathbb{R}$

3. $f(x) = 4x - 2, x \in \mathbb{R}$

4. $f(x) = x, x \in \mathbb{R}$

5. $f(x) = 1 - x, x \in \mathbb{R}$

6. $f(x) = \frac{3}{x}, x \in \mathbb{R}, x \neq 0$

7. $f(x) = \frac{1}{x+2}, x \in \mathbb{R}, x \neq -2$

8. $f(x) = \frac{1}{5-x} + 2, x \in \mathbb{R}, x \neq 5$.

3.3 Symmetry about the line y = x

Functions and their inverses have an interesting geometrical property as you will see below.

Activity 3   The graph of an inverse function

Below is a list of functions, each with its inverse. On graph paper, plot and draw the graph of a function, together with its inverse, on the same axes. Repeat this process for each function in the list. Use the same scale on both axes. Use values of x between -6 and $+6$, and y values between -12 and $+12$.

(a) $f(x) = 2x, f^{-1}(x) = \frac{1}{2}x$;

(b) $f(x) = x - 4, f^{-1}(x) = x + 4$;

(c) $f(x) = x^2, (x \in \mathbb{R}, x \geq 0), f^{-1}(x) = +\sqrt{x}, (x \in \mathbb{R}, x \geq 0)$.

Describe the relationship between the graph of each function and that of its inverse. You may find drawing the line $y = x$ on each pair of axes helpful.

In the previous Section 3.2, the inverse of the function

$f(x) = \frac{9}{5}x + 32$ was found to be $f^{-1}(x) = \frac{5}{9}(x - 32)$.

The figure opposite shows the graphs of these two functions on the same pair of axes. The dotted line is the graph $y = x$. These graphs illustrate a general relationship between the graph of a function and that of its inverse, namely that one graph is the **reflection** of the other in the line $y = x$.

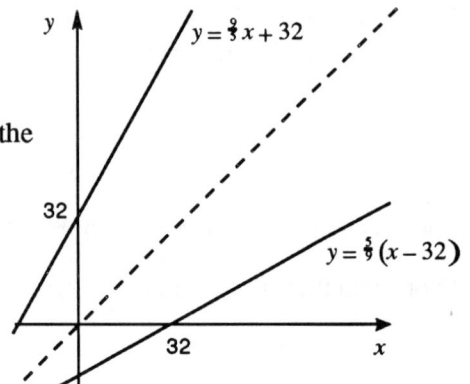

There is an interesting group of functions which have graphs that are symmetrical about the line $y = x$. The figure opposite shows such a graph, for the function $f(x) = \frac{1}{x}$. The graph of its inverse function will be the reflection of this curve in the dotted line, but as the curve is symmetrical about this line, the inverse function will be the same as the function itself.

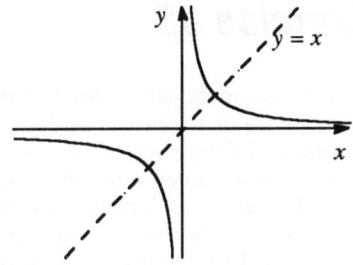

A function which is its own inverse is called a **self inverse** function. In this case, if $f(x) = \frac{1}{x}$, then $f^{-1}(x) = \frac{1}{x}$ too.

One final point that is worth noting is that
$$f\left(f^{-1}(x)\right) = f^{-1}\left(f(x)\right) = x.$$

For example, you have already seen that if $f(x) = \frac{9}{5}x + 32$,

then $\qquad f^{-1}(x) = \frac{5}{9}(x - 32).$

Now $\qquad f\left(f^{-1}(x)\right) = f\left(\frac{5}{9}(x - 32)\right)$

$$= \frac{9}{5}\left(\frac{5}{9}(x - 32)\right) + 32$$

$$= x - 32 + 32$$

$$= x,$$

and similarly for $f^{-1}\left(f(x)\right)$.

Activity 4

If $f(x) = 4x - 3$, find $f^{-1}(x)$ and check that
$$f\left(f^{-1}(x)\right) = f^{-1}\left(f(x)\right) = x$$

Exercise 3C

1. Use a graphic calculator or computer to make sketches of each of these functions. (You can plot and draw them if you do not have a calculator or computer at hand). Use values of x and y between -5 and $+5$ and on each pair of axes show the graph of the inverse function. You will find that superimposing the line $y = x$ is helpful. Use the same scales for x and y.

   (a) $f(x) = x + 3$ (b) $f(x) = x^3$ (c) $f(x) = -2x$

   (d) $f(x) = \dfrac{-1}{x}$ (e) $f(x) = 4 - x$ (f) $f(x) = \dfrac{1}{x^3}$.

2. For (b) and (c) in Question 1 above check that
   $$f\big(f^{-1}(x)\big) = f^{-1}\big(f(x)\big) = x.$$

3. Copy the graphs below and sketch the graphs of the inverse functions.

   (a)

   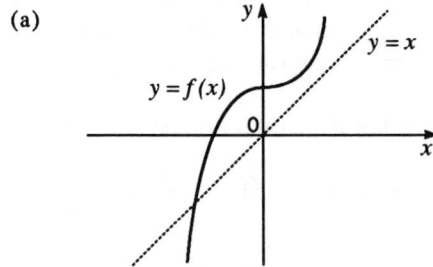

   (b)

   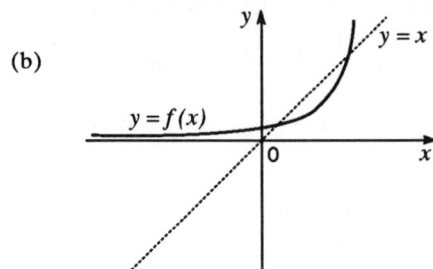

3.4   Functions with no inverse

In Section 2.1 you saw that for an algebraic rule, or formula, to be a function, the rule must map each member of the domain to one and only one member of the range. That is, it must give only one answer. So, for example, if $y^2 = x$ then it is not possible to express y as a function of x since $y = \pm\sqrt{x}$; that is, for $x = 4$, $y = \pm 2$ etc.

One value of x gives two values of y. However, $y = \sqrt{x}$, $x \geq 0$, is a function, as there is only one number that is the positive square root of x, for any real number x, which is greater than zero.

You can tell if a mapping is a function from its graph. If a line parallel to the y-axis crosses the graph in more than one place, the mapping is **not** a function. For instance, the figure opposite shows that the value $x = 2$ in the domain is mapped to three values in the range, so this is not the graph of a function.

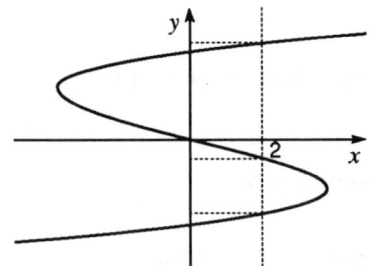

A **many to one** function will map different members of its domain to the same member of the range. An example of such a function is

$$f(x) = x^2, x \in \mathbb{R}.$$

For instance $f(3) = 3^2 = 9$ and $f(-3) = (-3)^2 = 9$. So two values in the domain, 3 and -3, are mapped to the same value in the range, namely 9.

Activity 5

Use a calculator or computer to draw the graph of $y = x^2$ for values of x between -3 and $+3$. Use the same scales on both axes. Draw on the same pair of axes the graph of the inverse mapping, by using the reflection property.

Does the graph represent a function?

If the domain of $f(x)$ is changed from $x \in \mathbb{R}$ to $x \in \mathbb{R}$, $x \geq 0$, make a sketch of $f(x)$ on new axes. Also show what the graph of the inverse must look like on the same pair of axes. Is this the graph of a function?

Activity 6

An even function has a graph which is symmetric about the y-axis. Draw some sketchs of functions which are even, like the one shown opposite. Then show on the same axes what the inverse mapping should look like.

A many to one function maps different values in the domain to the same value in the range. The graph of a many to one function can be crossed in more than one place by a line parallel to the x-axis **once**, as shown opposite.

The inverse of such a function must therefore map the same value to different members of the original function's domain. In the graph opposite, d must be mapped to a, b and c. This means that the inverse **cannot** be a function, (as it does not send every member of **its** domain to only one member of its range).

So a many to one function cannot have an inverse function.

If a function is to have an inverse, the function cannot be many to one - it must be **one to one**. That is, every member of its domain is mapped to its own unique member of the range. The graph of such a function can only be crossed by a line parallel to the x-axis **once**, as shown in the figure opposite.

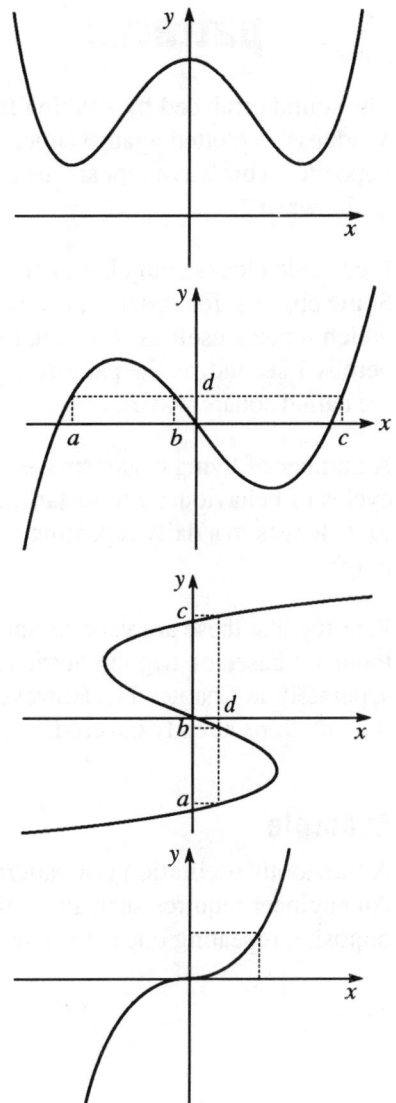

Exercise 3D

Use a graphic calculator or computer to sketch the graphs of these functions. Check whether a horizontal line can be drawn which crosses each graph in more than one place to decide whether each graph is one to one or many to one. For the functions which are one to one draw the graph of the inverse function in each case using the reflection relationship.

1. $f(x) = x^2 + 2, \ x \in \mathbb{R}$

2. $f(x) = x^3 + 2, \ x \in \mathbb{R}$

3. $f(x) = \dfrac{1}{x^2}, \ x \in \mathbb{R}, \ x \neq 0$

4. $f(x) = (x+1)^2, \ x \in \mathbb{R}$

5. $f(x) = x^3 + x^2 - 6x, \ x \in \mathbb{R}$

6. $f(x) = x^3 - 9x^2 + 27x - 27, \ x \in \mathbb{R}$

7. $f(x) = \dfrac{1}{x-2} + 1, \ x \in \mathbb{R}, \ x \neq 2$

8. $f(x) = 3, \ x \in \mathbb{R}$.

3.5 Modelling repeating patterns

The sound produced by a tuning fork is a wave shape when its 'loudness' is plotted against time. This is shown in the first graph opposite. This wave repeats itself after 0.02 seconds - the **period** of the wave.

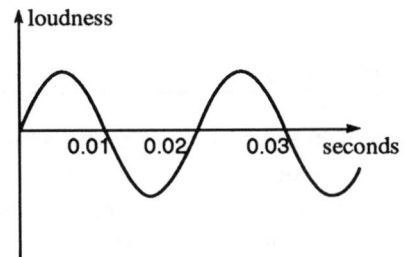

Electronic clocks count **beats** produced by an internal circuit. Some circuits, for instance, can be made to produce a voltage which repeats itself as shown in the second graph. The period here is 1 second, as the pattern is repeated after this time. These are called square waves.

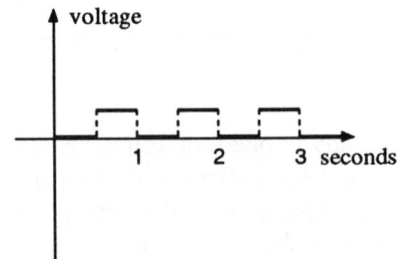

A number of living organisms, especially plants, exhibit repeating cycles of behaviour. For instance, a plant's stem may conduct sap to its leaves in a daily repeating pattern, as shown in the third graph.

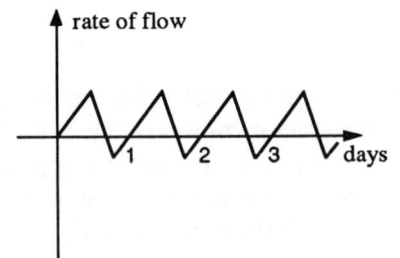

Patterns like these are very common. The most important among them are based on trigonometric functions, and these are covered separately in Chapter 11. However, some can be modelled with the functions already covered.

Example

A 'sawtooth' oscillation is a pattern which occurs in electronics. An engineer requires such an oscillation to have the shape shown opposite, repeating once every second.

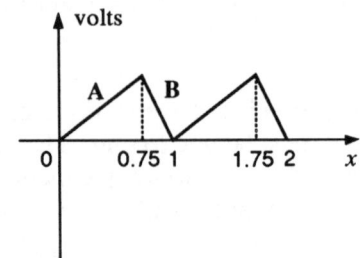

The equation of the line marked A is $y = \dfrac{8x}{3}$. The equation of the line marked B is $y = -8x + 8$. So the function which models the first 'tooth' is

$$f(x) = \begin{cases} \dfrac{8}{3}x, & 0 \le x \le 0.75 \\ -8x + 8, & 0.75 \le x \le 1 \end{cases}$$

The period of the function is 1 second, and this completes the definition of the function.

Example

The rate of flow of sap in a certain species of plant is thought to follow the pattern as shown opposite. A positive rate of flow means the flow is upwards.

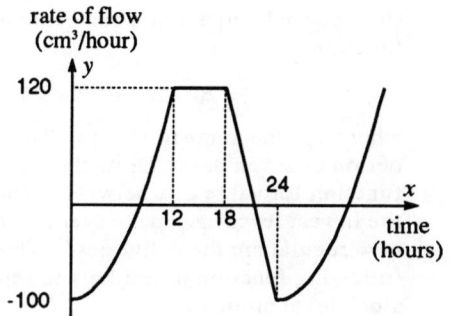

The section of the graph for values of x between 0 and 12 is the graph of

$$y = \frac{220}{144}x^2 - 100.$$

The part when x is between 12 and 18 is constant at 120, so its equation for these values of x is $y = 120$. The last part of the cycle, when x is between 18 and 24, is given by the graph of

$$y = -\frac{220}{6}x + 780.$$

So the function which models this pattern is

$$f(x) = \begin{cases} \dfrac{220}{144}x^2 - 100 & \text{for} \quad 0 \le x \le 12 \\ 120 & \text{for} \quad 12 \le x \le 18 \\ -\dfrac{220}{6}x + 780 & \text{for} \quad 18 \le x \le 24 \end{cases}$$

with a period of 24 hours.

Example

Sketch the graph $f(x) = \begin{cases} x & \text{for} \ 0 \le x < 2 \\ 0 & \text{for} \ 2 \le x \le 3 \end{cases}$

when $f(x)$ has a period of 3 units, for values of x between -3 and 6.

The solution is shown opposite.

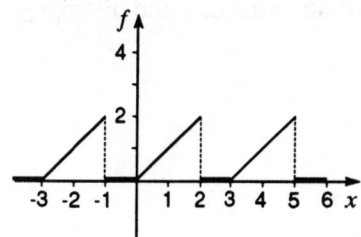

Exercise 3E

1. A square wave generator produces a voltage that is given by the function $f(x)$, where

$$f(x) = \begin{cases} 0 \text{ if } 0 \le x \le \frac{1}{2} \\ 1 \text{ if } \frac{1}{2} \le x \le 1 \end{cases}$$

and $f(x)$ has a period of 1 unit. Draw the graph for values of x between 0 and 3.

2. The stock level of coal at a small power station (in thousand tonne units) can be modelled by the function

$$f(x) = 20 - 2x \text{ for } 0 \le x \le 7$$

where x is the number of days. The function has a period of seven days. Draw the graph of the function for value of x between 0 and 21. What is the lowest level the stocks ever reach, and how regular are the deliveries? Why would the following function be unlikely to represent a stock level problem:

$$g(x) = 20 - 2x \text{ for } 0 \le x \le 10 ?$$

3. Draw the graph of $f(x)$ for values of x between 0 and 12, where

$$f(x) = \begin{cases} 1 - (x-1)^2 \text{ for } 0 \le x \le 2 \\ (x-3)^2 - 1 \text{ for } 2 \le x \le 4 \end{cases}$$

$f(x)$ has a period of 4 units. You may need to plot and draw the first 'cycle' of the pattern.

4. Plot and draw the graph of the function below, showing a full two periods of its cycle.

$f(x)$ has a period of 2 units. You may need to plot and draw the first 'cycle' of the pattern.

$$f(x) = \begin{cases} \dfrac{1}{2}x^2 \text{ for } 0 \le x \le 1 \\ \dfrac{1}{x} - \dfrac{1}{2} \text{ for } 1 \le x \le 2 \end{cases}$$

3.6 Case study

The marketing team of a company has to determine the marketing strategy for a recently introduced T-shirt. A major part of this strategy is the price for which it will be sold. The cheaper it is, the more it will sell. The team have found in trials in selected parts of the country that if the price is £5, then 200 will be sold in a month. If the price is £10, only 10 are sold.

The **assumption** is made that the number sold in a month is related to the selling price by the formula

$$n = 390 - 38x$$

where n is the number sold and x is the price. This corresponds to the straight **demand curve** shown opposite.

Using this **assumed** demand curve, it is necessary to find the price which will maximise the profit for the company, if each shirt costs £2 to manufacture and distribute.

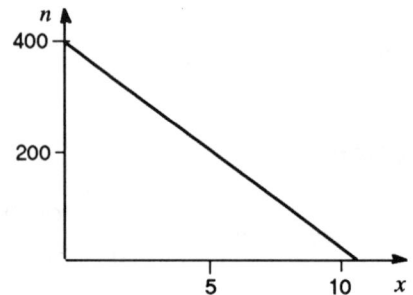

Activity 7

Use the example outlined above for this activity.

(a) Work out a formula for the total amount of money taken in sales each month (denote this by s), in terms of x. This will be the number sold each month, n, times the price of each one. You can use the demand curve formula $n = 390 - 38x$.

(b) Work out a formula for the cost, c, of producing n T-shirts at £2 per shirt, in terms of x. Again, you can use the demand curve formula for n, and you should multiply out any brackets before continuing.

(c) The profit made is the income from sales, s, minus the total cost of making n shirts, c, giving

$$p(x) = s - c$$

Use your answers to parts (a) and (b) above to find a formula for p in terms of x. Simplify the formula.

(d) Use a graph plotter to sketch the graph of the function $p(x)$, for values of x between 0 and 10. From this, estimate the best price (to the nearest pound) to maximise the profits, and work out the number n that should be produced each month.

(If you can magnify a part of the graph, you may be able to find the best price to the nearest 10p or 1p).

A major assumption was made by the team when they started to set up the model for the profit. Only two pieces of data, linking the number sold, n, with the price, x, were known, and it was assumed that these pieces of data formed part of the **straight** demand curve, $n = 390 - 38x$.

$$n = \frac{1900}{x} - 180$$

However, it is unlikely that the price and number of shirts sold really are related so simply. Two **possible** models for the demand curve are shown opposite. There are many others.

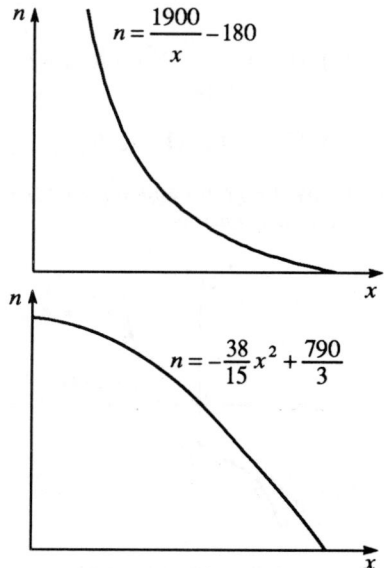

Activity 8

Using either, or both, of these models, find a new function for the profit, in a similar way to the earlier steps. Use a graph of this function to estimate the best prices with the new demand curve.

$$n = -\frac{38}{15}x^2 + \frac{790}{3}$$

What should the team do to decide which curve to use?

3.7 Miscellaneous Exercises

1. Using a graphic calculator or computer to make sketches of the following functions; decide whether each has an inverse function or not. If an inverse exists, find the algebraic rule for it, and state its domain.

   (a) $f(x) = 3x - 2, x \in \mathbb{R}$

   (b) $f(x) = 4 - 3x, x \in \mathbb{R}$

   (c) $f(x) = x^2 - 1, x \in \mathbb{R}$

   (d) $f(x) = x^2 - 1, x \in \mathbb{R}, x \geq 0$

   (e) $f(x) = (x - 3)^2, x \in \mathbb{R}$

   (f) $f(x) = \dfrac{1}{x} - 4, x \in \mathbb{R}, x \neq 0$

   (g) $f(x) = \dfrac{1}{(x+1)^2 - 2x}, x \in \mathbb{R}, x \neq \pm 1$.

2. Use a graphic calculator to sketch the following functions. Use the sketch to superimpose the graph of the inverse function in each case.

   (a) $f(x) = x^2 + 4, x \in \mathbb{R}, x > 0$

   (b) $f(x) = \dfrac{1}{x - 2} + 1, x \in \mathbb{R}, x \neq 2$

   (c) $f(x) = 4x - 1, x \in \mathbb{R}$.

3. Copy the graph and superimpose the graph of the inverse function.

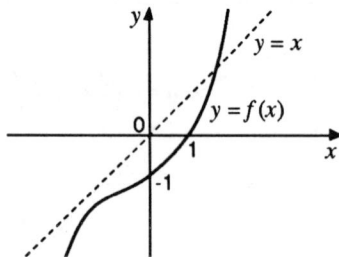

Show where the graph of $f^{-1}(x)$ cuts the x and y axes.

4. Copy the graph below, and superimpose the graph of $f(x - 2)$. Hence sketch the graph of $f^{-1}(x - 2)$.

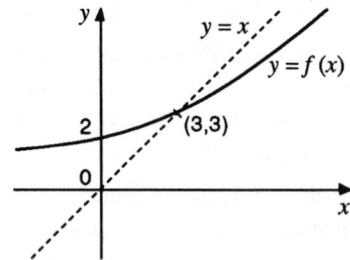

5. Plot and draw the following periodic functions, showing two full periods for each.

   (a) $f(x) = \begin{cases} x & \text{for } 0 \leq x \leq 2 \\ 4 - x & \text{for } 0 \leq x \leq 4 \end{cases}$

   (b) $f(x) = \begin{cases} x^2 & \text{for } 0 \leq x \leq 1 \\ 1 & \text{for } 0 \leq x \leq 2 \end{cases}$

   (c) $f(x) = 2 - \dfrac{x}{2}$ for $0 \leq x \leq 2$

   (d) $f(x) = x^5$ for $0 \leq x \leq 1$.

4 GRAPH TRANSFORMS

After studying this chapter you should

- be able to use appropriate technology to investigate graphical transformations;

- understand how complicated functions can be built up from transformations of simple functions;

- be able to predict the graph of functions after various transformations.

4.0 Introduction

You have already seen that the ability to illustrate a function graphically is a very useful one. Graphs can easily be used to explain or predict, so it is important to be able to quickly sketch the main features of a graph of a function. New technology, particularly graphic calculators, provides very useful tools for finding shapes but as a mathematician you will still need to gain the ability to understand what effect various transformations have on the graph of a function. First try the activity below **without** using a graphic calculator or computer.

Activity 1

You should be familiar with the graph of

$$y = x^2.$$

It is shown on the right. Without using any detailed calculations or technology, predict the shape of the graphs of the following

(a) $y = x^2 + 1$ (b) $y = 2x^2$ (c) $y = 3x^2$

(d) $y = (x-1)^2$ (e) $y = (x+1)^2$ (f) $y = \dfrac{1}{x^2}$ $(x \neq 0)$.

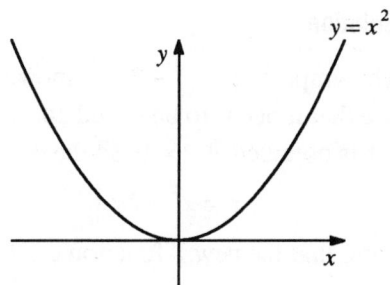

Now check your answers using a graphic calculator or computer.

4.1 Transformation of axes

Suppose $y = x^2$, then the graph of $y = x^2 + 2$ moves the curve up by two units as is shown in the figure opposite. For any x value, the y value will be increased by two units.

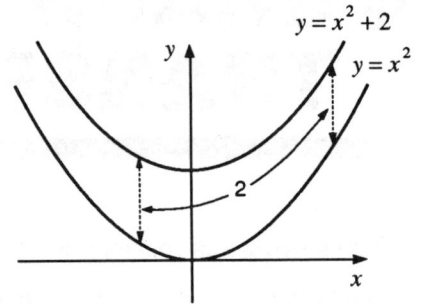

What does the graph of $y = x^2 + a$ look like?

What you are doing in the example above is equivalent to moving the x axes down by 2 units, which you can see by defining

$$Y = y - 2.$$

Then $Y = x^2$ and you are back to the original equation.

Describe the graph of $y = x^3 + 1$

This type of transformation

$$f(x) \mapsto f(x) + a$$

is called a **translation** of the graph parallel to the y-axis of a units.

Example

Find the value of a so that

$$y = x^2 - 2x + a$$

just touches the x-axis.

Solution

The graph of $y = x^2 - 2x$ is shown opposite. From this, you can see that it needs to be raised one unit, since its minimum value of -1 is obtained at $x = 1$. So the new equation will be

$$Y = x^2 - 2x + 1.$$

Note that the new Y function can be written as

$$Y = (x - 1)^2 \geq 0 \text{ for all } x$$

and equality only occurs when $x = 1$ (as illustrated).

As well as translations parallel to the y-axis, you can perform similar operations parallel to the x-axis.

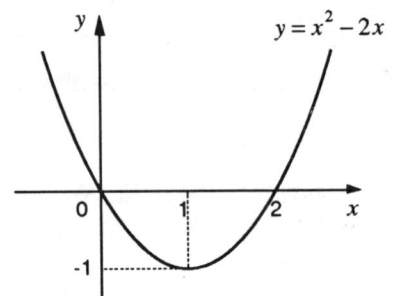

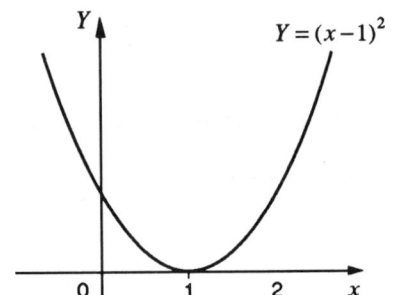

Activity 2   Translations parallel to the x-axis

Again use the familiar $y = x^2$ curve, but this time write it as

$$f(x) = x^2.$$

Evaluate $f(x-2)$. Sketch the graph of $y = f(x-2)$.

What is the relationship between this curve and the original.

If you know the shape of $y = f(x)$, what does the graph of $y = f(x-a)$ look like?

Example

The function $f(x)$ is defined by

$$f(x) = x^3 - 3x^2 + 3x - 1.$$

By considering $y = f(x+1)$, deduce the shape of the graph of $f(x)$.

Solution

$$f(x+1) = (x+1)^3 - 3(x+1)^2 + 3(x+1) - 1$$

$$(x+1)^2 = (x+1)(x+1) = x^2 + 2x + 1$$
$$(x+1)^3 = (x+1)(x+1)^2$$
$$= (x+1)(x^2 + 2x + 1)$$
$$= x^3 + 3x^2 + 3x + 1$$

$$f(x+1) = x^3 + 3x^2 + 3x + 1$$
$$- 3(x^2 + 2x + 1) + 3(x+1) - 1$$
$$= x^3 + x^2(+3-3) + x(3-6+3) + 1 - 3 + 3 - 1$$
$$= x^3$$

Hence $y = f(x-1) = x^3$ and this is illustrated opposite.

This means that $f(x)$ must also have this shape, but moved along one unit parallel to the x-axis.

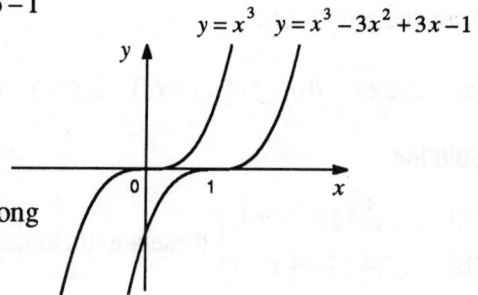

$y = x^3 \quad y = x^3 - 3x^2 + 3x - 1$

Activity 3

Using a graphic calculator or computer,

(a) illustrate the curves
$$y = x^3 \text{ and } y = x^3 - 3x^2 + 3x - 1$$

and hence verify the result in the sketch on the previous page;

(b) illustrate the curves
$$y = x^4 \text{ and } y = x^4 - 8x^3 + 24x^2 - 32x + 16$$

and deduce a simpler form to write the second function.

Use x range –2 to 4 and y range 0 to 10.

Exercise 4A

1. Without using a graph plotting device, draw sketches of
$$f(x), f(x+5), f(x)+5$$
for the following functions

   (a) $f(x) = 2x - 1$

   (b) $f(x) = x^2 - 1$

   (c) $f(x) = (x-1)^2$.

2. Use a graph plotting device to illustrate the graphs of
$$f(x) = x^2, \ g(x) = x^2 + 2x + 2.$$

   Hence or otherwise write $g(x)$ in the form
$$f(x+a) + b$$
by finding the constants a and b.

3. If $f(x) = \frac{1}{x}$, sketch the graphs of

   (a) $f(x)$ (b) $f(x-1)$ (c) $f(x-1)+1$.

4.2   Stretches

In this section you will be investigating the effect of stretching either the y- or x-axis.

Example

For the function
$$y = f(x) = x + 1$$

draw the graphs of

(a) $f(2x)$ (b) $f(\frac{1}{2}x)$ (c) $2f(x)$ (d) $\frac{1}{2}f(x)$.

Solution

(a) $f(2x) = 2x + 1$ ⎫
(b) $f(\frac{1}{2}x) = \frac{1}{2}x + 1$ ⎬ these are illustrated opposite

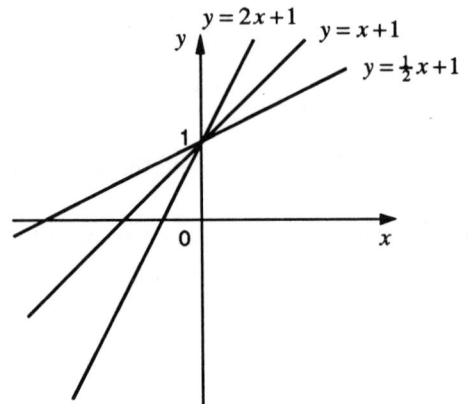

(c) $2f(x)=2(x+1)$
(d) $\frac{1}{2}f(x)=\frac{1}{2}(x+1)$ } again illustrated opposite

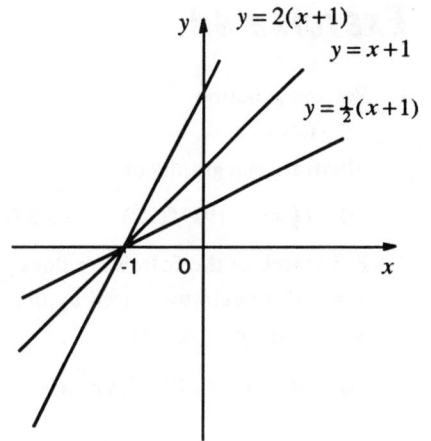

You should be beginning to get a feel for what the various types of transformations do, and the next activity will help you to clarify your ideas.

Activity 4   Stretches

For the function

$$y = f(x) = x^2 + 1$$

draw the graphs of

(a) $2f(x)$ (b) $\frac{1}{2}f(x)$ (c) $f(2x)$ (d) $f(\frac{1}{2}x)$.

Use a graph plotting device to help you if you are not sure of what the graphs look like.

The example and the activity have shown you that

$y = \alpha\, f(x)$ is a stretch, parallel to the y-axis, by a factor α

$y = f(\alpha\, x)$ is a stretch, parallel to the x-axis, by a factor $\frac{1}{\alpha}$

Example

If $f(x) = \dfrac{1}{x}$, illustrate   (a) $2f(x)$ (b) $f(\frac{1}{2}x)$.

Solution

(a) $2f(x) = \dfrac{2}{x}$;   this is illustrated opposite.

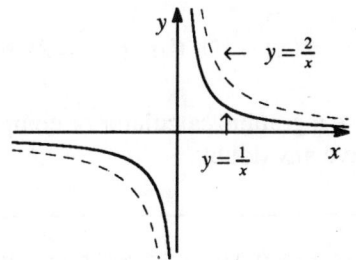

(b) $f(\frac{1}{2}x) = \dfrac{1}{(\frac{1}{2}x)} = \dfrac{2}{x}$;

which is identical to $2f(x)$

For this rather special function, a stretch of factor α parallel to the y-axis is identical to a stretch of factor α parallel to the x-axis.

Why are the two transformations identical for the function $y = \frac{1}{x}$?

Exercise 4B

1. For the function
$$f(x) = 2x - 1$$
illustrate the graphs of

   (a) $f(\tfrac{1}{2}x)$ (b) $f(2x)$ (c) $2f(x)$ (d) $\tfrac{1}{2}f(x)$.

2. For which of the following does the function $y = f(x)$ remain unaltered by the transformation $y = \frac{1}{\alpha}f(\alpha x)$?

   (a) $f(x) = x$ (b) $f(x) = x + 1$

   (c) $f(x) = x^2$ (d) $f(x) = \frac{2}{x}$.

3. For the function $y = f(x)$, shown below, sketch the curves defined by

   (a) $y = f(\tfrac{1}{2}x)$ (b) $y = 2f(x)$.

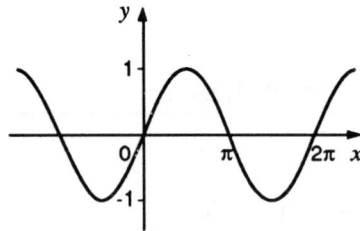

4.3 Reflections

If $f(x) = x + 1$, then the graph of $-f(x) = -(x + 1)$ is seen to be a **reflection** in the x-axis.

On the other hand
$$f(-x) = -x + 1$$
can be seen to be a **reflection** in the y-axis.

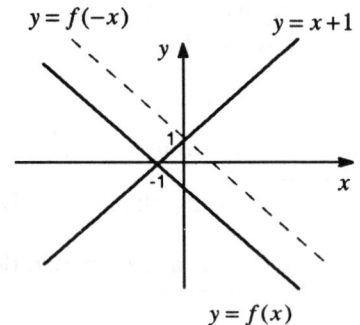

Activity 5   Reflections

For each of the functions below sketch (i) $-f(x)$ (ii) $f(-x)$

(a) $f(x) = x^2$ (b) $f(x) = 2x + 1$ (c) $f(x) = x^3$ (d) $f(x) = \dfrac{1}{x}$.

Use a graphic calculator or computer to check your answers if you have any doubt.

It is now possible to combine various transformations.

Example

If the graph of $y = f(x)$ is shown opposite, illustrate the shape of $y = 2f(-x) + 3$.

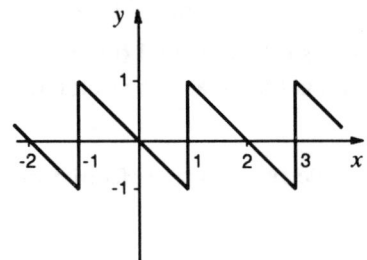

Solution

To find $f(-x)$, you reflect in the y-axis to give the graph opposite.

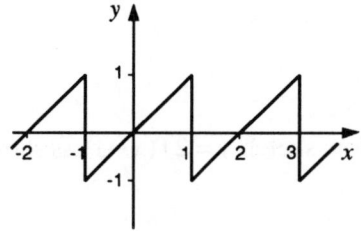

The sketch of $2f(-x)$ is shown opposite.

This is a stretch of factor 2 parallel to the y-axis.

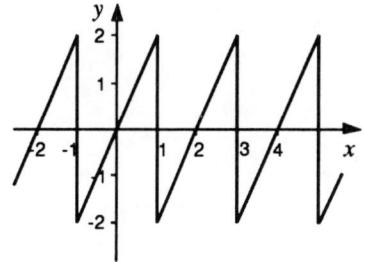

Finally adding 3 to each value gives the graph shown opposite,

$$y = 2f(-x) + 3.$$

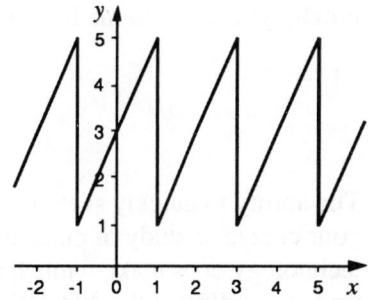

Example

Sketch the graph of $y = \dfrac{2}{x-1} + 2$.

Solution

Of course, you could find its graph very quickly using a graphic calculator or computer. It is, though, instructive to build up the sketch starting from a simple function, say

$$f(x) = \frac{1}{x}$$

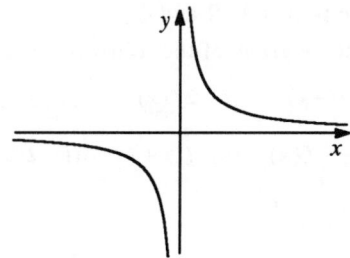

and performing transformations to obtain the required function.

In terms of f, you can write

$$y = 2f(x-1) + 2.$$

So you must first sketch $y = f(x-1)$ as shown opposite.

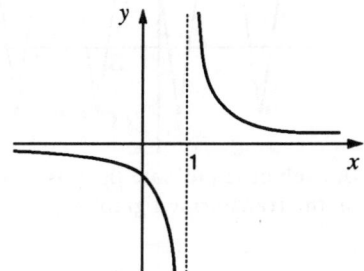

Now sketch $y = 2f(x-1)$ as shown opposite.

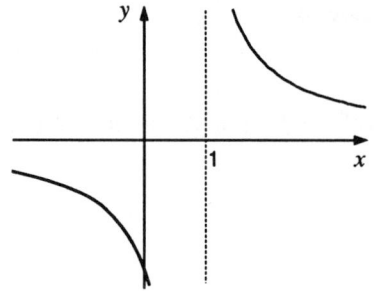

Finally you add 2 to the function to give the sketch opposite.

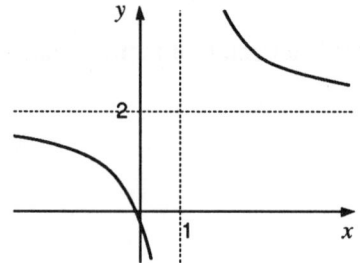

The ability to quickly sketch curves will be very useful throughout your course of study of mathematics. Although modern technology does make it much easier to find graphs, the process of **understanding** both what various transformations do and how more complex functions can be built up from a simple function is crucial for becoming a competent mathematician.

Exercise 4C

1. The graph below is a sketch of $y = f(x)$, showing three points A, B and C.

   Sketch a graph of the following functions

   (a) $f(-x)$ (b) $2f(x)$ (c) $f(\frac{x}{3})$

   (d) $-\frac{1}{2}f(x)$ (e) $f(x+3)$ (f) $2f(x+1)$

   (g) $f(x)+5$.

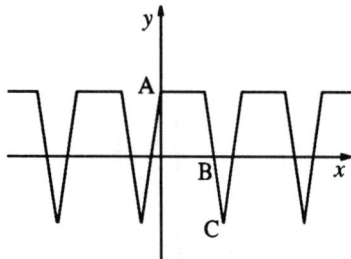

   In each case indicate the position of A, B and C on the transformed graphs.

2. Using the functions $f(x) = x^2$, $g(x) = \frac{1}{x}$

   show how each of the following functions can be expressed in terms of f or g. Hence sketch these graphs.

   (a) $y = 2x^2 + 1$ (b) $y = 4 - x^2$

   (c) $y = \dfrac{1}{(x+4)} + 2 \ (x \neq -4)$ (d) $y = -\dfrac{2}{x} + 1 \ (x \neq 0)$

   (e) $y = x^2 + 2x + 42$.

4.4 Miscellaneous Exercises

1. The function $y = f(x)$ is illustrated below.

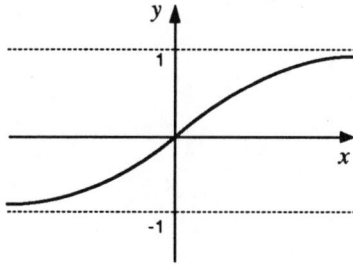

Sketch the following functions:

(a) $y = -f(x)$ (b) $y = f(-x)$

(c) $y = f(x) + 1$ (d) $y = 2f(x)$.

2. Express each of the following functions in terms of either $f(x) = x^2$ or $g(x) = \frac{1}{x}$.

   (a) $y(x) = 4x^2 + 1$

   (b) $y(x) = 1 - \frac{1}{(x+1)}$ $(x \neq -1)$.

3. Sketch the graph of

$$f(x) = \frac{1}{x^3} \quad (x \neq 0).$$

Show that

$$f(-x) = -f(x).$$

What does this tell you about the function?

5 SOLVING PROBLEMS

Objectives

After studying this chapter you should

* have gained experience in formulating several types of problem in mathematical terms;

* appreciate how algebra can be used to solve problems;

* have improved your fluency in handling a variety of algebraic expressions;

* be able to solve several types of equations and inequalities.

5.0   Introduction

A more traditional title for this chapter might have been 'Elementary Algebraic Techniques', however, this would not tell the whole story. It is certainly true that you will have the opportunity to practice and develop your algebraic skills here, but perhaps the more important aim is to emphasise algebra as an effective problem-solving aid.

Algebra cannot solve every problem. Even where it can, there are sometimes alternatives. Nevertheless, algebra is an efficient way of communicating ideas and formulating problems; once a problem has been expressed in algebraic form, finding the answer to the problem is often much simpler.

5.1   Setting up and solving linear problems

Do not be put of by the simplicity of this problem.

> A coal merchant charges £8 per tonne of coal plus a fixed delivery charge of £5. How much coal can be delivered for £50?
>
> What is the answer if the merchant delivers only in 100 kg bags?

The purpose of beginning with such an easy problem is to demonstrate the general method of solving problems algebraically. Algebraic solutions should always fall into three stages:

Stage 1   Translate the problem into a mathematical problem.

Stage 2   Solve the mathematical problem.

Stage 3   Translate the answer back into the terms of the original problem.

The simple problem above could be solved algebraically like this:

Let the number of tonnes delivered be x.

Total cost of x tonnes $= £(5+8x)$.

But the total cost $= £50$,

so $5+8x = 50$.

$\left.\right\}$ **Stage 1**

$5+8x = 50$

$\Rightarrow \qquad 8x = 45$

$\Rightarrow \qquad x = 5.625$

$\left.\right\}$ **Stage 2**

$x = 5.625$,

so the amount of coal that can be delivered is 5.625 tonnes or 5 tonnes 625kg.

$\left.\right\}$ **Stage 3**

(Check the answer : $5+(8\times 5.625)= 5+45 = 50$.)

The last part of the problem could be put slightly differently. If the merchant delivered only in 100 kg bags then an answer of 5.625 tonnes is inadmissible, and clearly the required answer, by common sense, is 5.6 tonnes.

For the mathematics to come up with the proper answer the problem needs to be set up as an inequation, or inequality, rather than as equation.

Let the number of bags delivered be x.

Total cost of x bags $= £(5+0.8x)$

Total cost must not exceed £50, and x must be a natural number (i.e. a positive whole number):

$$5 + 0.8x \leq 50, \ x \in \mathbb{N} \quad \text{(the set of natural numbers)}$$

$$\Rightarrow \quad 0.8x \leq 45, \ x \in \mathbb{N}$$

$$\Rightarrow \quad x \leq 56.25, \ x \in \mathbb{N}$$

The last line suggests directly that the largest possible value of x must be 56.

Since x represents the number of 100 kg bags the answer to the problem must now now be **5.6 tonnes**.

Note also the importance of choosing the right domain for x. The inequality combined with the condition '$x \in \mathbb{N}$' yields the correct solution.

Below are three further examples. Before reading the solutions you should try to solve the problems by yourself, or in a group; then compare your solutions to the versions given in the text.

Make sure you set out each solution clearly. Try to follow the format of the worked examples given above.

Example

A coach travels along the M3 from Winchester to London. It sets off at 10.30 am and travels at a constant 65 mph. A car makes the same journey, travelling 10 mph faster but leaving 5 minutes later. When does the car overtake the coach?

Solution

Let t represent the number of minutes since the coach left Winchester, and let T be the number of minutes since the car left Winchester.

$$\text{Distance travelled by coach} = \frac{65t}{60}$$

$$\text{Distance travelled by car} = \frac{75T}{60}$$

The car left 5 minutes later than the coach, so when $T = 0, t = 5$.

Hence $T = t - 5$.

The car overtakes the coach when they have travelled the same distance.

$$\frac{65}{60}t = \frac{75}{60}T$$

$$\Rightarrow \quad 65t = 75(t-5)$$

$$\Rightarrow \quad 13t = 15(t-5)$$

$$\Rightarrow \quad 13t = 15t - 75$$

$$\Rightarrow \quad 2t = 75$$

$$\Rightarrow \quad t = 37.5$$

So the car overtakes the coach $37\frac{1}{2}$ minutes after the coach's departure, i.e. at $11.07\frac{1}{2}$ am.

Example

Tickets for a pantomime cost £8 for adults and £5 for children. A party of 25 pays £143 in total. How many adults were there in the party?

Solutions

Let the number of adults be x. Then the number of children must be $25 - x$.

Cost of x adult tickets and $(25-x)$ children tickets

$$= £\left[8x + 5(25-x)\right]$$

But the total cost was £143, so

$$8x + 5(25-x) = 143$$

$$\Rightarrow \quad 3x + 125 = 143$$

$$\Rightarrow \quad 3x = 18$$

$$\Rightarrow \quad x = 6$$

Hence there were **6 adults** in the party.

(Check: 6 adult tickets and 19 child tickets cost $6 \times 8 + 19 \times 5 = 143$.)

Example

Alan has a non-interest current account in his bank from which mortgage repayments of £250 are made monthly. At the beginning of the year the account contains £3000. Barbara has a similar account: her monthly payments are £370 and her account contains £4150 at the start of the year. After how many payments does Alan's account contain more money than Barbara's?

Solution

Let the number of payments be n. After n payments:

Alan's account contains $£(3000 - 250n)$

Barbara's account contains $£(4150 - 370n)$

Alan's account has more money than Barbara's.

$$\Rightarrow \quad 3000 - 250n > 4150 - 370n, n \in \mathbb{N}$$
$$\Rightarrow \quad -250n > 1150 - 370n, n \in \mathbb{N}$$
$$\Rightarrow \quad 120n > 1150, n \in \mathbb{N}$$
$$\Rightarrow \quad n > 9.583, n \in \mathbb{N}$$

The last line suggests that the lowest possible value of n is 10.
Hence Alan's account first has more money than Barbara's after
10 payments.

Activity 1   Using simultaneous equations

The solution to the second example might have begun like this:
Let the party consist of x adults and y children.

$$\text{Total cost} = 8x + 5y = 143$$

$$\text{Total number of people} = x + y = 25$$

Solve these simultaneous equations and check that the answer is
the same as before.

(If you need to, revise how to solve simultaneous equations.)

Similarly the first example might have been solved as the answer
to these simultaneous equations.

$$\left. \begin{array}{l} 13t = 15T \\ t - 5 = T \end{array} \right\}$$

Check that the answers obtained satisfy these questions.

All the problems in this section are called **linear** as they contain
terms in x and y but not in x^2, y^2, xy, etc.

Activity 2   Solving inequalities

This solution of the inequality $3000 - 250n > 4150 - 370n$, given below (which occurs in the solution to the third example) is wrong. Find the mistake.

$$3000 - 250n > 4150 - 370n, \quad n \in \mathbb{N}$$
$$\Rightarrow \quad -1150 - 250n > -370n, \qquad n \in \mathbb{N}$$
$$\Rightarrow \qquad -1150 > -120n, \qquad n \in \mathbb{N}$$
$$\Rightarrow \qquad 9.583... > n, \qquad n \in \mathbb{N}$$
$$\Rightarrow \qquad n < 9...$$

Exercise 5A below contains some more examples for you to try. The exercise begins with a brief opportunity to revise the solution of linear equations and inequalities.

Exercise 5A

1. Solve these equations. If the answers are not exact, give them to 3 s.f.

   (a) $1400 - 3a = 638$

   (b) $\dfrac{4(b+45)}{5} = 24$

   (c) $16c + 24 = 21c - 481$

   (d) $6(35 - d) = d + 19$

   (e) $\dfrac{13.2e - 10.7}{4} = 6.1$

   (f) $\frac{1}{2}(3f + 11) = 6f - 47$

2. Solve the following pairs of simultaneous equations

   (a) $\begin{cases} 5p + 3q = 612 \\ p + q = 150 \end{cases}$

   (b) $\begin{cases} 13m - 5n = 434 \\ m - n = 50 \end{cases}$

   (c) $\begin{cases} 9k + 6l = 306 \\ 5k - 2l = 154 \end{cases}$

   (d) $\begin{cases} y = 2x - 4 \\ y = 5x + 11 \end{cases}$

   (e) $\begin{cases} p = 2q - 30 \\ q = 6p + 16 \end{cases}$

   (f) $\begin{cases} u + 3v + 10 = 0 \\ 3u = 2v + 1 \end{cases}$

3. Solve these inequalities

   (a) $15a + 365 \le 560$

   (b) $61000 - 103b < 62648$

   (c) $\frac{2}{3}(11c - 70) > 8, c \in \mathbb{N}$

   (d) $5(d + 6.3) > 2d + 3.2$

   (e) $-\left(\dfrac{187 + 65e}{3} \right) \le 12e - 500$

   (f) $16.3(f - 5) \ge 4(7.9 - 3.2f), f \in \mathbb{N}$

4. A school is looking to furnish two new classrooms. Tables cost £11 each and chairs £6. Each table will have two chairs except the two teachers' tables which will have one chair each. The total budget is £770.

   Assuming the budget is spent exactly, how many chairs and tables will the school buy?

5. An amateur operatic society needs to buy 20 vocal scores for their latest show. Hardback copies cost £14.50, paperback copies £9.20. £250 has been put aside to buy these scores. What is the maximum number of hardback copies they can buy and satisfy their needs?

6. At 6.00 pm a house's hot water tank contains only 10 litres of water, while the cold tank contains 100 litres. The cold water tank is emptying at a rate of 12 litres/min, half of this amount flowing into the hot tank. Water from the mains supply flows into the cold water tank at 3 litres/min. After how many minutes will the two tanks contain equal amounts of water?

7. Andrea is trying to save money on electricity bills. At present she estimates her family uses 8000 unit of electricity per year at 6.59 pence per unit. They also have to pay an annual standing charge of £36.28

   She is told that installing an 'Economy 7' system might save money. This means that off-peak units are charged at 2.63 pence per unit and the rest at 6.59p, but there is an additional annual standing charge of £10.04. She wants to know how many of the 8000 units would have to be off-peak units in order to save at least £50 per year. Solve her problem.

5.2 Revision

Section 5.1 concentrated more on the process of formulating problems in algebraic terms than on the algebra itself. Hence the algebra was uncomplicated. However, this is often not true, so this section is intended as an opportunity to revise some algebraic techniques you may have studied before. The worked examples should serve as a reminder of them before you tackle Exercise 5B.

Example

Multiply out (i.e. write without brackets):

   (a) $p(p-7q)$ (b) $2mn(m^2-3n^2)$
   (c) $(x+3)(x-5)$ (d) $(k^2-2l)(k+l)$.

Solution

(a) $p(p-7q)=p^2-7pq$

(b) $2mn(m^2-3n^2)=2m^3n-6mn^3$

(c) $(x+3)(x-5) = x(x-5)+3(x-5)$

$$= x^2-5x+3x-15$$

$$= x^2-2x-15$$

(d) $(k^2-2l)(k+l) = k^2(k+l)-2l(k+l)$

$$= k^3+k^2l-2kl-2l^2.$$

Example

Factorise (i.e. simplify using brackets) :

   (a) $2u^2+6u$ (b) ab^2-3a^3b.

Solution

(a) $2u^2+6u=2u(u+3)$

(b) $ab^2-3a^3b=ab(b-3a^2)$.

Example

Simplify the expression

$$(p+3)(p-1)-(p+2)(p-3).$$

Solution

$$(p+3)(p-1)-(p+2)(p-3)$$
$$= (p^2+2p-3)-(p^2-p-6)$$
$$= p^2+2p-3-p^2+p+6$$
$$= 3p+3$$
$$= 3(p+1).$$

Example

Factorise these :

(a) x^2+5x+6 (b) $x^2-9x+20$

(c) $x^2-3x-40$ (d) $x^2+2x-99$.

Solution

All these expressions are quadratic and will factorise to the form $(x+m)(x+n)$. Now

$$(x+m)(x+n) = x^2+(m+n)x+mn,$$

so to find the appropriate values of m and n compare $x^2+(m+n)x+mn$ with the quadratic required.

(a) x^2+5x+6:

$m+n=5$ and $mn=6$, so m and n must be 2 and 3, and
$$x^2+5x+6 = (x+2)(x+3)$$

(b) $x^2-9x+20$:

Two numbers are required whose **product** is +20 and whose **sum** is −9. From the list on the right the required numbers are clearly −4 and −5, and
$$x^2-9x+20 = (x-4)(x-5)$$

$$20 = 1\times20 = (-1)\times(-20)$$
$$= 2\times10 = (-2)\times(-10)$$
$$= 4\times5 = (-4)\times(-5)$$

(c) $x^2-3x-40$

Similarly from the list of factors of 40 the factorisation must be $(x-8)(x+5)$

(d) $x^2+2x-99 = (x+11)(x-9)$.

Example

Factorise

(a) x^2-9 (b) $4a^2-9b^2$

(c) $2x^2-9x+4$ (d) $3x^2-2x-16$

(e) $3x^2+6x-24$.

Solution

(a) and (b) are the difference of two squares. You may recall that

$$p^2 - q^2 = (p+q)(p-q)$$

and so, by comparison,

(a) $x^2 - 9 = (x+3)(x-3)$

(b) $4a^2 - 9b^2 = (2a)^2 - (3b)^2$

$$= (2a+3b)(2a-3b)$$

(c) and (d) are more difficult. Some methods are given below, but the technique is still best learned through experience.

(c) $2x^2 - 9x + 4$ must factorise to the form $(2x+m)(x+n)$. Now $(2x+m)(x+n) = 2x^2 + (m+2n)x + mn$. So the product of m and n must be 4, but notice that the n is doubled in the $(m+2n)x$ term. Looking at the list opposite the factorisation will be $(2x-1)(x-4)$.

$$4 = 1 \times 4 = (-1) \times (-4)$$
$$= 2 \times 2 = (-2) \times (-2)$$

(d) $3x^2 - 2x - 16$.

Similar techniques as before, but note that in multiplying out $(3x+m)(x+n)$ the 'n' is multiplied by '$3x$' to help give a term $(m+3n)x$.

$$3x^2 - 2x - 16 = (3x-8)(x+2)$$

(e) looks similar to (d), but notice that all the terms have a factor of 3, so

$$3x^2 + 6x - 24 = 3(x^2 + 2x - 8)$$
$$= 3(x+4)(x-2).$$

Exercise 5B

1. Multiply out :

   (a) $2(5x+7y)$ (b) $a(a-b)$

   (c) $3lm(6l-5m)$ (d) $p^2(2p-3q^2+1)$

   (e) $h^3k^3(2hk+3hk^2)$.

2. Multiply out :

   (a) $(x+1)(x+2)$ (b) $(x+3)(x-5)$

   (c) $(x-8)(x-2)$ (d) $(x+6)(x-5)$

   (e) $(x-10)(x+7)$.

3. Multiply out :

   (a) $(2x-7)(x-3)$ (b) $(2x+3)(x+1)$

   (c) $(3x+20)(x-8)$ (d) $(a-2b)(2a+b)$

   (e) $(2m^2-n)(m+5n^2)$.

4. Multiply out :

   (a) $(x+1)^2$ (this is not x^2+1)

   (b) $(p-3)^2$ (c) $(x-10)^2$

   (d) $(x+a)^2$ (a is any number)

   (e) $(2x-5)^2$.

5. Simplify these :

    (a) $2a(a-b)+b(2a-b)$

    (b) $(y-2)(y+3)+(2y+5)(y+1)$

    (c) $(n+4)^2-(n-1)(n+7)$

    (d) $5(m-6)(m+2)-2(2m+3)(m-7)$

    (e) $(u+2v)(u-v)-(4u-v)(u+3v)$.

6. Factorise these expressions :

    (a) $6x+15$ (b) u^2-3u

    (c) $3p^2+24p$ (d) $12a^2b^3-6ab^2$

    (e) $5x^2y+15y-35$.

7. Factorise these :

    (a) x^2+6x+8 (b) x^2+x-30

    (c) $x^2-7x+10$ (d) x^2+5x+4

    (e) $x^2-3x-70$ (f) $x^2-10x+9$

    (g) $x^2+6x-16$ (h) $x^2-5x-84$.

8. Factorise these :

    (a) x^2-16 (b) x^2-25x

    (c) $2x^2+7x+3$ (d) $2x^2-7x-4$

    (e) $3x^2-11x-20$ (f) $2x^2-4x-6$

    (g) $3x^2-4x-20$ (h) $100x^2-64$.

9. Factorise these :

    (a) $5x^2+3x-2$ (b) $4x^2+5x-6$

    (c) $4x^2-4x-3$ (d) $6x^2+15x-36$

    (e) $6x^2+5x-25$ (f) $12x^2-7x-10$.

5.3 Setting up and solving quadratic equations

Activity 3   A problem of surface area

The diagram shows an open-topped box with a square base. The sides of the box are 3 cm high.

The box is to be made from a total of 160 cm² of card. What size must the square be? Try to set out an algebraic solutions as in the previous section.

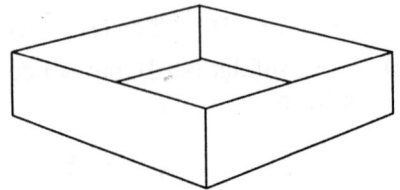

Solving the above problem algebraically leads to a quadratic equation, that is, an equation involving a quadratic function. You should have already covered how to solve this type of equation. The following worked example and activities should help to refresh your memory.

Example

Solve the equation $x(x+6)=16$.

Solution

$$x(x+6)=16$$
$$\Rightarrow \quad x^2+6x=16$$
$$\Rightarrow \quad x^2+6x-16=0.$$

$x^2 + 6x - 16$ can be factorised to give $(x+8)(x-2)$. Hence the equation becomes

$$(x+8)(x-2) = 0$$
$$\Rightarrow \quad x = -8 \text{ or } 2.$$

Example

Solve the equation $3x^2 + 10x = 25$.

Solution

The procedure is exactly the same, though the factorisation is more difficult.

$$3x^2 + 10x = 25$$
$$\Rightarrow \quad 3x^2 + 10x - 25 = 0$$
$$\Rightarrow \quad (3x - 5)(x + 5) = 0$$
$$\Rightarrow \quad x = \tfrac{5}{3} \text{ or } -5.$$

Activity 4   Solutions and graphs

(a) Sketch the graph of $y = (x+9)(x-3)$
    At what points does the curve cross the horizontal axis?

(b) At what point will the graph of $y = (x+m)(x+n)$ cross the horizontal axis?

(c) Sketch the graph of $y = (2x-1)(x+2)$. Explain why the curve crosses the x-axis where it does.

(d) Sketch the graphs of these functions all on one diagram, without using any electronic aids.

$$y = (x-6)(x+2)$$
$$y = 2(x-6)(x+2)$$
$$y = \tfrac{1}{3}(x-6)(x+2)$$
$$y = -(x-6)(x+2)$$

(e) Give a possible equation for the quadratic graph on the right. How many possibilities are there? What else would you need to know to narrow these possibilities down to one?

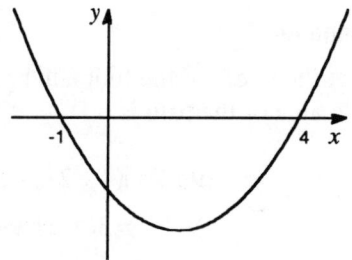

Activity 5   Match the graphs to the functions

Below are eight quadratic functions, numbered 1 to 8, and five
graphs, lettered A to E. Each graph corresponds to one of the
functions. Decide which function goes with which graph. Draw
sketches of the graphs of the functions that are not used.

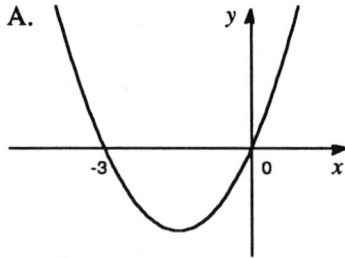

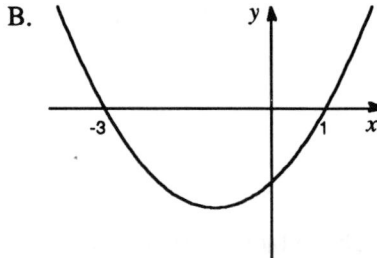

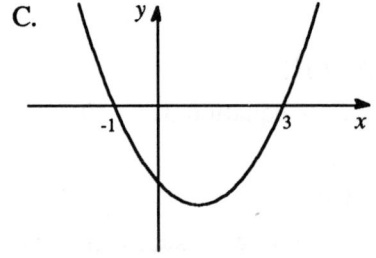

A.

B.

C.

1. $x^2 + 3x$

2. $(x-1)(x+3)$

3. $(x+17)(x-15)$

D.

E.

4. $(x+2)(x+3)$

5. $(x-17)(x+15)$

6. $x^2 - 5x + 6$

7. $(x+1)(x-3)$

8. $x(x-3)$.

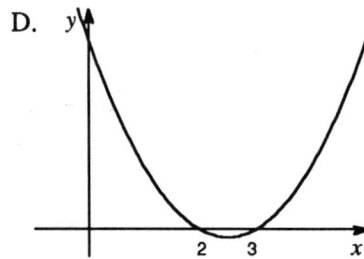

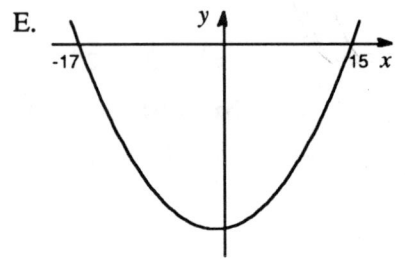

Quadratic equations often arise when solving problems connected
with area. (Can you think why?) When a quadratic equation does
present itself, it is important to bear in mind the domain of the
variable concerned: in general, quadratic equations have two
solutions, but it is not necessarily true that both of them are
solutions to the original problem, as you will see in the next
example.

Example

Joe wishes to make a gravel path around his rectangular pond. The
path must be the same width all the way round, as shown in the
diagram. The pond measures 4 m by 9 m and he has enough gravel
to cover an area of 48 m². How wide should the path be?

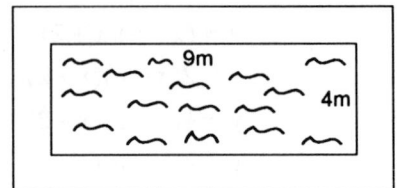

Solution

Let the width of the footpath be x metres. The diagram shows that
the area of the path is

$$(9+2x)(4+2x) - 36$$
$$= 36 + 26x + 4x^2 - 36$$
$$= 4x^2 + 26x.$$

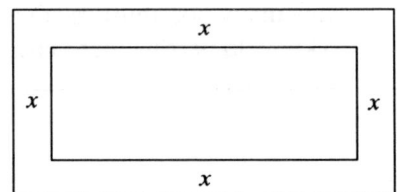

Since the required area is 48 m², then $4x^2 + 26x = 48$.

To this equation must be added the condition that $x > 0$.

$$4x^2 + 26x = 48, \qquad x > 0$$
$$\Rightarrow \quad 2x^2 + 13x - 24 = 0, \qquad x > 0$$
$$\Rightarrow \quad (2x - 3)(x + 8) = 0, \qquad x > 0$$
$$\Rightarrow \quad x = \tfrac{3}{2} \text{ or } -8, \qquad x > 0$$
$$\Rightarrow \quad x = \tfrac{3}{2}.$$

Hence the width of the path must be 1.5m, since the other value, $x = -8\text{m}$, is unrealistic, and does not satisfy the condition $x > 0$.

Exercise 5C

1. Solve these quadratic equations by factorising

   (a) $x(x+3) = 4$ (b) $x(x+5) = 50$

   (c) $x^2 = x + 72$ (d) $x^2 + 77 = 18x$

   (e) $x^2 + 50x + 96 = 0$ (f) $x^2 - 50x + 600 = 0$

   (g) $2x^2 - 13x + 15 = 0$ (h) $x(2x-1) = 21$

   (i) $3x^2 - 6x = 45$ (j) $3x^2 - 14 = x$.

2. A triangular flowerbed is to be dug in the corner of a rectangular garden. The area is to be 65 m². The owner wants the length BC of the bed to be 3 m longer than the width AB. Find what the length and width should be.

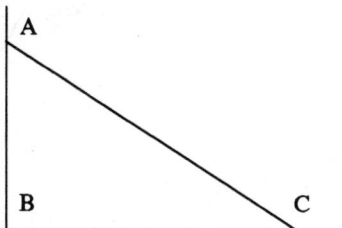

3. A window is designed to be twice as wide as it is high. It is to be made up of three parts: a central, fixed part and two identical rectangles on either side of width 0.5 metres. Find the height of the window, if the central part is to have an area of 3 m².

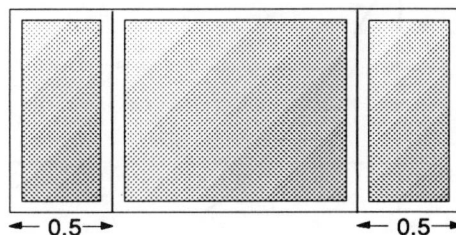

5.4   Approximate solutions

In practical problems which lead to solving quadratic equations, it is not always possible to factorise the equation.

Activity 6   A fencing problem

50 m of fencing is being used to enclose a rectangular area, with a long straight wall as one of the sides of the rectangle. What dimensions give an area of 150 m²?

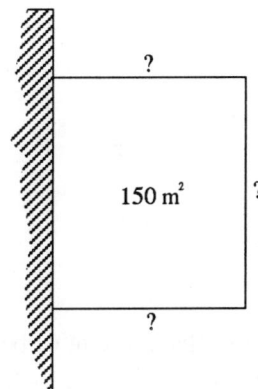

As Activity 6 will have illustrated, not all quadratic equations can be solved by factorisation. Another example is the equation

$$x^2 - 15x + 40 = 0$$

which cannot be factorised.

You may know how to deal with such equations already - an algebraic method will be discussed in section 5.5 - but you can use a combination of graphical or trial and improvement methods to find approximate solutions.

The **graphical method** can be undertaken with the help of a graphic calculator or with pencil and graph paper. Either way, draw the graph of

$$y = x^2 - 15x + 40$$

and find where it crosses the horizontal axis.

If you are doing this by hand you will need to be accurate; if you are using a computer or calculator, greater accuracy can be obtained by zooming in on each solution in turn.

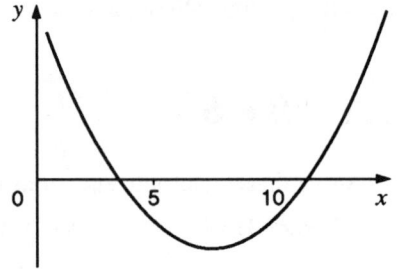

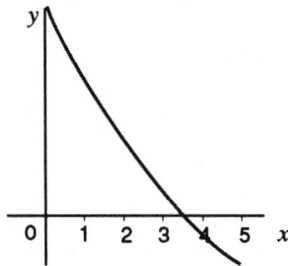

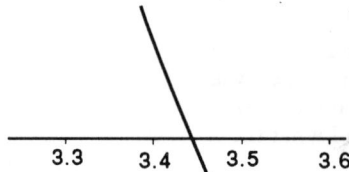

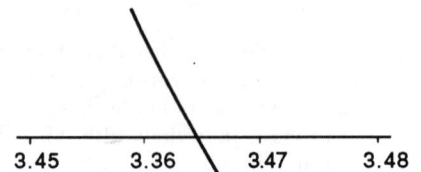

You may be familiar with the **trial and improvement** technique. It is best explained by example. The results of successive trials can be set out in table.

x	$x^2 - 15x + 40$
0	40
1	26
2	14
3	4
4	−4

This shows that there must be a solution between $x = 3$ and $x = 4$, probably near 3.5. Now look more closely between $x = 3$ and $x = 4$.

$$\left.\begin{array}{c|c} 3.4 & 0.56 \\ 3.5 & -0.25 \end{array}\right\} \text{ root between } x = 3.4 \text{ and } 3.5$$

$$\left.\begin{array}{c|c} 3.46 & 0.0716 \\ 3.47 & -0.0091 \end{array}\right\} \begin{array}{l} \text{root between } x = 3.46 \text{ and } 3.47 \\ \text{being closer to } 3.47 \end{array}$$

The solution is 3.47 to 3 s.f. Greater accuracy can be achieved by continuing the process as long as necessary. This, of course, is only one solution. The process must be repeated to find the other. (Try it yourself.)

Activity 7

The quadratic $x^2 + 10x - 20$ cannot be factorised using whole numbers but can be factorised approximately using decimals. Find its factors correct to 2 d.p.'s.

Does a quadratic equation always have precisely two solutions?

Activity 8 Computer program

Devise a computer program to solve a quadratic equation by trial and improvement.

Exercise 5D

1. Use a graphical approach to solve these equations to 3s.f.

   (a) $x(2x-3) = 9$ (b) $x^2 + x = 14$

   (c) $2x(20-x) = 195$ (d) $300 - x^2 = 6x$.

2. Solve these equations to 3s.f. using trial and improvement.

   (a) $x(x+3) = 8$ (b) $7x - x^2 = 5$

   (c) $x^2 + 8x + 11 = 0$ (d) $x(x+25) = 8000$.

3. The distance across the diagonal of a square field is 50 m shorter than going round the perimeter. To the nearest metre, find the length of the side of the field.

4. An architect decides that the smallest room in the house needs a window with area 4000 cm².

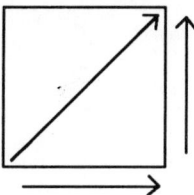

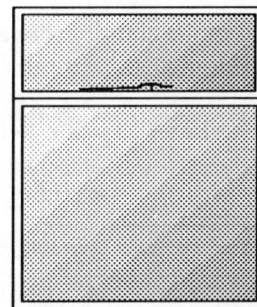

   The window will be in two parts: a square part at the bottom, and a rectangular part above it hinged at the top and with height 25 cm. Find the width the window must be to satisfy these requirements. Answer to the degree of accuracy you think is appropriate to the problem.

5.5 Investigating quadratic functions

Graphical and numerical methods are of great importance in solving equations. Their big advantages are that they can be used to solve practically any equation, and solve them to any degree of accuracy. However it can be a slow process, and where quadratic equations cannot be solved by factoring there is another method of solution that is often faster.

Consider the equation $x^2 - 4x - 3 = 0$.

Activity 9   Completing the square

(a) Another way of expressing the function $x^2 - 4x - 3$ is to write it in the form

$$(x - p)^2 - 9.$$

Multiply out this expression and equate it to $x^2 - 4x - 3$ to find the values of p and q.

(b) Having put the function in this new form, now solve the equation $x^2 - 4x - 3 = 0$.

Remember that you should obtain two solutions.

(c) Solve the equations

       (i) $x^2 - 2x - 1 = 0$, and

      (ii) $x^2 - 3x - 5 = 0$

by a similar method.

The method used in the activity above is called **completing the square**. Any quadratic can be written in this form. For example,

$$2x^2 - 12x + 15 = 2\left(x^2 - 6x + \tfrac{15}{2}\right)$$
$$= 2\left((x - 3)^2 - 9 + \tfrac{15}{2}\right)$$
$$= 2\left((x - 3)^2 - \tfrac{3}{2}\right).$$

> $(x-3)^2$: the 3 comes from half 'x'coefficient
>
> -9 : introduced to cancel out 3^2

This quadratic $2x^2 - 12x + 15 = 0$ can now be solved since

$$2\left((x - 3)^2 - \tfrac{3}{2}\right) = 0 \;\Rightarrow\; (x - 3)^2 - \tfrac{3}{2} = 0$$
$$(x - 3)^2 = \tfrac{3}{2}$$
$$x - 3 = \pm\sqrt{\tfrac{3}{2}}$$
$$x = 3 \pm \sqrt{\tfrac{3}{2}}.$$

Example

By completing the square, solve

   (a) $x^2 + 6x + 4 = 0$

   (b) $x^2 - 5x + 1 = 0$

   (c) $5x^2 - 6x - 9 = 0$.

Solutions

(a) $x^2 + 6x + 4 = (x+3)^2 - 9 + 4$

$$= (x+3)^2 - 5$$

Hence the equation $x^2 + 6x + 4 = 0$ can be rewritten

$$(x+3)^2 - 5 = 0$$

$\Rightarrow \qquad (x+3)^2 = 5$

$\Rightarrow \qquad x+3 = \pm\sqrt{5}$

$\Rightarrow \qquad x = -3 \pm \sqrt{5}$

$$= -5.24 \text{ or } -0.764 \text{ to } 3\,\text{s.f.}$$

(b) $x^2 - 3x + 1 = (x-1.5)^2 - (1.5)^2 + 1$

$$= (x-1.5)^2 - 1.25.$$

Hence $x^2 - 3x + 1 = 0$ can be re-written

$$(x-1.5)^2 - 1.25 = 0$$

$\Rightarrow \qquad (x-1.5)^2 = 1.25$

$\Rightarrow \qquad x - 1.5 = \pm\sqrt{1.25}$

$\Rightarrow \qquad x = 1.5 \pm \sqrt{1.25}$

$$= 0.382 \text{ or } 2.62 \text{ to } 3\,\text{s.f.}$$

(c) $5x^2 - 6x - 9 = 5[x^2 - 1.2x - 1.8]$

$$= 5[(x-0.6)^2 - (0.6)^2 - 1.8]$$

$$= 5[(x-0.6)^2 - 2.16].$$

Hence $5x^2 - 6x - 9 = 0$ can be re-written

$$(x-0.6)^2 - 2.16 = 0$$

$\Rightarrow \qquad x = 0.6 \pm \sqrt{2.16}$

$$= -0.870 \text{ or } 2.07 \text{ to } 3\,\text{s.f.}$$

The technique of completing the square has wider applications than just solving equations. It is, for example, useful in helping to sketch curves of quadratic functions.

Look, for example, at the function $5x^2 - 6x - 9$ (see (c) above). Completing the square, as suggested by Activity 9, can be seen as the splitting up of a quadratic function into several simpler ones. In this case, $f(x) = 5x^2 - 6x - 9$ is equivalent to the transformation

$$a : x \mapsto x - 0.6$$

followed by $b : x \mapsto x^2$

followed by $c : x \mapsto x - 2.16$

followed by $d : x \mapsto 5x.$

These four functions gradually build up the expression required:

$$a(x) = x - 0.6$$
$$ba(x) = (x - 0.6)^2$$
$$cba(x) = (x - 0.6)^2 - 2.16$$
$$dcba(x) = 5[(x - 0.6)^2 - 2.16]$$
$$= 5x^2 - 6x - 9.$$

Activity 10

In earlier chapters you found out how simple graphs were affected by simple transformations. Explain why the graph of
$y = 5x^2 - 6x - 9$ looks like the sketch opposite.

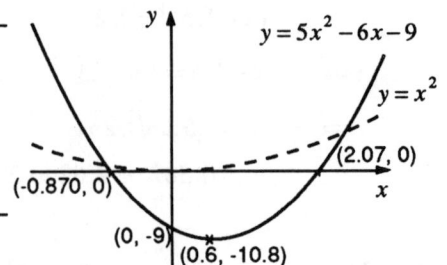

Activity 11 Transformations of $y = x^2$

(a) Using the graph of $y = x^2$, sketch the graphs of these functions, to the same scale.

   (i) $y = (x - 3)^2 + 2$

   (ii) $y = (x + 5)^2 - 8$

   (iii) $y = 3[(x - 1)^2 + 1]$

   (iv) $y = x^2 - 8x + 10$

   (v) $y = 2x^2 + 5x - 3.$

Do not use any graph-plotting package except to check your answers. Label the crucial points in the same way as the sketch above.

(b) Explain how the minimum point of the quadratic curve $y = x^2 - 10x + 20$ can be found without having to draw the curve itself.

Activity 12

On the right is a sketch of the graph of

$$y = (x - p)^2 - q.$$

Copy it and mark on

(a) the co-ordinates of the points marked with crosses

(b) the line of symmetry of the graph, with its equation

(c) the distances marked with arrows, $\longleftrightarrow$.

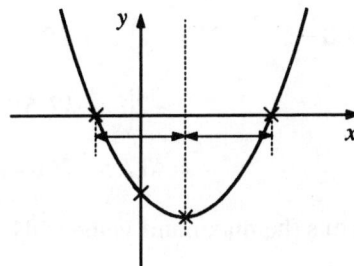

Activity 13   Maximum values

(a) Plot the graphs of the functions

$$y = 4 + 2x - x^2$$
$$y = 10 - 5x - 2x^2$$

and give an explanation for their general shape.

(b) Show that

$$4 + 2x - x^2 = 5 - (x - 1)^2$$

and rewrite $10 - 5x - 2x^2$ in a similar way.

(c) What are the maximum values of these two functions? How can you tell without drawing a graph?

As you can see another application of completing the square is that of finding the maximum or minimum values of quadratic functions.

Example

Find the minimum value of the function $x^2 + 3x + 7$.

Solution

$$x^2 + 3x + 7 = (x+1.5)^2 + 4.75$$

So the minimum value must be 4.75, which occurs when $x = -1.5$

Example

Find the value of x that maximises the function $100 + 50x - 2x^2$ and find this maximum value.

Solution

$$100 + 50x - 2^x = -2\left[x^2 - 25x - 50\right]$$

$$= -2\left[(x - 12.5)^2 - 206.25\right]$$

$$= 412.5 - 2(x - 12.5)^2.$$

Thus the maximum value is 412.5 achieved when $x = 12.5$.

Exercise 5E

1. Solve the quadratic equations to 3 s.f.

    (a) $a^2 - 4a - 7 = 0$ (b) $b^2 - 12b + 30 = 0$

    (c) $c^2 - c - 1 = 0$.   (d) $2d^2 + 20d = 615$

    (e) $2e^2 - 8e + 1 = 0$ (f) $5f^2 + 4f = 13$

    (g) $53 + 30g - g^2 = 0$ (h) $1225 + 46h - 2h^2 = 0$.

2. (a) What value of x maximises the function $x^2 + 40x + 25$?

    (b) What is the minimum value taken by the function

    $$x^2 + 9x + 14?$$

    (c) What is the maximum value of $46 - 24x - x^2$?

    (d) What is the maximum value of $25 - 60x - 2x^2$ and what value of x makes this function a maximum?

3. 100 m of fencing is used to make an enclosure as shown in the diagram, jutting out from the corner of a rectangular building.

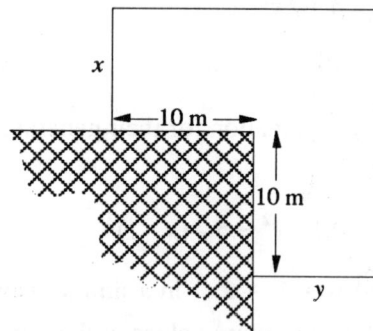

If x and y are as marked, show that $y = 40 - x$ and hence

    (a) find x if the area of the enclosure is to be 600 m²;

    (b) find the maximum area that can be enclosed.

5.6 Quadratic solution formula

Completing the square is a technique that can be used to solve any quadratic equation. In many cases it easy to use. However, in practice quadratic equations can sometimes have awkward coefficients and 'completing the square' leads to some unhappy arithmetic; for example,

$$7.8x^2 - 11.2x - 4.9 = 0$$

$$\Rightarrow \quad 7.8[x^2 - 1.4359x - 0.6282] = 0 \text{ (working to 4 d.p.'s)}$$

$$\Rightarrow \quad (x - 0.7179)^2 - 1.1437 = 0$$

$$\Rightarrow \quad x = 0.7179 \pm \sqrt{1.1437}$$

$$= -0.351 \text{ or } 1.79 \text{ to } 3 \text{ s.f.}$$

For cases like this there is a formula which will enable you to calculate the answer more quickly.

Activity 14   Finding a formula

(a) If $x^2 + bx + c = 0$, show that $\left(x + \dfrac{b}{2}\right)^2 + c - \left(\dfrac{b}{2}\right)^2 = 0$ and

hence that

$$x = -\frac{b}{2} \pm \sqrt{\frac{b^2}{4} - c} \ .$$

Can you see why this formula is the same as

$$x = \frac{-b \pm \sqrt{b^2 - 4c}}{2} \ ?$$

(b) Now extend this process to solve the general quadratic equation

$$ax^2 + bx + c = 0 \ .$$

The two solutions to $ax^2 + bx + c = 0$ can be shown to be given by

$$x = \frac{-b \pm \sqrt{b^2 - 4ac}}{2a}$$

Example

Solve the equation $7.8x^2 - 11.2x - 4.9 = 0$.

Solution

Use the formula with $\begin{cases} a = 7.8 \\ b = -11.2 \\ c = -4.9 \end{cases}$

Solutions are $x = \dfrac{11.2 \pm \sqrt{(-11.2)^2 - 4 \times (7.8) \times (-4.9)}}{2 \times 7.8}$

$= \dfrac{11.2 \pm \sqrt{278.32}}{15.6}$

$= -0.351$ or 1.79 to 3 s.f.

Activity 15

(a) Complete the square for the function $x^2 - 6x + 10$. Sketch its graph, labelling the minimum point clearly.

(b) Try to solve the equation $x^2 - 6x + 10 = 0$ using the formula. What happens? Explain your answer graphically.

(c) Try to solve the equation $x^2 - 12x + 36 = 0$ using the formula.

Use a graph to explain what happens.

(d) What general statements can you make about quadratic equations?

The quantity $(b^2 - 4ac)$ is clearly of importance in the behaviour of quadratics. It is referred to as the **discriminant** of the function.

What happens if $b^2 - 4ac = 0$ or $b^2 - 4ac < 0$?

Exercise 5F

1. Use the formula $\dfrac{-b \pm \sqrt{b^2 - 4ac}}{2a}$ to solve these equations to 3 s.f.

   (a) $0.7a^2 - 0.2a - 0.8 = 0$

   (b) $17b^2 + 12b - 1 = 0$

   (c) $2.05c^2 = 4.79c + 12.26$

   (d) $1500d^2 - 17\,000d = 100\,000$

   (e) $\dfrac{1}{2}e^2 - \dfrac{1}{4}e + \dfrac{1}{45} = 0$

   (f) $2\pi f^2 = 150f - 400$.

2. Find the discriminant of each of the following equations. How many solutions does each have? (You are not expected to find the solutions.)

   (a) $x^2 + x + 10 = 0$

   (b) $x^2 + 10x - 1 = 0$

   (c) $3x^2 - x - 15 = 0$

   (d) $4x^2 - 36x + 81 = 0$

   (e) $1.72x^2 + 5.71x + 4.68 = 0$

   (f) $9x^2 + 60x + 100 = 0$.

3. A stone is hurled vertically upwards at a speed of 21 metres per second. Its height above the ground in metres is given by the formula
   $$2 + 21t - 5t^2.$$

   (a) After how many seconds does the stone hit the ground?

   (b) What is the maximum height reached by the stone?

4. The surface area of a circular cylinder of radius r and height h is $2\pi r(r + h)$ What radius, to 3 s.f., will give a cylinder of height 15 cm a surface area of 600 cm^2.

5.7 Equations with fractions

You may at this moment be using A4 paper. Paper sizes in the 'A' series have a special property.

If you fold a piece of A4 paper in half you get two pieces of size A5 (see diagram).

If you put two A4 sheet together side by side, then the resulting size is A3.

Two A3 sheets side by side make an A2 sheet and so on.

All the rectangles in the 'A' series are similar, that is, the ratio

$$\frac{\text{shortest side}}{\text{longest side}}$$

is the same for each size. But what is this ratio?

The problem can be solved algebraically. Let r denote the required ratio, and let l stand for the length of a sheet of A4 paper. The width is therefore rl.

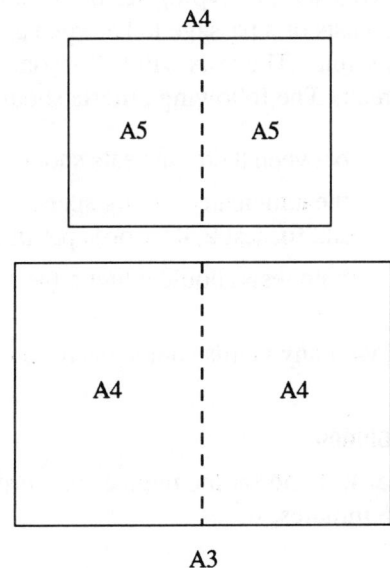

Cutting the sheet in half makes an A5 sheet, with the same ratio. For the A5 sheet:

$$\text{shortest side} = \tfrac{1}{2}l \qquad \text{longest side } = rl.$$

Hence $\qquad \dfrac{\frac{1}{2}l}{rl} = r, \ 0 < r < 1.$

This is an equation involving an algebraic fraction. The best way to solve it is to multiply both sides by rl, the denominator of the fraction, so that the fraction disappears. This leaves

$$\tfrac{1}{2}l = r^2 l, \ 0 < r < 1$$

$$\Rightarrow \quad \tfrac{1}{2} = r^2, \ 0 < r < 1$$

$$\Rightarrow \quad r = +\sqrt{\tfrac{1}{2}} = 0.707 \text{ to 3 s.f.}$$

The answer must now be validated.

Measure a piece of A4 paper to check this answer.

Algebraic fractions often turn up when solving problems connected with ratio or rates. Here is another example.

Example

Two tests for a typing certificate are being designed. Each test consists of a passage to be typed as quickly and accurately as possible. The tests will follow on, one after the other, without a break. The following criteria should also be used:

- between them the tests should last 25 minutes;
- the anticipated typing speed for test 1 is 36 words per minute and for test 2, 45 words per minute;
- both tests should contain the same number of words.

How many words should each test contain?

Solution

Let W stand for the number of words in each test. The total time is 25 minutes, so

$$\frac{W}{45} + \frac{W}{36} = 25,$$

which must be solved to find W.

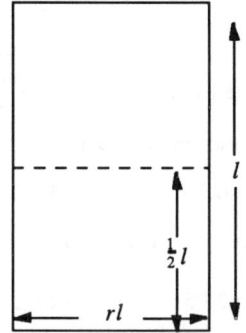

In any equation involving fractions an efficient way to start is to multiply through by the denominators so that no fractions are left. In doing this, take care to multiply each separate term by the same number. So to solve the equation above you first multiply the equation by 45, to give

$$45 \times \frac{W}{45} + 45 \times \frac{W}{36} = 45 \times 25$$

$$\Rightarrow \quad W + \frac{45}{36}W = 1125$$

$$\Rightarrow \quad W + \frac{5}{4}W = 1125.$$

Now multiply both sides by 4:

$$4W + 5W = 4500$$

$$\Rightarrow \quad 9W = 4500$$

$$\Rightarrow \quad W = 500.$$

Example

Solve the equation $\frac{x}{2} = \frac{10}{x+1}$.

Solution

$$\frac{x}{2} = \frac{10}{x+1}$$

$$\Rightarrow \quad x = \frac{20}{x+1} \quad \text{(multiply both sides by 2)}$$

$$\Rightarrow \quad x(x+1) = 20 \quad \text{(multiply both sides by } (x+1))$$

$$\Rightarrow \quad x^2 + x - 20 = 0$$

$$\Rightarrow \quad (x+5)(x-4) = 0$$

$$\Rightarrow \quad x = -5 \text{ or } 4.$$

Example

If $\frac{24}{x} + \frac{30}{x-1} = 10$, find x.

Solution

$$\frac{24}{x} + \frac{30}{x-1} = 10$$

$\Rightarrow \quad 24 + \dfrac{30x}{x-1} = 10x \qquad$ (multiply both sides by x)

$\Rightarrow \quad 24(x-1) + 30x = 10x(x-1) \quad$ (multiply both sides by $(x-1)$))

$\Rightarrow \quad 24x - 24 + 30x = 10x^2 - 10x$

$\Rightarrow \quad 10x^2 - 64x + 24 = 0$

$\Rightarrow \quad 5x^2 - 32x + 12 = 0$

$\Rightarrow \quad (5x-2)(x-6) = 0$

$\Rightarrow \quad x = \frac{2}{5}$ or 6.

Activity 16   Spot the deliberate mistakes

These two solutions contain deliberate errors; spot them, correct them, and find the right answers.

(a) $\dfrac{24}{x+1} = \dfrac{x}{6}$

$\Rightarrow \quad \dfrac{24}{1} = \dfrac{x^2}{6}$

$\Rightarrow \quad 6 \times 24 = x^2$

$\Rightarrow \quad 144 = x^2$

$\Rightarrow \quad x = \pm 12.$

(b) $\dfrac{8}{x} + \dfrac{10}{x-3} = 3$

$\Rightarrow \quad 8 + \dfrac{10x}{x-3} = 3$

$\Rightarrow \quad 8(x-3) + 10x = 3$

$\Rightarrow \quad 18x - 24 = 3$

$\Rightarrow \quad x = 1.5.$

Exercise 5G

1. Solve the equations

    (a) $\dfrac{x}{7}+\dfrac{2x}{3}=17$ (b) $\dfrac{x}{9}-\dfrac{x}{12}=1$

    (c) $\dfrac{x}{8}=\dfrac{18}{x}$ (d) $\dfrac{3}{x+1}=\dfrac{1}{2}$

    (e) $\dfrac{7}{x+1}=\dfrac{8}{x-2}$ (f) $\dfrac{2x-3}{5}=\dfrac{27}{x}$

    (g) $\dfrac{x+1}{3}=\dfrac{18}{x-2}$ (h) $\dfrac{6}{x+2}+\dfrac{4}{x-2}=1$

    (i) $\dfrac{18}{2x-1}-\dfrac{15}{x}+1=0$ (j) $\dfrac{x+1}{4}-\dfrac{20}{x-5}=0$.

2. Foolscap paper obeys the following property.

    If the sheet is folded to make a square and a rectangle, the rectangular part is similar to the foolscap sheet itself, that is, for both rectangles the ratio $\frac{\text{longest side}}{\text{shortest side}}$ is the same. Find this ratio.

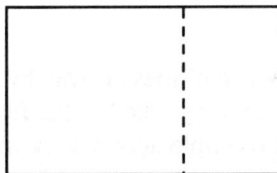

3. 'Petrol is 75p per gallon more expensive now that it was in 1981. You can buy as much petrol now for £132 as you could for £87 in 1981'.
    Assuming these statements are true, find the cost of petrol in 1981.

4. Sandra buys some 'fun sized' chocolate bars as prizes for her son's birthday party. She spends £16.80 on them in total.

    She recalls that she spent exactly the same amount of money on them for last year's party, but this year the same amount bought 20 bars fewer, since the price had gone up by 2p a bar.

    What is the current cost of a 'fun-sized' chocolate bar?

5.8   Inequalities

Here is a variation on a problem posed in Activity 6 of Section 5.4

Activity 17   A quadratic inequality

A farmer uses 50 m of fencing to form a rectangular enclosure with one side against a wall as shown opposite. The area of the enclosure must be at least 250 m².

Show that $x^2-25x+125\le 0$, where x metres is the width of the enclosure.

Find the range of values of x that satisfy this inequality. (You might find it helpful to draw a graph of the function)

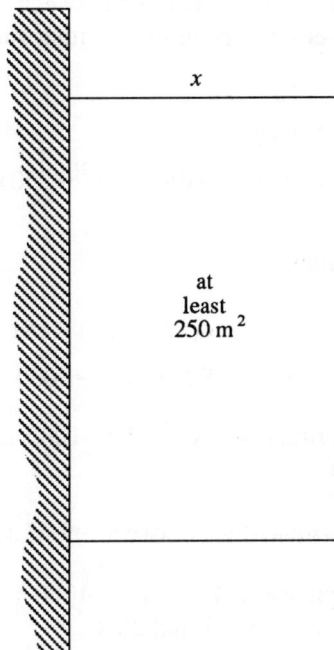

Quadratic inequalities can be solved by a mixture of algebraic and graphical means. For example, to solve $2x^2 + 7x - 5 > 0$, you might first consider the graph of the function $y = 2x^2 + 7x - 5$, shown on the right. The solution to the inequality is clearly the two regions $x < a$ and $x > b$, where a and b are the two solutions to the quadratic equation $2x^2 + 7x - 5 = 0$.

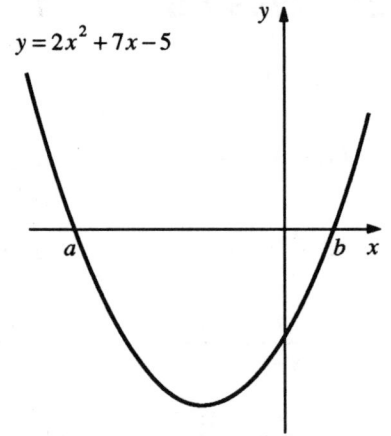

The values of a and b can be found either by completing the square or by using the formula. This gives $a = 4.11$ and $b = 0.608$. Hence the solution to the equality is $x < -4.11$ or $x > 0.608$.

This solution could have been found without drawing a graph at all. Once you know where the quadratic function is zero there are two possibilities.

Either - the function is positive between these two values, negative elsewhere;

Or - the function is negative between these two values, positive elsewhere.

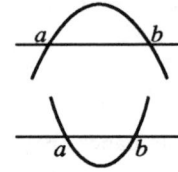

You can quickly discover which of these is true by choosing a number between a and b and seeing whether the function is positive or negative. In the example above, look to see what happens at $x = 0$:

$$x = 0 \Rightarrow y = 2x^2 + 7x - 5 = -5$$

so function is negative between $x = -4.11$ and $x = 0.608$ and the required solution is either side of these values.

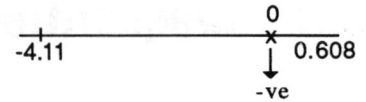

Example

Solve the inequality $4(x^2 - 250) \leq 65x$

Solution

$$4(x^2 - 250) \leq 65x$$
$$\Rightarrow \quad 4x^2 - 65x - 1000 \leq 0.$$

The function $4x^2 - 65x - 1000$ is zero at $x = -9.65$ and 25.9 to 3 s.f.

Choose a value between -9.65 and 25.9, say $x = 0$.

When $x = 0$, $4x^2 - 65x - 1000 = -1000$ so the function is negative between -9.65 and 25.9.

The required solution is between these two values:

$$-9.65 \leq x \leq 25.9.$$

Two alternative methods are illustrated in the next two activities.

Activity 18

(a) Consider the inequality $x^2 > 9$. Why is the answer $x > 3$ incomplete? What is the full solution?

(b) Now consider the inequality

$$x^2 - 6x - 11 \geq 0.$$

Complete the square and use this form of the function to solve the inequality directly.

Activity 19

In the inequality $2x^2 - 7x - 39 < 0$ the quadratic function can be factorised. Rewrite the inequality in the form

$$(\quad)(\quad) < 0.$$

If the product of two brackets is negative, as required by this inequality, what can you say about the two factors?

Hence solve the inequality.

Two further worked examples are given below, one for each of these alternative methods. After that there follows an exercise for you to practice these techniques.

Example

Solve $x^2 + 8x + 9 < 0$.

Solution

$x^2 + 8x + 9$ cannot be factorised, but it is easy to complete the square.

$$x^2 + 8x + 9 < 0$$
$$\Rightarrow \quad (x+4)^2 - 7 < 0$$
$$\Rightarrow \quad (x+4)^2 < 7$$
$$\Rightarrow \quad -\sqrt{7} < x + 4 < +\sqrt{7}$$
$$\Rightarrow \quad -\sqrt{7} - 4 < x < +\sqrt{7} - 4.$$

Example

Solve $x^2 - 3x - 28 \geq 0$.

Solution

$x^2 - 3x - 28$ can be factorised

$$x^2 - 3x - 28 \geq 0$$
$$\Rightarrow \quad (x - 7)(x + 4) \geq 0.$$

The two factors can be either both be positive or both be negative, as the diagram suggests, so the solution must be

$$x \leq -4, x \geq 7.$$

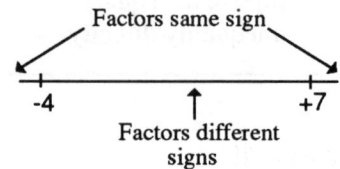

Exercise 5H

1. Solve these by factorising

   (a) $x^2 - 5x - 66 \geq 0$

   (b) $3x^2 + 1 < 4x$

   (c) $(x - 7)(x + 5) \geq 28$

   (d) $x^2 + 110x + 3000 > 0$.

2. Solve these inequalities by completing the square.

   (a) $x^2 + 2x > 4$

   (b) $x^2 - 8x + 10 \leq 0$

   (c) $x^2 + 24x + 100 < 0$

   (d) $x(x - 100) \geq 2000$.

3. Solve these inequalities by whatever method you choose.

   (a) $x^2 - 7x \leq 60$

   (b) $x^2 + 12x + 28 > 0$

   (c) $x(50 - 3x) < 200$

   (d) $6x^2 + 11x \leq 350$

   (e) $\dfrac{14x - 3x^2}{9} \geq 1$

   (f) $0.7x^2 - 3.9x + 2.5 < 0$.

4. In a right-angled triangle the hypotenuse is more than double the shortest side. The third side of the triangle is 2 cm longer than the shortest side. What values can the shortest side take? (Remember the domain when setting up the problem).

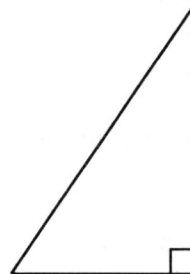

5.9   Modulus sign

The use of the modulus sign will be introduced through this problem.

> Two cars are driving in opposite directions on the motorway. One starts from London at a constant speed of 75 mph; the other starts at the same time from Leeds, 198 miles away, at a constant 60 mph.
>
> Both cars are equipped with twoway radios but need to be within 10 miles of each other to make contact. When can they contact one another?

If t is the number of minutes after the journeys start then

distance of car 1 from London

$$= \frac{75t}{60} \text{ miles}$$
$$= \frac{5t}{4} \text{ miles.}$$

Distance of car 2 from London $= (198 - t)$ miles.

Their distance apart is found by subtracting one distance from the other, i.e. distance apart in miles is

$$= 198 - t - \frac{5t}{4}$$
$$= 198 - \frac{9t}{4}.$$

So, for example, after one hour ($t = 60$) the cars are 63 miles apart. But when $t = 120$, the formula gives -72. This means that the cars are 72 miles apart, but that car 2 is now closer to London than car 1 i.e. the cars have crossed.

There is an agreed convention for getting round this problem, and that is to write

$$\text{distance apart} = \left| 198 - \frac{9t}{4} \right|$$

where the two vertical lines mean ' the **modulus** of ' or 'the **absolute value** of '.

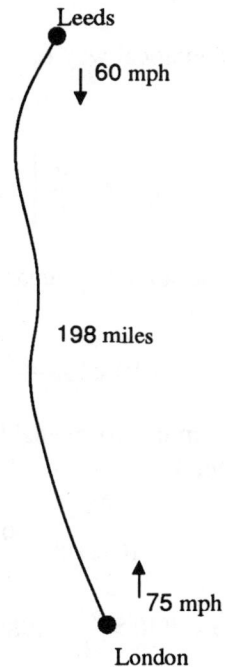

The modulus sign causes any minus sign to be disregarded. So,

$$\text{when } t = 60, \text{ distance apart } = |\ 63\ | = 63 \text{ miles,}$$
$$\text{when } t = 120, \text{ distance apart } = |\ -72\ | = 72 \text{ miles.}$$

The mathematical way of expressing the original problem is to find t when

$$\left| 198 - \frac{9t}{4} \right| < 10.$$

This can be solved by regarding it as short for two separate inequalities.

$$-10 < 198 - \frac{9t}{4} < 10.$$

Writing it in this form enables two inequalities to be solved at the same time;

$$-10 < 198 - \frac{9t}{4} < 10$$

$$\Rightarrow \quad 10 > \frac{9t}{4} - 198 > -10$$

$$\Rightarrow \quad 208 > \frac{9t}{4} > 188$$

$$\Rightarrow \quad 92\frac{4}{9} > t > 83\frac{5}{9}.$$

Hence the two cars are within 10 miles of one another between $83\frac{5}{9}$ and $92\frac{4}{9}$ minutes after the start of their journeys.

Example

Solve

(a) $\left| 40\left(1 - \dfrac{x}{7} \right) \right| < 5$

(b) $|\ 8t - 11| \geq 13.$

Solution

(a) $\left| 40\left(1 - \dfrac{x}{7} \right) \right| < 5$

$$\Rightarrow \quad -5 < 40\left(1 - \frac{x}{7}\right) < 5$$

$$\Rightarrow \quad -\frac{1}{8} < 1 - \frac{x}{7} < \frac{1}{8}$$

$$\Rightarrow \quad \frac{1}{8} > \frac{x}{7} - 1 > -\frac{1}{8}$$

$$\Rightarrow \quad 1\frac{1}{8} > \frac{x}{7} > \frac{7}{8}$$

$$\Rightarrow \quad \frac{63}{8} > x > \frac{49}{8} \text{ or } 7.875 > x > 6.125.$$

(b) $|8t - 11| \geq 13$

$$\Rightarrow \quad 8t - 11 \geq 13 \text{ or } \qquad 8t - 11 \leq -13$$

$$\Rightarrow \quad 8t \geq 24 \qquad \text{or} \qquad 8t \leq -2$$

$$\Rightarrow \quad t \geq 3 \qquad \text{or} \qquad t \leq -\tfrac{1}{4}.$$

Example

Laila wants some leaflets printed. She gets two quotes:

'Bulkfast' say they will charge a fixed charge of £60 plus 5p per copy. 'Smallorder' printers quote a £25 fixed charge plus 9p per copy.

Find a formula for the difference in price if Laila decides to have n copies printed. She decides that a difference of £5 or less is 'negligible' and, if this is the case, that quality of product will be the deciding factor between the two firms. For what values of n is the difference 'negligible'?

Solution

For n copies 'Bulkfast' will charge £$(60 + 0.05n)$ and 'Smallorder' £$(25 + 0.08n)$.

Difference in price $= |(60 + 0.05n) - (25 + 0.08n)|$

$$= |35 - 0.03n|$$

For the difference to be negligible

$$|35 - 0.03n| \le 5, \qquad n \in \mathbb{N}$$
$$\Rightarrow \quad -5 \le 35 - 0.03n \le 5, \qquad n \in \mathbb{N}$$
$$\Rightarrow \quad 5 \ge 0.03n - 35 \ge -5, \qquad n \in \mathbb{N}$$
$$\Rightarrow \quad 40 \ge 0.03n \ge 30, \qquad n \in \mathbb{N}$$
$$\Rightarrow \quad 1333.33 \ge n \ge 1000, \qquad n \in \mathbb{N}.$$

So the difference is negligible when n is between 1000 and 1333 copies inclusive.

Example

Solve $\left| x^2 + 4x - 1 \right| < 2$.

Solution

$$\left| x^2 + 4x - 1 \right| < 2 \Rightarrow -2 < x^2 + 4x - 1 < 2.$$

This can be solved as two separate inequalities $x^2 + 4x - 3 < 0$

and $x^2 + 4x + 1 > 0$ and the results combined.

A quicker technique is to complete the square;

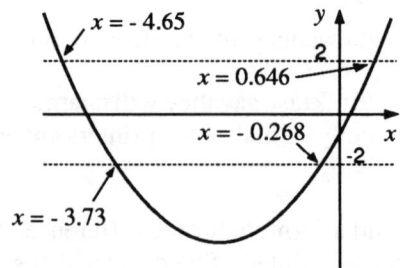

$$-2 < (x + 2)^2 - 5 < 2$$
$$\Rightarrow \quad 3 < (x + 2)^2 < 7$$
$$\Rightarrow \quad \sqrt{3} < x + 2 < \sqrt{7} \qquad \text{or} \qquad -\sqrt{3} > x + 2 > -\sqrt{7}$$
$$\Rightarrow \quad \sqrt{3} - 2 < x < \sqrt{7} - 2 \qquad \text{or} \qquad -\sqrt{3} - 2 > x > -\sqrt{7} - 2$$
$$\Rightarrow \quad -0.268 < x < 0.646 \qquad \text{or} \qquad -3.73 > x > -4.65.$$

Activity 20   Properties of the modulus function

Which of these statements below are true? x and y stand for any numbers.

(a) $|x + y| = |x| + |y|$ (b) $|x - y| = |x| - |y|$

(c) $|xy| = |x||y|$

(d) $\left| \dfrac{x}{y} \right| = \dfrac{|x|}{|y|}$ (e) $(|x|)^2 = x^2$.

Activity 21   Graphs involving the modulus function

(a) Sketch the graph of $y = x$. What will the graph of $y = |x|$ look like?

(b) Sketch the graph of $y = 2x - 5$ and hence the graph of $y = |2x - 5|$

(c) Sketch the graph of $y = x^2 - 7x + 5$ and $y = |x^2 - 7x + 5|$.

(d) The graph of $y = f(x)$ is shown on the right.

Sketch $y = |f(x)|$ and $y = f(|x|)$.

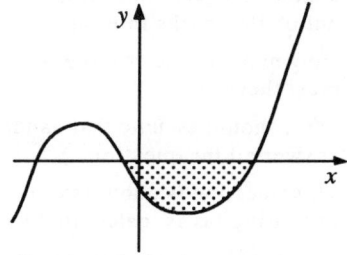

Exercise 5I

1. Solve these inequalities :

   (a) $|5a + 3| \le 2$

   (b) $|10 - b| < 6$

   (c) $|15 - 4c| \ge 3$

   (d) $\left| \dfrac{2d}{3} + 7 \right| \le 11$

   (e) $\left| 2(5 + \dfrac{12e}{5}) \right| < 16$

   (f) $\left| \dfrac{4 - 9f}{13} \right| \le 10$

2. Use a graph to solve :

   (a) $|12x^2 + 7x - 10| < 20, x \in \mathbb{N}$

   b) $|100 - 5x - x^2| < 50, x \in \mathbb{N}$.

3. Solve

   (a) $|x^2 - 8x + 10| < 3$

   (b) $|x^2 + 2x - 6| > 6$

   (c) $|x^2 + 10x + 18| < 8$.

4. The formula $F = \frac{9}{5}C + 32$ is used to convert degrees Celsius (C) into degrees Fahrenheit (F). Often, however, the formula $F = 2C + 30$ is used to make the arithmetic easier. Over what temperature range (in °C) is the approximate formula correct

   (a) to within 5°F (b) to within 2°F.

5. Di and Vi decide to race each other over 100 metres. Vi is a faster sprinter than Di. Vi's average time for the 100 metres is 12.5 seconds whilst Di's is 16 seconds. It is agreed that, while Vi will run the full distance, Di will be given a head start of d metres.

   (a) Suppose they run the race at the same speed as their average times suggest. How many seconds apart will they cross the finishing line?

   (b) They decide to set d so that no more than 1 second is expected between them at the finish. Find over what values d could range.

5.10 Miscellaneous Exercises

1. In setting an exam paper, it is agreed that short questions will carry 4 marks each and long questions 13 marks. The paper must contain 16 questions altogether and must be out of 100 marks in total.

   How many of the two types of question must there be?

   (You should assume that candidates have to answer all the questions.)

2. This question is about taxation. At the time of writing tax is calculated like this:

   - first £3005 earned in a year is not taxed.

   - income between £3005 and £20 700 is taxed at 25%;

   - all income over £20 700 is taxed at 40%.

   Two brothers compare their 1990/91 tax bills. Peter notices that, although his income was twice as much as Jonathan's, he paid 2.8 times as much tax. How much money did Peter earn? (Assume that neither brother receives any allowance other than the £3005 mentioned above.)

3. Staff at a factory are working overtime to get a job finished quickly. It is decided that continuous supervision is needed. Six supervisors are chosen, each to work a 6-hour shift, with an equal overlap time between each shift. The job will not take less than 32 hours and must be finished within 35 hours. How long could the overlap between shifts be?

4. (a) Clair needs to get to work by 9.00 am. She finds that if she leaves home at 7.00 am, the journey takes 1 hour, but that every two minutes after 7 o'clock adds another minute to her travelling time. At what time must she leave home in order to get to work exactly on time?

   (b) Her journey home takes 1 hour if she leaves at 4.30 pm. Every 4 minutes later than this adds 3 minutes to her travelling time. To the nearest minute, when must she leave work to arrive home at 6.00 pm?

5. Economists often talk about a firm's total cost function. This function relates the total cost C to the level of output Q units.

   Suppose a firm's total cost function is

   $$C = 4Q^2 + 100Q + 16000,$$

   find the values of $Q(\in \mathbb{N})$ for the total cost to be less than £70 000.

6. An office manager employs the following method of buying Christmas presents for staff. He buys a consignment of turkeys, all at the same price. 15 of them are reserved for his colleagues and the rest are resold when the price has risen by £2 per turkey above what he paid. He then sells the reserved turkeys at whatever price he needs to break even.

   This year he had to pay £1200 for the original consignment, and his colleagues paid exactly half the original price for their birds. How many turkeys did he buy?

7. A light plane travels from London to Inverness, a distance of 450 miles, at an average speed of 200 mph. Another plane travels the same route but gets there half an hour quicker. How fast was the second plane travelling?

8. Two cars travel 150 miles along the motorway. On average, one car travels 10 mph faster than the other, and completes the journey 15 minutes before the other one. What speed was the slower one doing, to the nearest mile per hour?

9. In a cycle race the contestants ride from A to B and back. Both outward and return journeys are the same distance, but the return journey is mostly downhill; average speeds on the return leg are 8 km/h faster than on the outward.

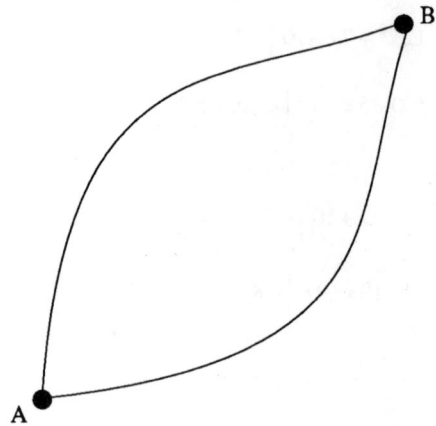

Neville covers the 60 km course in $2\frac{1}{2}$ hours. Work out his average speed for the first half of the course, to the nearest km/h.

10.

The hands on the clock move smoothly and continually. In the above picture the minute hand is exactly covering the hour hand. What time is it? (NOT 4.20)

11. The shape below is known as a pentagram. Find x.

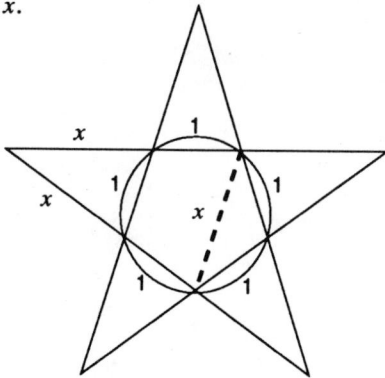

12. The diagram shows the elevation of a garden shed, the roof of which overhangs to the level of the top of the window. The distance AB is 30 cm longer than the length of the overhang, marked x in the diagram. Find x.

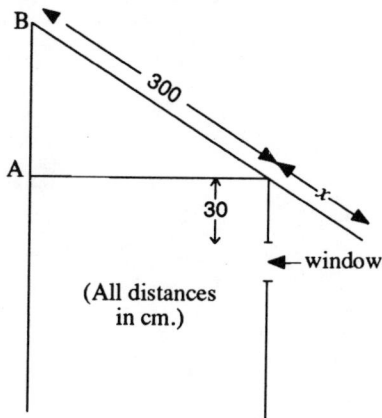

(All distances in cm.)

13. A fighter and a spy plane are travelling on perpendicular paths that cross at the point C. At 1.10 pm the fighter is 100 miles due west of C and travelling east at 400 mph. At the same time the spy plane is 60 miles due south of C and travelling north at 300 mph.

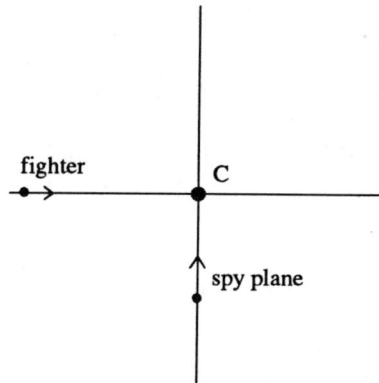

(a) If t is the time in hours after 1.10 pm, write down formulae for the distances of both planes from C at time t.

(b) For how many minutes are the two planes within 50 miles of each other?

14. A rectangular paddock is 3 times longer than it is wide. The width is increased by 20 m. This has the effect of doubling the area. Find the original dimensions of the paddock.

15. Solve these:

(a) $|x| = |x+1|$

(b) $\dfrac{12}{x+1} > \dfrac{x}{6}$

(c) $x^4 - 13x^2 + 36 = 0$

(d) $x - 6\sqrt{x} + 8 = 0$

(e) $x^3 + 2x^2 - 48x = 0$

(f) $x^2 - |x| - 2 = 0$.

6 EXTENDING ALGEBRA

Objectives

After studying this chapter you should

*   understand techniques whereby equations of cubic degree and higher can be solved;

*   be able to factorise polynomials;

*   be able to use the remainder theorem;

*   appreciate the nature of algebra as 'generalised arithmetic' and be able to use this idea when manipulating algebraic fractions.

6.0   Introduction

This chapter is different from the others in this book; although it begins with a realistic example, much of the rest does not have any immediate relevance to the outside world. Its purpose is to extend the algebra you already know.

This does not make it a dull chapter. On the contrary, there are many people who enjoy abstract mathematics as being interesting in its own right. Over the centuries mathematical techniques have developed partly in response to problems that needed to be solved, but there have always been those who have pondered on the less applicable side of the subject purely out of interest and curiosity.

6.1   The cubic equation

How can an open-topped box with volume 500 cm³ be made from a square piece of card 20 cm by 20 cm?

To make such a box, a net can be made by cutting off equal squares from each corner, as shown in the diagram.

Suppose each smaller square has side x cm. The dimensions of the box will then be x by $(20 - 2x)$ by $(20 - 2x)$. For the volume to be 500 cm³, x must satisfy the equation

$$x(20 - 2x)^2 = 500, \ 0 \le x \le 10$$

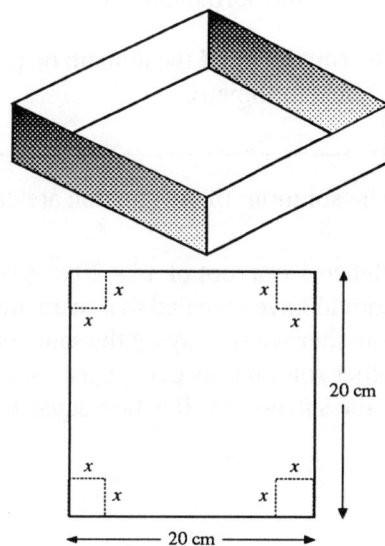

When the expression has been multiplied out, the highest power of x in this equation is x^3; accordingly this equation is called a **cubic equation**.

As with a quadratic equations, it is good practice to reduce this equation to the form $f(x) = 0$.

$$x(400 - 80x + 4x^2) = 500$$

$$\Rightarrow \quad 4x^3 - 80x^2 + 400x - 500 = 0$$

$$\Rightarrow \quad x^3 - 20x^2 + 100x - 125 = 0, \ 0 \le x \le 10$$

Again by comparison with quadratic equations, the next thing you might want to do would be to try to factorise it. With a cubic function, however, this is not easy.

Fortunately one solution to the problem can be seen 'by inspection'. A little experimentation reveals that $x = 5$ will fit the bill. Is this the only solution? Remember that for quadratic equations where there was one answer there was almost always another.

One method of solution is to draw the graph of the cubic function $x^3 - 20x^2 + 100x - 125$ and see where it crosses the x-axis.

Activity 1   Using a graph

Sketch the graph of the function

$$f(x) = x^3 - 20x^2 + 100x - 125$$

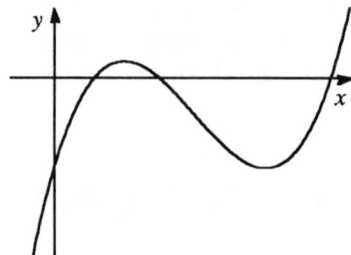

Verify that $f(5) = 0$ and find any other values of x where the curve crosses the horizontal axis.

Bearing in mind the domain of x, solve the original problem to the nearest millimetre.

The solutions of an equation are called the **roots** of an equation.

Hence 5 is a root of $x^3 - 20x^2 + 100x - 125 = 0$ and Activity 1 should have revealed that there are two other roots as well. Another way of saying the same thing is to refer to 5 and the two other solutions as being **zeros** of the function; in other words, these values make that function equal zero.

$$\boxed{\alpha \text{ is a root of } f(x) = 0 \iff f(\alpha) = 0}$$

How could the cubic equation have been solved without the need to draw the graph? Activity 2 gives a clue.

Activity 2   Zeros and factors

You may find it useful to use a graphic calculator.

(a) Sketch the graph of the function $x^2 + 2x - 143$. What is the factorised form of this function? How can you tell by looking at the graph?

(b) Now sketch the graph of the cubic function $(x+4)(x-1)(x-3)$. How does the form of the function correspond to its zeros?

(c) Sketch the graph of the function $x^3 - 79x + 210$. Hence write this function as the product of three linear factors.

(d) Can you factorise $x^3 - 5x^2 + 3x + 4$ by looking at its graph?

(e) Suggest a possible equation for the curve on the right.

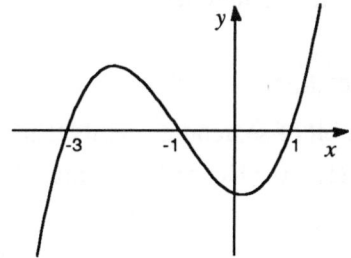

Now look again at the equation

$$x^3 - 20x^2 + 100x - 125 = 0$$

$x = 5$ is known to be a zero of the function $x^3 - 20x^2 + 100x - 125$ and so it can be deduced, from your work on Activity 2 that $(x-5)$ is a factor.

If $(x-5)$ is a factor, how does the rest of the factorisation go? It is evident that $x^3 - 20x^2 + 100x - 125 = (x-5) \times$(quadratic function). Moreover, the quadratic must be of the form

$$x^2 + bx + 25$$

Finding b is not quite so obvious but can be done as follows :

$$x^3 - 20x^2 + 100x - 125 = (x-5)(x^2 + bx + 25)$$

To make $-20x^2$ in the cubic equation:

$$-20 = -5 + b \implies b = -15$$

To make $+100x$ in the cubic equation:

$$100 = 25 - 5b \implies b = -15$$

Hence $x^3 - 20x^2 + 100x - 125 = (x-5)(x^2 - 15x + 25)$

The solution to the cubic equation is as follows

$$x^3 - 20x^2 + 100x - 125 = 0$$

$\implies$ $\quad (x-5)(x^2 - 15x + 25) = 0$

$\implies$ $\quad$ either $x - 5 = 0$ or $\quad x^2 - 15x + 25 = 0$

$\implies$ $\quad x = 1.9, 5$ or 13.1 to 1 d.p.

Example

Find all solutions of $x^3 - 3x^2 - 33x + 35 = 0$

Solution

A bit of searching reveals that $x = 1$ is a root of this equation, since $f(1) = 0$. Hence $(x-1)$ must be a factor, and you can write

$$x^3 - 3x^2 - 33x + 35 = (x-1)(x^2 + bx - 35)$$

To find b : the cubic equation contains $-33x$, so

$$-33 = -b - 35$$

$\implies$ $\quad b = -2$

$\implies$ $\quad x^3 - 3x^2 - 33x + 35 = (x-1)(x^2 - 2x - 35)$

In this case, the quadratic factor itself factorises to

$$(x-7)(x+5)$$

so

$$x^3 - 3x^2 - 33x + 35 = (x-1)(x-7)(x+5)$$

The original equation can thus be re-written

$$(x-1)(x-7)(x+5) = 0$$

$\implies$ $\quad x = -5, 1$ or 7

Example

Solve $x^3 - 14x - 15 = 0$

Solution

$x = -3$ is a root of this equation. Hence $(x+3)$ is a factor.

$$x^3 - 14x - 15 = (x+3)(x^2 + bx - 5)$$

To find b: cubic equation has $0x^2$, so

$$0 = b + 3$$
$$\Rightarrow \quad b = -3$$
$$\Rightarrow \quad x^3 - 14x - 15 = (x+3)(x^2 - 3x - 5)$$

The original equation thus reads

$$(x+3)(x^2 - 3x - 5) = 0$$
$$\Rightarrow \quad \text{either } x + 3 = 0 \text{ or } x^2 - 3x - 5 = 0$$
$$\Rightarrow \quad x = -3, -1.19, \text{ or } 4.19, \text{ to 3 s.f.}$$

Activity 3   Missing roots

(a)   Find a linear factor of $x^3 - 4x^2 - 2x + 20$ and the corresponding quadratic factor. Hence find all the solutions of $x^3 - 4x^2 - 2x + 20 = 0$.

(b)   Illustrate your answer by means of a sketch graph.

Exercise 6A

1. (a) Work out the missing quadratic factors :

   (i) $x^3 - 3x^2 - 6x + 8 = (x-4)(...)$

   (ii) $x^3 + 8x^2 + 12x - 9 = (x+3)(...)$

   (iii) $2x^3 - x^2 - 117x - 324 = (2x+9)(...)$

   (b) Use your answers to (a) to find all the roots of

   (i) $x^3 - 3x^2 - 6x + 8 = 0$

   (ii) $x^3 + 8x^2 + 12x - 9 = 0$

   (iii) $2x^3 - x^2 - 117x - 324 = 0$

2. Explain how you know that

   (a) $(x-3)$ is a factor of $x^3 - 2x^2 + x - 12$

   (b) $(x+5)$ is a factor of $2x^3 + 6x^2 - 23x - 15$

   (c) $(2x-1)$ is a factor of $4x^3 + 2x^2 + 8x - 5$.

3. Find all the roots of these equations

    (a) $x^3 - 5x^2 + 6x - 2 = 0$

    (b) $x^3 + 3x^2 - 46x = 48$

    (c) $2x^3 - x^2 - 18x + 9 = 0$

4. Four identical 'square corners' are cut from a square piece of card measuring 10 cm by 10 cm. The resulting net will make an open topped box with volume 64 cm³. Find the size of the squares that must be removed.

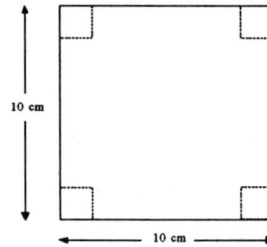

6.2   No simple solution

You should by now have asked the question : "What happens if no simple solution can be found?" The process you have used depends crucially on first being able to find a root. You may wonder whether, as with quadratic equations that cannot be factorised, there is a formula that will give all the roots automatically.

The answer is that a systematic method of finding roots of cubics does exist. The bad news, however, is that

(a)   it is a long-winded method which is seldom used;

(b)   it involves complicated maths.

A practical solution is to use either a graphical method or trial and improvement. Both methods can be time-consuming and both depend on first knowing approximately where the roots are.

For example, to solve the equation $x^3 - 100x^2 + 2000x - 1500 = 0$ it is a help to work out a few values of the function first. Let the function be labelled $f(x)$.

A table of values is shown on the right. The arrows show where $f(x)$ changes from positive to negative or vice versa. Hence the zeros of $f(x)$ must occur

> between 0 and 10
>
> between 20 and 30
>
> between 70 and 80

Trial and improvement, or graphs drawn in the correct regions, eventually give answers of 0.78, 26.40 and 72.82 to 2 d.p.

x	$f(x)$
-30	-178500
-20	-89500
-10	-32500
0	-1500 ←
10	9500
20	6500 ←
30	-4500
40	-17500
50	-26500
60	-25500
70	-8500 ←
80	30500

Exercise 6B

1. (a) Explain how you can tell that the function
   $f(x) = x^3 + x^2 - x + 5$ has a zero between
   $x = -3$ and $x = -2$.

   (b) Find this value to 3 s.f.

2. Solve these cubic equations to 3 s.f.

   (a) $2x^3 - 150x^2 + 75000 = 0$

   (b) $x^3 + 4x^2 - 32x - 100 = 0$

6.3 Factor theorem

Quadratics and cubics are particular examples of **polynomial functions** : a quadratic function is a polynomial of degree 2 : a cubic has degree 3. In general a **polynomial of degree** n has the form

$$a_n x^n + a_{n-1} x^{n-1} + \ldots + a_2 x^2 + a_1 x + a_0$$

where a_0, a_1, ... a_n are real numbers which are not all zero. When solving cubic equations, the spotting of a root leads immediately to a simple linear factor : e.g.

$$3 \text{ is a zero of } x^3 - 5x^2 + 11x - 15$$

$\Rightarrow \quad (x - 3) \text{ is a factor of } x^3 - 5x^2 + 11x - 15$

$\Rightarrow \quad x^3 + 5x^2 + 11x - 15 = (x - 3) \times (\text{quadratic})$

The same technique can be applied to polynomials of any degree. For example,

$$2 \text{ is a zero of } x^4 + 3x^2 - 17x + 6$$

$\Rightarrow \quad (x - 2) \text{ is a factor of } x^4 + 3x^2 - 17x + 6$

$\Rightarrow \quad x^4 + 3x^2 - 17x + 6 = (x - 2) \times (\text{cubic})$

In general

If $P(x)$ is a polynomial of degree n and has a zero at $x = \alpha$, that is $P(\alpha) = 0$, then $(x - \alpha)$ is a factor of $P(x)$ and

$$P(x) = (x - \alpha)Q(x)$$

where $Q(x)$ is a polynomial of degree $(n-1)$

This is known as the **factor theorem.**

Example

Show that $(x-5)$ is a factor of the polynomial

$$x^5 - 4x^4 - x^3 - 21x^2 + 25.$$

Solution

Denote the polynomial by $P(x)$.

$$\begin{aligned} P(5) &= 5^5 - 4 \times 5^4 - 5^3 - 21 \times 5^2 + 25 \\ &= 3125 - 2500 - 125 - 525 + 25 \\ &= 0 \end{aligned}$$

Hence 5 is a zero of $P(x)$ and so by the factor theorem, $(x-5)$ is a factor.

Activity 4   Fractions as roots

(a)   Show that $x = \frac{1}{2}$ is a zero of $2x^3 - 3x^2 - 3x + 2$.

(b) $2x^3 - 3x^2 - 3x + 2$ has two linear factors other than $x = \frac{1}{2}$. Find them. Check that all three factors multiply together to give the original cubic.

(c)   Find the quadratic expression missing from this statement :

$$2x^3 - 3x^2 - 3x + 2 = (2x - 1)(\ldots\ldots\ldots)$$

(d)   Suppose the fraction $\dfrac{p}{q}$ is a root of a polynomial. What could the associated factor be?

*(e) Suppose $b, c, d \in \mathbb{Z}$. Can the equation
$$2x^3 - 3x^2 - 3x + 2 = (2x - 1)(\ldots\ldots\ldots)$$
have fractions as roots, not counting integers? Explain your answer.

Activity 5   Tips for root-spotting

(a)   Just by looking at the equation $x^3 + 3x^2 + 3x + 2 = 0$ it is possible to deduce that, if there is an integer root, it can only be ± 1 or ± 2. Why? What is the root?

(b)   If there is an integer root of
$x^4 + 3x^3 + x^2 + 2x - 3 = 0$ then what could it be? For example, how can you tell that neither $+2$ nor -2 could be a root?

Exercise 6C

1. Show that
    (a) $(x-1)$ is a factor of $x^4 + 3x^3 - 2x^2 + 5x - 7$

    (b) $(x+1)$ is a factor of
    $$x^5 - 7x^4 - 8x^3 + 2x^2 + x - 1$$

    (c) $(x-3)$ is a factor of $x^4 - 2x^3 - 7x - 6$

    (d) $(x-2)$ is a factor of $2x^6 + 5x^4 - 27x^3 + 8$

    (e) $(x+10)$ is a factor of
    $$5x^5 + 23x^4 - 269x^3 - 90x + 100$$

    (f) $(x-8)$ is a factor of
    $$x^{10} - 9x^9 + 8x^8 - x^2 + x + 56$$

2. Three of the polynomials below have a linear factor from the list on the right. Match the factors to the polynomials. The 'odd one out' does have a simple linear factor, but not one in the list. Find the factor.

A: $x^4 - 5x^2 + 3x - 2$	1: $x+1$
B: $x^5 - 4x^4 + 2x^3 - 8x^2 - x + 4$	2: $x-2$
C: $x^3 - 4x^2 - 6x - 1$	3: $x+3$
D: $x^3 - 12x^2 - 43x + 6$	

3. Find a linear factor for each of these polynomials.

    (a) $x^3 + 4x^2 + x - 6$

    (b) $x^4 - 6x - 4$

    (c) $x^4 - 6x^3 - 8x^2 + 6x + 7$

    (d) $2x^4 + x^3 - 6x^2 - x + 1$

    (e) $3x^3 + 10x^2 - 11x + 2$

6.4 Solving higher order equations

The next activity shows how you can tackle higher order equations using the factor theorem.

Activity 6   Solving a quartic equation

The equation $x^4 - 4x^3 - 13x^2 + 4x + 12 = 0$ is an example of a quartic equation, a polynomial equation of degree 4.

Find a simple linear factor of the quartic, and use this factor to complete a statement of this type :
$$x^4 - 4x^3 - 13x^2 + 4x + 12 = (x - \alpha) \times (\text{cubic})$$

Hence find all the solutions of the quartic equation and sketch the graph of $y = x^4 - 4x^3 - 13x^2 + 4x + 12$ without the help of a calculator or computer.

Activity 6 shows how the factor theorem can help solve higher order equations. However, it may have taken you some time! Probably the longest part was finding the cubic factor. When the same process is applied to, say,

$$x^6 - x^4 + 17x - 14 = (x+2) \text{ (polynomial of degree 5)}$$

it will take longer still.

Fortunately there are ways of finding the polynomial factor which are more efficient. Two methods will be introduced here. You may, later on, like to reflect on how they are essentially the same method expressed differently.

The first method involves juggling with coefficients, and is best demonstrated by example.

Consider the quartic $x^4 + 5x^3 + 7x^2 + 6x + 8$. $x = -2$ is a zero of this function and hence $(x+2)$ is a factor. The cubic factor can be found thus :

$$x^4 + 5x^3 + 7x^2 + 6x + 8$$

$$= x^3(x+2) + 3x^3 + 7x^2 + 6x + 8$$

$$= x^3(x+2) + 3x^2(x+2) + x^2 + 6x + 8$$

$$= (x^3 + 3x^2)(x+2) + x(x+2) + 4x + 8$$

$$= (x^3 + 3x^2 + x)(x+2) + 4(x+2)$$

$$= (x^3 + 3x^2 + x + 4)(x+2)$$

Hence the cubic factor is $x^3 + 3x^2 + x + 4$.

Example

Find the missing expression here :

$$x^4 - 3x^3 + 5x^2 + x - 4 = (x-1)(\ldots\ldots)$$

Solution

$$x^4 - 3x^3 + 5x^2 + x - 4$$

$$= x^3(x-1) - 2x^3 + 5x^2 + x - 4$$

$$= x^3(x-1) - 2x^2(x-1) + 3x^2 + x - 4$$

$$= (x^3 - 2x^2)(x-1) + 3x(x-1) + 4x - 4$$

$$= (x^3 - 2x^2 + 3x)(x-1) + 4(x-1)$$

$$= (x^3 - 2x^2 + 3x + 4)(x-1)$$

so the missing expression is $x^3 - 2x^2 + 3x + 4$.

The second method is sometimes called 'long division of polynomials'. The statement

$$x^4 + 5x^3 + 7x^2 + 6x + 8 = (x+2)(\text{cubic})$$

can be written instead in this form

$$\frac{x^4 + 5x^3 + 7x^2 + 6x + 8}{x+2} = \text{cubic polynomial}$$

The cubic polynomial is thus the result of dividing the quartic by $(x+2)$. This division can be accomplished in a manner very similar to long division of numbers.

$$
\require{enclose}
\begin{array}{r}
x^3 + 3x^2 + x + 4 \\
x+2 \enclose{longdiv}{x^4 + 5x^3 + 7x^2 + 6x + 8} \\
\underline{x^4 + 2x^3} \\
3x^3 + 7x^2 \\
\underline{3x^3 + 6x^2} \\
x^2 + 6x \\
\underline{x^2 + 2x} \\
4x + 8 \\
\underline{4x + 8} \\
0
\end{array}
$$

So $x^4 + 5x^5 + 7x^2 + 6x + 8 = (x+2)\left(x^3 + 3x^2 + x + 4\right)$,

the same answer is obtained as on the previous page by the 'juggling' method.

Activity 7   Long division

(a) When $x = 10$, the quotient $\dfrac{x^4 + 5x^3 + 7x^2 + 6x + 8}{x+2}$ becomes

the division sum $15\ 768 \div 12$. Evaluate this by long division (**not** short division). Discuss the resemblance between your sum and the algebraic long division above. Try putting other values of x into the quotient (e.g. $x = 9$, $x = -5$).

(b) Evaluate $15\ 899 \div 13$ by long division. What is the corresponding algebraic long division?

Example

Evaluate the quotient $\dfrac{x^4 - 5x^3 - 2x^2 + 25x - 3}{x - 3}$.

Solution

$$
\begin{array}{r}
x^3 - 2x^2 - 8x + 1 \\
x-3 \overline{\smash{\big)}\, x^4 - 5x^3 - 2x^2 + 25x - 3} \\
\underline{x^4 - 3x^3} \\
-2x^3 - 2x^2 \\
\underline{-2x^3 + 6x^2} \\
-8x^2 + 25x \\
\underline{-8x^2 + 24x} \\
x - 3 \\
\underline{x - 3} \\
0
\end{array}
$$

The quotient is therefore $x^3 - 2x^2 - 8x + 1$.

Example

Work out the missing cubic factor in this statement:
$$4x^4 - 11x^2 + 15x - 18 = (2x - 3)(\ldots\ldots)$$

Solution

Note that there are no terms in x^3. To simplify the division, a term $0x^3$ is included.

$$
\begin{array}{r}
2x^3 + 3x^2 - x + 6 \\
2x-3 \overline{\smash{\big)}\, 4x^4 + 0x^3 - 11x^2 + 15x - 18} \\
\underline{4x^4 - 6x^3} \\
6x^3 - 11x^2 \\
\underline{6x^3 - 9x^2} \\
-2x^2 + 15x \\
\underline{-2x^2 + 3x} \\
12x - 18 \\
\underline{12x - 18} \\
0
\end{array}
$$

The missing cubic factor is therefore $2x^3 + 3x^2 - x + 6$.

Exercise 6D

1. Find these quotients by long division

   (a) $7982 \div 26$

   (b) $22149 \div 69$

   (c) $45694 \div 134$

   (d) $55438 \div 106$

2. Use the 'juggling' method to find the missing factors

   (a) $x^4 + 8x^3 + 17x^2 + 12x + 18 = (x+3)(......)$

   (b) $x^4 - 5x^3 - x^2 + 25 = (x-5)(......)$

   (c) $x^3 + 4x^2 - 8 = (x+2)(......)$

3. Use long division to find the missing factors

   (a) $x^4 + 6x^3 + 9x^2 + 5x + 1 = (x+1)(......)$

   (b) $x^4 + 7x^3 - 39x - 18 = (x+6)(......)$

   (c) $2x^3 - 4x^2 - 7x + 14 = (x-2)(......)$

   (d) $9x^5 + 9x^4 - 16x^3 + 11x + 2 = (3x+2)(......)$

4. Evaluate these quotients

   (a) $\dfrac{x^3 + 6x^2 - 6x + 7}{x^2 - x + 1}$

   (b) $\dfrac{2x^4 + 5x^3 - 5x - 2}{x^2 + 3x + 2}$

   (c) $\dfrac{x^4 - 6x^3 + 4x^2 - 6x + 3}{x^2 + 1}$

   (d) $\dfrac{x^4 + 2x^3 - 2x - 4}{x^3 - 2}$

6.5 Factorising polynomials

In the next examples, you will see how to factorise polynomials of degrees higher than two.

Example

Factorise the quartic $x^4 - 4x^3 - 7x^2 + 34x - 24$ as fully as possible and hence solve the equation $x^4 - 4x^3 - 7x^2 + 34x - 24 = 0$.

Solution

$x = 1$ is a zero of the quartic, so $(x-1)$ is a factor.

Long division yields

$$x^4 - 4x^3 - 7x^2 + 34x - 24 = (x-1)(x^3 - 3x^2 - 10x + 24)$$

Now factorise the cubic; since $x = 2$ is a zero of the cubic, $(x-2)$ is a factor.

Long division gives

$$x^3 - 3x^2 - 10x + 24 = (x-2)(x^2 - x - 12)$$

The quadratic $x^2 - x - 12$ factorises easily to give $(x-4)(x+3)$

The full factorisation of the original quartic is therefore

$$(x-1)(x-2)(x-4)(x+3)$$

and the solutions to the equation are thus

$$x = -3, 1, 2 \text{ and } 4.$$

Example

Solve the equation $x^4 + 3x^3 - 11x^2 - 19x - 6 = 0$.

Solution

First, factorise the quartic as far as possible.

$x = -1$ is a zero so $(x + 1)$ is a factor.

By long division

$$x^4 + 3x^3 - 11x^2 - 19x - 6 = (x+1)(x^3 + 2x^2 - 13x - 6)$$

$x = 3$ is a zero of the cubic so $(x - 3)$ is a factor.

$$x^3 + 2x^2 - 13x - 6 = (x - 3)(x^2 + 5x + 2)$$

The quadratic $x^2 + 5x + 2$ has no straightforward linear factors, but the equation $x^2 + 5x + 2 = 0$ does have solutions, namely -4.56 and -0.438.

The quartic thus factorises to $(x + 1)(x - 3)(x^2 + 5x + 2)$ yielding solutions $x = -4.56, -1, -0.438$ and 3.

Exercise 6E

1. Solve the following quartic equations :

   (a) $x^4 + 8x^3 + 14x^2 - 8x - 15 = 0$

   (b) $x^4 + 8x^3 - 13x^2 - 32x - 36 = 0$

2. Solve the quintic equation

   $$x^5 - 8x^3 + 6x^2 + 7x - 6 = 0$$

6.6 Remainders

In the previous section you saw how to divide a polynomial by a factor. These ideas will be extended now to cover division of a polynomial by an expression of the form $(x - \alpha)$.

If $(x - \alpha)$ is not a factor, there will be a remainder.

Activity 8

(a) Carry out the long division $\dfrac{x^3+5x^2-2x+1}{x+1}$

(b) If you substitute $x=9$ in this quotient it becomes $1117 \div 10$. Carry out the division. Repeat with $x=10,11,12$. Do not use a calculator. Comment on your answers and how they correspond to part (a).

As with numbers, long divisions of polynomials often leave remainders. For example $(x+3)$ is **not** a factor of x^3+6x^2+7x-4, and so long division will yield a remainder.

$$
\begin{array}{r}
x^2+3x-2 \\[4pt]
x+3\overline{\smash{\big)}\,x^3+6x^2+7x-4} \\[2pt]
\underline{x^3+3x^2} \\[2pt]
3x^2+7x \\[2pt]
\underline{3x^2+9x} \\[2pt]
-2x-4 \\[2pt]
\underline{-2x-6} \\[2pt]
2
\end{array}
$$

One way of expressing this might be to write

$$\frac{x^3+6x^2+7x-4}{x+3}=x^2+3x-2,\ \text{rem }2$$

but the normal method is to write either

$$x^3+6x^2+7x-4=(x^2+3x-2)(x+3)+2$$

or $\quad\dfrac{x^3+6x^2+7x-4}{x+3}=x^2+3x-2+\dfrac{2}{x+3}$

Activity 9   Remainders

(a) Substitute any positive integer for x in the quotient
$$\frac{x^3 + 6x^2 + 7x - 4}{x + 3}.$$ Verify by division that the remainder is 2.
Try some negative values of x (except -3) and comment.

(b) Find the remainder when $x^3 + 6x^2 + x - 7$ is divided by $x + 1$.

(c) Find the remainder for the division
$(x^3 + x^2 - 4x + 8) \div (x - 2)$.

Exercise 6F

1. Find the remainders

(a) when $x^2 - 15x + 10$ is divided by $(x - 5)$

(b) when $x^3 + 4x^2 - 7x + 10$ is divided by $(x + 3)$

2. Use Question 1 to complete these statements

(a) $\dfrac{x^3 + 4x^2 - 7x + 10}{x + 3} = (\ldots\ldots) + \dfrac{\ldots\ldots}{x + 3}$

(b) $x^2 - 15x + 10 = (\ldots\ldots)(x - 5) + \ldots\ldots$

6.7   Extending the factor theorem

Activity 10

(a) How can you tell that $(x - 1)$ is not a factor of $x^3 - 7x + 10$?

(b) When the division is carried out a statement of the form

$$x^3 - 7x + 10 = (x - 1)Q(x) + R$$

will result, where $Q(x)$ is a quadratic function and R the remainder. Without doing the division, calculate R.
(Hint : choose a suitable value of x to substitute in the equation above.)

(c) Without doing the division, calculate the remainder when
$x^4 + 3x^3 - 5x + 10$ is divided by $x + 2$.

Activity 10 illustrates the result known as the **remainder theorem**.

If $P(x)$ is a polynomial of degree n then

$$P(x) = (x - \alpha)Q(x) + R$$

where $Q(x)$ is a polynomial of degree $n-1$ and $R = P(\alpha)$.

The useful fact that $R = P(\alpha)$ can be demonstrated simply by considering what happens to the equation $P(x) = (x - \alpha)Q(x) + R$ when $x = \alpha$ is substituted into it.

Does this provide a proof of the remainder theorem?

Activity 11   Division by quadratics

Carry out the algebraic division

$$\frac{x^4 + 3x^3 - 10x^2 - 26x + 28}{x^2 + 2x - 3}$$

Can you suggest a remaindertheorem for division by quadratics? Can you generalise to division by any polynomial?

Exercise 6G

1. Work out the remainder when

   (a) $x^2 + 5x - 7$ is divided by $(x-2)$;

   (b) $x^4 - 3x^2 + 7$ is divided by $(x+3)$;

   (c) $5x^3 + 6x^2 + 2x - 3$ is divided by $(x+5)$.

2. When the quadratic $x^2 + px + 1$ is divided by $(x-1)$ the remainder is 5. Find p.

3. $x^2 + px + q$ divides exactly by $(x-5)$ and leaves remainder -6 when divided by $(x+1)$. Find p and q.

4. Find the linear expressions which yield a remainder of 6 when divided into $x^2 + 10x + 22$.

6.8 Rational numbers and functions

Algebra is often called 'generalised arithmetic'. Some of the work in this chapter, particularly that on factors and remainders, should have emphasised the link between algebra and arithmetic, between letters and numbers.

This link is useful when dealing with rational functions. A rational function is another name for an algebraic fraction, such as

$$\frac{x}{x+2} \quad \text{or} \quad \frac{x^2+y^2}{xy}$$

Similarly, a rational number is any number that can be expressed as a fraction, for example,

$$1\tfrac{2}{5} \qquad -\tfrac{3}{7} \qquad 1.2 \qquad -8$$

Many arithmetic processes have their algebraic equivalent. A number of examples are given below.

1. Common factors

Algebra	Arithmetic
The fraction $\dfrac{x^2-2x-3}{x^2+6x+5}$ can be divided to $\dfrac{x-3}{x+5}$ since both numerator and denominator have a factor $x+1$.	The fraction $\dfrac{21}{77}$ can be cancelled down to $\dfrac{3}{11}$ since both numerator and denominator can be divided by 7.

2. Improper fractions

Algebra	Arithmetic
$\dfrac{x^2-1}{x+2}$ is an improper fraction and can be written $(x-2)+\dfrac{3}{x+2}$	$\dfrac{8}{5}$ is an improper fraction and can be written $1+\dfrac{3}{5}$, or $1\tfrac{3}{5}$

Algebraically, an improper fraction is one in which the degree (i.e. the power of x) of the numerator is greater than or equal to that of the denominator.

e.g. $\dfrac{4x-1}{x+5}$ is improper and can be written $4 - \dfrac{21}{x+5}$

3. Adding and subtracting fractions

Algebra	**Arithmetic**
To write $\dfrac{1}{x} + \dfrac{3}{x+5}$ as a single fraction, write both as fractions with the same denominator. The lowest common multiple of x and $(x+5)$ is $x(x+5)$, so the sum can be written	To write $\dfrac{1}{2} + \dfrac{3}{7}$ as a single fraction, write both as fractions with the same denominator. The lowest common multiple of 2 and 7 is 14, so the sum can be written
$\dfrac{x+5}{x(x+5)} + \dfrac{3x}{x(x+5)} = \dfrac{4x+5}{x(x+5)}$	$\dfrac{7}{14} + \dfrac{6}{14} = \dfrac{13}{14}$

The lowest common multiple of two numbers is not always their product, either in algebra or in arithmetic.

$\dfrac{3}{2x} - \dfrac{1}{x^2+x}$	$\dfrac{3}{4} - \dfrac{1}{6}$
The LCM of $2x$ and x^2+x is $2x(x+1)$. The subtraction can be written	The LCM of 4 and 6 is 12, so the subtraction can be written
$\dfrac{3(x+1)}{2x(x+1)} - \dfrac{2}{2x(x+1)} = \dfrac{3x+1}{2x(x+1)}$	$\dfrac{9}{12} - \dfrac{2}{12} = \dfrac{7}{12}$

4. Multiplication and division

Algebra	**Arithmetic**
$\left(\dfrac{x}{3y}\right) \times \left(\dfrac{y+1}{x^2-1}\right) = \dfrac{x(y+1)}{3y(x^2-1)}$	$\dfrac{2}{15} \times \dfrac{4}{3} = \dfrac{2 \times 4}{15 \times 3} = \dfrac{8}{45}$

Sometimes dividing can simplify the multiplication.
Here is an example:

$$\left(\frac{x}{x+1}\right) \times \left(\frac{x+2}{x^2}\right) \qquad \bigg| \qquad \frac{3}{4} \times \frac{5}{9}$$

$$= \left(\frac{1}{x+1}\right) \times \left(\frac{x+2}{x}\right) \qquad \bigg| \qquad = \frac{1}{4} \times \frac{5}{3}$$

$$= \frac{x+2}{x(x+1)} \qquad \bigg| \qquad = \frac{5}{12}$$

Finally here is an example of division

$$\left(\frac{x+1}{x^2+2x-8}\right) \div \left(\frac{x}{x^2+3x-4}\right) \qquad \bigg| \qquad \frac{4}{7} \div \frac{3}{14}$$

$$= \left(\frac{x+1}{x^2+2x-8}\right) \times \left(\frac{x^2+3x-4}{x}\right) \qquad \bigg| \qquad = \frac{4}{7} \times \frac{14}{3}$$

$$= \left(\frac{x+1}{x-2}\right) \times \left(\frac{x-1}{x}\right) \qquad \bigg| \qquad = \frac{4}{1} \times \frac{2}{3}$$

$$= \frac{(x+1)(x-1)}{x(x-2)} \qquad \bigg| \qquad = \frac{8}{3} \left(\text{or } 2\frac{2}{3}\right)$$

Activity 12   Some classic mistakes

(a) The following are attempts at simplifications. Can you give the correct versions of the LHS in each case?

$$\frac{1}{x} + \frac{1}{y} = \frac{2}{x+y}, \quad \frac{x+2}{x+4} = \frac{x+1}{x+2}, \quad \frac{x-2}{x+5} = -\frac{2}{5}$$

(b) In physics, resistors, R_1 and R_2 in parallel are equivalent to a single resistor, R, where

$$\frac{1}{R} = \frac{1}{R_1} + \frac{1}{R_2}$$

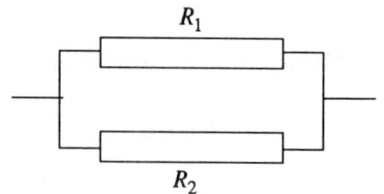

Explain why this is not the same as $R = R_1 + R_2$ and find the correct formula for R.

Exercise 6H

1. Simplify these where possible.

   (a) $\dfrac{3y}{5xy}$ (b) $\dfrac{p}{p^2-2p}$

   (c) $\dfrac{t-10}{t+2}$ (d) $\dfrac{n^2-n-2}{(n-2)^2}$

   (e) $\dfrac{q^2+1}{q}$ (f) $\dfrac{4x^2-36}{x^2+5x+6}$

   (g) $\dfrac{20a^2(b+1)}{8a^3b}$ (h) $\dfrac{2y^2+y-1}{y^2+10y+9}$

2. Find the LCMs of

   (a) x and $2y$

   (b) p^2 and $2p$

   (c) x^2-x and x

   (d) $3y^2z$ and $4yz^2$

   (e) $(2x-4)$ and $(x+3)$

   (f) $(x+1)(x-2)$ and $(x-3)(x+1)$

3. Write these as single fractions

   (a) $\dfrac{4}{a}+\dfrac{5}{b}$ (b) $\dfrac{7}{p}-\dfrac{3}{2p}$ (c) $\dfrac{3}{mn}-\dfrac{4}{n}$

   (d) $\dfrac{10}{xy}+\dfrac{14}{3x^2}$ (e) $\dfrac{2}{x}+\dfrac{3}{x^2}$ (f) $\dfrac{5}{x+2}-\dfrac{3}{x}$

   (g) $\dfrac{6}{p^2}+\dfrac{10}{p(p-1)}$ (h) $\dfrac{3r}{r-1}+\dfrac{2}{r+2}$

   (i) $\dfrac{16p}{q^2}-\dfrac{8q}{p^2}$ (j) $\dfrac{x}{2x+3}+\dfrac{6x}{x+2}$

4. Write these as single fractions. Simplify where possible.

   (a) $\dfrac{x}{y}\times\dfrac{p}{q}$ (b) $\dfrac{m}{n}\times\dfrac{3}{mn}$ (c) $\dfrac{10}{y}\div\dfrac{15}{y^2}$

   (d) $\dfrac{a^2b}{c}\div abc$ (e) $\left(\dfrac{x+1}{x-2}\right)\times\left(\dfrac{(x-2)^2}{x^2-1}\right)$

   (f) $\left(\dfrac{x^2+10x+16}{x-2}\right)\div\left(\dfrac{x^2+4x-32}{x^2-x-2}\right)$

   (g) $\left(\dfrac{x^2+5x}{x^2-9x+18}\right)\times\left(\dfrac{x-6}{x^2+11x+30}\right)\times\left(\dfrac{x-3}{x}\right)$

   (h) $\left(\dfrac{4p^2}{p^2-16}\right)\div\left(\dfrac{2p}{p-4}\right)$

6.9   Rationals and irrationals

Until about the 5th century AD it was firmly believed that whole numbers and their ratios could be used to describe any quantity imaginable. In other words, that the set Q of rational numbers contained every number possible. Gradually, though, mathematicians became aware of 'incommensurable quantities', quantities that could not be expressed as the ratio of two integers. Such numbers are called **irrational numbers**.

The origin of this concept is uncertain, but one of the simplest examples of an irrational number arises from Pythagoras' Theorem. This theorem gives the length of the diagonal of the unit square as $\sqrt{2}$.

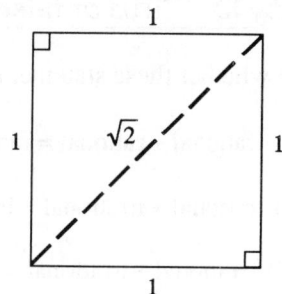

The proof that $\sqrt{2}$ is irrational is one of the most famous proofs in all mathematics. It employs a technique called 'reductio ad absurdum' (reduction to the absurd) the proof begins by assuming that $\sqrt{2}$ is rational and then shows that this assumption leads to something impossible (absurd). See if you can follow the reasoning:

Suppose $\sqrt{2}$ is rational.

This means that $\sqrt{2} = \frac{p}{q}$, where p and q are integers with no common factor, and $q \neq 0$.

$$\sqrt{2} = \frac{p}{q} \implies \frac{p^2}{q^2} = 2 \implies p^2 = 2q^2$$

$\implies p^2$ is even $\implies p$ is even.

Hence p can be written as $2r$, where r is an integer.

Since p is even, q must be odd.

$$p^2 = 2q^2 \implies 4r^2 = 2q^2 \implies q^2 = 2r^2$$

$\implies q^2$ is even $\implies q$ is even.

q is thus seen to be both odd **and** even, which is impossible. The original assumption that $\sqrt{2}$ is rational must therefore be false.

Hence $\sqrt{2}$ is irrational.

Other examples of rationals are π and any square root like $\sqrt{5}$. Still more examples can be constructed from these, e.g. $\sqrt{2}+1, 6\pi$, etc. The set of rational **and** irrational numbers is called the set of real numbers, denoted by $\mathbb{R}$.

Activity 13   True or false?

Decide whether these statements are true or false

(a) (i)   rational + rational = rational

   (ii)   rational + irrational = irrational

   (iii) irrational + irrational = irrational

(b) (i) rational × rational = rational

   (ii) rational × irrational = rational

   (iii) irrational × irrational = rational

(c) 'Between any two rational numbers there is another rational number'.

(d) 'Between any two irrational numbers, there is another irrational number'.

A different way of looking at rational and irrational numbers comes from considering equations.

To solve any linear equation involving integer coefficients, the set of rationals is sufficient.

$$\text{e.g.} \quad 71x + 1021 = 317 \Rightarrow x = -\frac{704}{71}$$

However, Q is **not** sufficient to solve every polynomial of degree 2 and higher. While some do have rational solutions,

$$\text{e.g.} \quad x^2 - 7x + 12 = 0 \Rightarrow x = 3 \text{ or } 4$$

in general, they do not.

$$\text{e.g.} \quad x^2 - 3x + 1 = 0 \Rightarrow x = \frac{1}{2}(3 \pm \sqrt{5}).$$

In fact, there is still a further class of numbers; these are the **transcendental** numbers. If you are interested, find out what these are by consulting a mathematical dictionary.

Irrational expressions like $\sqrt{5}$ are called **surds**. Surds cannot be expressed as ratios of natural numbers.

Activity 14   Handling surds

(a) Use a calculator to verify that $\sqrt{8} = 2\sqrt{2}$. Explain why this is true.

(b) What surd can be written as $3\sqrt{2}$?

(c) Generalise (b) to $p\sqrt{2}$, where $p \in \mathbb{N}$.

(d) Express $\sqrt{12}$ as a multiple of $\sqrt{3}$.

   Use a calculator to check your answer.

A method of manipulating surds is used to 'rationalise' the denominators in expressions like

$$\frac{1}{\sqrt{5}+1}$$

To rationalise a denominator means literally to turn an irrational denominator into a rational one.

Activity 15   How to rationalise a denominator

(a)   What is $(\sqrt{5}+1)(\sqrt{5}-1)$?

(b)   Multiply the fraction $\left(\dfrac{1}{\sqrt{5}+1}\right)$ by $\left(\dfrac{\sqrt{5}-1}{\sqrt{5}-1}\right)$.

(c)   Explain why $\dfrac{1}{\sqrt{5}+1}=\dfrac{\sqrt{5}-1}{4}$.

(d)   What is $(\sqrt{10}-2)(\sqrt{10}+2)$?

(e)   Write $\dfrac{3}{(\sqrt{10}-2)}$ as a fraction with a rational denominator.

Example

(a) $\sqrt{20}=\sqrt{4\times5}=\sqrt{4}\times\sqrt{5}=2\sqrt{5}$

(b) $\sqrt{216}=\sqrt{4\times54}=2\sqrt{54}$

But $\sqrt{54}$ can itself be simplified :
$2\sqrt{54}=2\sqrt{9\times6}=2\sqrt{9}\times\sqrt{6}=6\sqrt{6}$

Example

(a) $\dfrac{5}{\sqrt{6}-1}=\left(\dfrac{5}{\sqrt{6}-1}\right)\left(\dfrac{\sqrt{6}+1}{\sqrt{6}+1}\right)=\dfrac{5(\sqrt{6}+1)}{5}=\sqrt{6}+1$

(b) $\dfrac{1}{\sqrt{11}+\sqrt{7}}=\left(\dfrac{1}{\sqrt{11}+\sqrt{7}}\right)\left(\dfrac{\sqrt{11}-\sqrt{7}}{\sqrt{11}-\sqrt{7}}\right)=\dfrac{\sqrt{11}-\sqrt{7}}{4}$

Activity 16   Fence posts

The diagram shows a circular field cut in half by the diameter AB. The owner of the field wants to build two fences, one round the circumference of the circle, the other across the diameter. Fence posts are placed at A and B and further posts spaced equally along the diameter.

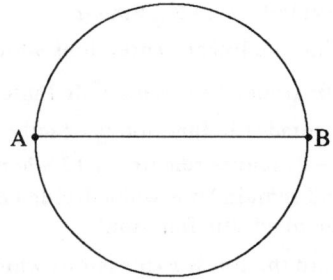

Explain why the owner must use a different equal spacing when fence posts are put around the circumference. ('Spacing' is to be taken to mean 'distance round the circumference').

Activity 17 $\sqrt{3}$

Prove that $\sqrt{3}$ is irrational. Using the same method of proof as was used for $\sqrt{2}$.

Can the same method be used to 'prove' that $\sqrt{4}$ is irrational?

Exercise 6I

1. Write down an irrational number
    (a) between 3 and 4;
    (b) between 26 and 27;
    (c) between -6 and -5.

2. Simplify the surds
    (a) $\sqrt{52}$ (b) $\sqrt{75}$ (c) $\sqrt{120}$ (d) $\sqrt{245}$

3. Use a similar technique to simplify
    (a) $\sqrt[3]{16}$ (b) $\sqrt[3]{54}$ (c) $\sqrt[4]{48}$

4. Rationalise the denominators in these expressions
    (a) $\dfrac{1}{\sqrt{2}-1}$ (b) $\dfrac{3}{\sqrt{21}-3}$ (c) $\dfrac{2}{\sqrt{5}-\sqrt{2}}$

    (d) $\dfrac{3}{\sqrt{2}}$ (e) $\dfrac{5}{\sqrt{14}-2}$

6.10 Miscellaneous Exercises

1. Find a linear factor for each of these polynomials :
    (a) $x^4 - 3x^3 - 10x^2 - x + 5$
    (b) $x^5 + 3x^4 + x^3 + 5x^2 + 12x - 4$

2. Solve these equations :
    (a) $x^4 - 3x^3 - 10x^2 - x + 5 = 0$
    (b) $x^4 + 2x^3 - 67x^2 - 128x + 192 = 0$
    (c) $x^4 + 2x^3 - 5x^2 + 6x - 24 = 0$

3. Copy and complete these identities :
    (a) $x^3 - 7x^2 - x - 6 = (x - 7)(\ \dots\ ) + \dots$
    (b) $\dfrac{x^5 + 5x^4 - 3x^2 + 2x + 1}{x + 2} = \dots + \dfrac{\dots}{x + 2}$

4. What is the remainder when
    (a) $x^4 - 5x^2 + 12x - 15$ is divided by $x + 3$;
    (b) $x^3 - 5x^2 - 21x + 7$ is divided by $x - 10$?

5. $x^3 + ax^2 + 5x - 10$ leaves remainder 4 when divided by $x+2$. Find a.

6. Find the linear expressions which leave remainder 14 when divided into $x^2 - 5x - 10$.

7. A quadratic function is exactly divisible by $x-2$, leaves remainder 12 when divided by $x+1$ and remainder 8 when divided by $x-3$. What is the quadratic function?

8. Find the linear expressions which leave remainder -8 when divided into $x^3 - 12x^2 + 17x + 22$.

9. Simplify these, where possible

   (a) $\dfrac{x^2 - 1}{2(x+1)}$ (b) $\dfrac{6a^3}{2a + a^2}$

   (c) $\dfrac{m^2 - 4}{2m}$ (d) $\dfrac{3x^2 + 5x - 2}{3x + 6}$

   (e) $\dfrac{x^3 - 1}{x - 1}$

10. Write these as single fractions in their simplest form.

   (a) $\dfrac{6p}{5} + \dfrac{1}{10p}$ (b) $\dfrac{3}{n^2} - \dfrac{5}{4n}$

   (c) $\dfrac{5}{x-5} + \dfrac{2}{x+2}$ (d) $\dfrac{7}{x^2 - 25} - \dfrac{3}{x+5}$

   (e) $\dfrac{a}{b} \div \dfrac{2a}{5}$ (f) $\dfrac{x^2}{6} \times \dfrac{2}{x(x+5)}$

11. (a) Find the values of p, q and r in these identities

   (i) $\dfrac{x+4}{x(x+2)} \equiv \dfrac{p}{x} + \dfrac{q}{x+2}$

   (ii) $\dfrac{5x-1}{x^2 - 1} \equiv \dfrac{p}{x+1} + \dfrac{q}{x-1}$

   (iii) $\dfrac{2}{x(x+1)(x+2)} \equiv \dfrac{p}{x} + \dfrac{q}{x+1} + \dfrac{r}{x+2}$

   (b) Write these functions as the sum or difference of fractions

   (i) $\dfrac{4x-3}{x(x-3)}$ (ii) $\dfrac{2x+19}{(x+2)(x+5)}$

   (iii) $\dfrac{x^2 + 9}{x(x-1)(x+3)}$

12. (a) The hot tap on its own takes $7\frac{1}{2}$ minutes to fill a bath. The cold tap on its own takes 5 minutes. How long does it take both taps to fill the bath?

   (b) Generalise the procedure adopted in (a). Suppose the hot tap takes h minutes and the cold tap c minutes. Find a formula for the time taken by both taps.

13. A motorist stops for a break exactly one-third of the way along a long route. The average speed for the first third of the distance is u; the average speed for the remaining two-thirds of the distance is v. What is the average speed for the whole journey?

14. An equilateral triangle has sides of length a.

   Show that the area is $\dfrac{a^2 \sqrt{3}}{4}$.

   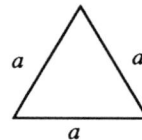

15. (a) Show that the area of this shape is $x^2(4 + \sqrt{15})$.

   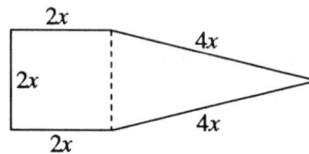

   (b) Given that the area is actually equal to 4, show that x is equal to $2\sqrt{(4 - \sqrt{15})}$.

16. For each of the diagrams below find a formula for the area in terms of x and the value of x if the numerical value of the area is as shown. Express your answers in the same form as those in question 15. Simplify all surds and rationalise all denominators.

   (a)

   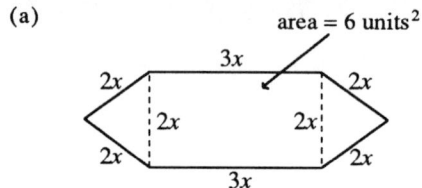

   (b)

   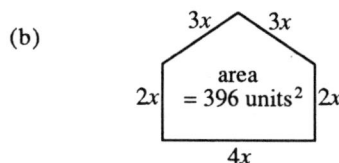

(c)

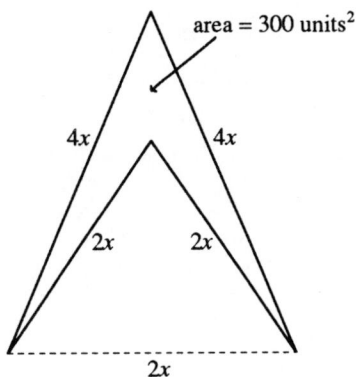

area = 300 units^2

4x   4x

2x   2x

2x

(d)

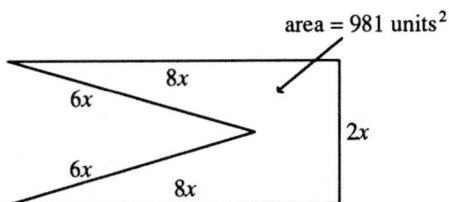

area = 981 units^2

6x   8x

      2x

6x

8x

17. (a) The cubic $x^3 + 2x^2 - x - 7$ is divided by the quadratic $(x-2)(x+3)$. The quotient is the linear function $l(x)$ and the remainder is expressed in the form $A(x-2) + B(x+3)$. Hence

$$x^3 + 2x^2 - x - 7 \equiv$$
$$(x-2)(x+3)L(x) + A(x-2) + B(x+3).$$

Put $x = 2$ into this identity and hence find the value of B. Find A in a similar way. Hence express the remainder in the form $\alpha x + \beta$ where α and β are constants.

(b) Use a similar technique to find the remainder when

(i) $x^4 + 2x^3 - x^2 + 7$ is divided by $x(x-2)$;

(ii) $x^5 - 3x^4 + 2x^3 + 5x - 10$ is divided by $(x+1)(x-4)$.

18. (a) Show that

(i) $(p+4q)$ is a factor of $p^2 + 2pq - 8q^2$ and find the other linear factor;

(ii) $(x-2y)$ is a factor of $x^3 + x^2 y - 7xy^2 + 2y^3$ and find the quadratic factor;

(iii) $(a+3b)$ is a factor of $2a^3 + 7a^2 b - 9b^3$ and find the two other linear factors.

(b) Factorise these expressions completely

(i) $x^2 - 2ax - 3a^2$

(ii) $p^2 + 10pq - 24q^2$

(iii) $a^3 - 2a^2 b - 11ab^2 + 12b^3$

19. Factorise the expression

$$16x^5 - 81x$$

as completely as possible.

7 STRAIGHT LINES

Objectives

After studying this chapter you should

- be familiar with the equation of a straight line;
- understand what information is needed to define a straight line;
- appreciate the significance of the gradient of a straight line;
- be able to draw by eye a line of best fit and find its equation;
- be able to solve simple linear inequality problems;
- be able to find the distance between two points in the xy-plane and the coordinates of the midpoint.

7.0   Introduction

In solving problems, a great deal of effort is often made to find and use a function which 'models' the situation being studied. The easiest type of functions to use and work with are those whose graphs are straight lines. These are called **linear** functions.

You will see how linear models can be constructed and then used for predicting in the following activity. For this activity you will need:

> spring, stand and set of weights.

Activity 1   Spring extensions

The extension of a spring held vertically depends on the weight fixed on the free end. Measure the extension, x, in cm, of the spring beyonds its natural length for a variety of weights, w, in grams.

(a) Plot the data points on a graph of x against w.

(b) Draw a straight line as accurately as possible through the data points.

(c) Assuming that the straight line has an equation of the form

$$x = \alpha w$$

for some constant α, find a point on your line and use it to find the value of α.

(d) Use your model to predict the extension for various weights; test your model by obtaining further experimental data.

7.1 Gradients

The figure shows a straight line - the graph increases its 'heights' by equal amounts for equal increases of the horizontal variable. This is because the slope or gradient is constant. Remember the gradient is defined as

$$\text{gradient} = \frac{\text{increase in } y}{\text{increase in } x}$$

When the graph is curved as shown in the figure opposite, then the gradient is no longer constant.

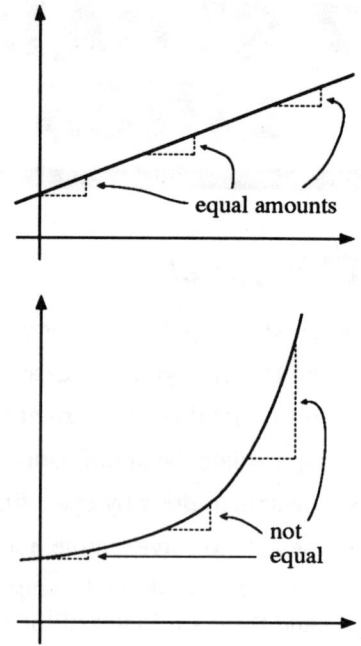

Example

A company offers a personal life insurance so that your life is insured against accidental death for

£250 000 for a payment of £50 per year

or

£100 000 for a payment of £25 per year.

If other rates are 'pro-rata' (in proportion) determine

(a) the amount of life insurance that can be obtained for a yearly payment of
     (i) £10   (ii) £80

(b) the yearly payment for life insurance of amount
     (i) £160 000   (ii) £550 000

Solution

Define

x = yearly payment in £'s

y = amount of insurance in £1000's

The two pieces of information can now be written as

$y = 250$ when $x = 50$,

$y = 100$ when $x = 25$.

These are illustrated on the graph opposite.

A graphical approach would be to draw the straight line between the two data points, extending in each direction beyond the points. The line could then be used to find corresponding values of yearly payment and amount of insurance.

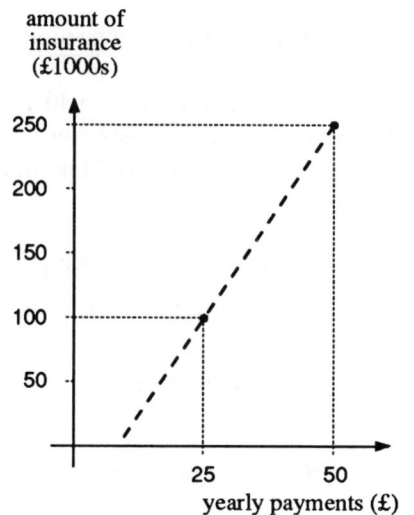

Algebraically, the relationship between x and y can be written as

$$y = mx + c$$

Since the two points, (25, 100) and (50, 250), satisfy this equation

$$250 = 50\,m + c \qquad\qquad (1)$$

$$100 = 25\,m + c \qquad\qquad (2)$$

You can solve these equations for m and c by first substituting (2) in (1).

$$250 - 100 = 50m + c - (25m + c)$$

$$= 50m - 25 + (c - c)$$

$$\Rightarrow \quad 150 = 25m$$

$$\Rightarrow \quad m = 6$$

Substituting m in (1) now gives

$$C = 250 - 50 \times 6 = -50$$

(You should now check that (2) is in fact satisfied by $m = 6$ and $c = -50$)

So the life insurance system can be modelled by the equation

$$\boxed{y = 6x - 50} \qquad\qquad (3)$$

(a) (i) $x = 10 \Rightarrow y = 6 \times 10 - 50 = 10$ or £10 000 life insurance.

(ii) $x = 50 \Rightarrow y = 6 \times 80 - 50 = 430$ or £430 000 life insurance.

(b) It is easier to make x the subject of equation (3)

Now $\qquad 6x = y + 50$

$$\Rightarrow \quad x = \frac{1}{6}(y + 50)$$

(c) (i) $y = 160 \Rightarrow x = \dfrac{1}{6}(160 + 50) = 35$

(ii) $y = 550 \Rightarrow x = \dfrac{1}{6}(550 + 50) = 100$

In the example above, it is worth noting that the value of m, namely 6, represents the gradient of the line. It shows that any increase of £1 in the yearly payment results in an extra insurance of £6000.

Can you suggest why $x \neq 0$ when $y = 0$?

The real bonus of this algebraic approach is that the model equation, $y = 6x - 50$, can be used to solve any problem related to this life insurance system.

Activity 2   Ski-passes

At a particular ski resort in Switzerland, ski-passes are advertised at the following two rates:

> 6 day pass for 38 Swiss Francs
> 13 day pass for 80 Swiss Francs.

Assuming that other days are charged at the same rate, find using algebraic methods a linear model to describe this situation. Use the model to find the cost of a ski pass for

(a) 3 days   (b) 30 days

Exercise 7A

1. The volume of 20 g of the metal Lithium was measured by a chemist and was found to be 37 cm^3. Use the fact that 0 g of Lithium must have a volume of 0 cm^3 to draw a graph of weight on the vertical axis against volume on the horizontal axis. Find the gradient of the graph, and hence state the density of Lithium in g cm^{-3}.

2. A car overtakes a lorry. At the start of the manoeuvre, the car is travelling at a speed of 13.4 ms^{-1} (30 mph). Five seconds later, after passing the lorry, it is travelling at a speed of 22 ms^{-1} (50 mph). Draw a graph of the speed, v, of the car against the time, t. Put v on the vertical axis, using metres per second, and t on the horizontal axis, using seconds. You may assume the graph is a straight line. Find the gradient of the graph, and so find the rate at which the speed of the car has increased, in metres per second per second.

3. An electronic scale works by measuring a current that is produced when the weighing pan is pressed down. The heavier the weight, the greater the current. To 'calibrate' it, the manufacturers need to find the rate at which current increases as the weight increases. The scale is tested with four different known weights, and the results are shown in this table.

Weight (grams)	100	200	500	1000
Current (mA)	25	45	105	205

Plot these values with the weight on the horizontal axis. Draw a straight line graph through the points, and so find the gradient of the line. Hence state the rate at which the current

(in milliamps, mA) is increasing as the weight increases.

4. The speed of a train pulling into a station decreases from 11.2 ms^{-1} to 0 ms^{-1} in 15 seconds. Draw a straight line graph showing this information, with the speed on the vertical axis. By finding the gradient of the line, find the rate at which the train decelerates, in ms^{-2}

5. A petrol pump works at a rate of 20 litres per minute. Draw a graph showing the volume of petrol pumped on the vertical axis against the time taken in seconds on the horizontal axis. You should choose scales from 0 to 40 litres and from 0 to 2 minutes. Use your graph to find the time required to pump 35 litres.

*6. A bank advertises loans, giving monthly repayments for various amounts of money lent, in a table :

Amount borrowed (£)	500	1000	2000
Monthly repayment (£)	53	106	212

Draw a graph showing this data assuming that it is a straight line. Put the monthly repayments on the vertical axis. Find the gradient of the graph. What does this gradient represent?

7. Boats are hired at the following rates

> 2 hours for £11
> 5 hours for £20

Assuming charges for other times are pro-rata, develop a linear model to describe this relationship and use it to find the hire charges for

(a) $1\frac{1}{2}$ hours   (b) 3 hours   (c) 12 hours

7.2 Equation of a straight line

One of the advantages that linear functions have is that their equations can all be written in a simple form. When the equation of a straight line is known, it is possible to find its gradient and the point it crosses the vertical axis immediately, without drawing. Also, it is possible to write down the equation of the line once the gradient and any single point on the line is known.

Remember that the equation of any straight line can be written in the form

$$y = mx + c$$

where m and c are constant.

This is illustrated opposite; m is the value of the gradient of the line, whilst c is the length of the intercept on the y-axis (since when $x = 0$, $y = m \times 0 + c = c$)

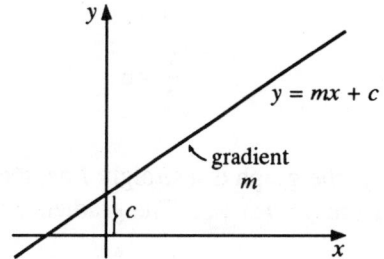

Example

A straight line with a gradient of -2 passes through the point $(4, -1)$. Find the equation of the line, and draw its graph.

Solution

As the graph is a straight line, its equation can be written in the form

$$y = mx + c$$

As the gradient is -2, m in this equation must be -2, and

$$y = -2x + c$$

The point $(4, -1)$ lies on this line, so its x and y coordinates must satisfy the equation. That is, the coordinates $x = 4$ and $y = -1$ can be substituted to make the equation true.
So

$$-1 = -2 \times 4 + c$$
$$-1 = -8 + c$$
$$\Rightarrow \quad 7 = c$$

So the equation is given by

$$y = -2x + 7$$

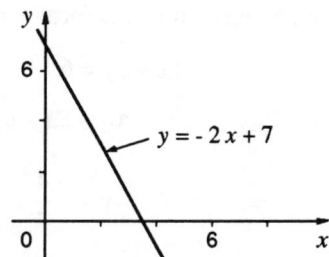

Example

A straight line passes through the points A, $(1, 0)$ and B, $(3, 6)$. Find the gradient of the line, and its equation.

Solution

The figure opposite shows a sketch of the line. The gradient can be calculated using this formula:

$$\text{gradient} = \frac{y_B - y_A}{x_B - x_A} \quad \left(\frac{\text{differences in } y\text{'s}}{\text{differences in } x\text{'s}}\right)$$

$$= \frac{6-0}{3-1} \quad \text{since } y_B = 6, \ y_A = 0, \ x_B = 3 \text{ and } x_A = 1$$

$$= \frac{6}{2} = 3$$

As the graph is a straight line, the equation can be written in the form $y = mx + c$. The gradient is 3, so this is the value of m :

$$y = 3x + c$$

To find c, the coordinates of a point on the line must be substituted into the equation. The coordinates of point A $(x = 1, \ y = 0)$ are used here ;

$$0 = 3 \times 1 + c$$

$$0 = 3 + c$$

$$\Rightarrow \quad c = -2$$

The equation is $y = 3x - 3$.

Example

The graph $4x + 3y - 6 = 0$ is a straight line. Find its gradient, and the intercept with the y-axis.

Solution

To read the gradient and intercept of a line from its equation it must be written in the form $y = mx + c$ first. Now

$$4x + 3y = 6$$

$$\Rightarrow \qquad 3y = -4x + 6$$

$$\Rightarrow \qquad y = -\frac{4}{3}x + \frac{6}{3}$$

$$= -\frac{4}{3}x + 2$$

If this equation is compared with $y = mx + c$, it can be seen that the gradient is $-\frac{4}{3}$ and the intercept with the y-axis is 2.

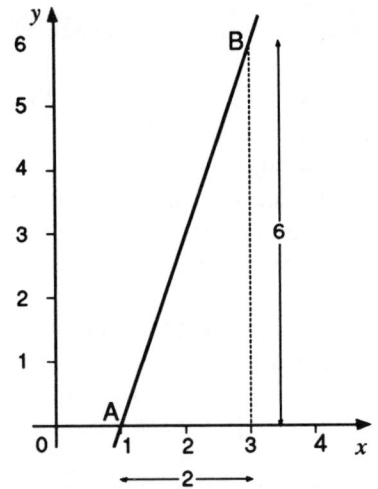

Activity 3

Draw the following lines on the same axes

(a) $2y + x = 0$ (b) $4y = 1 - 2x$ (c) $6y + 3x = 1$

What do you notice about these lines?

Exercise 7B

1. Write down the gradient and its intercept with the y-axis of each of these lines

    (a) $y = 3x - 1$ (b) $y = -4x - 3$

    (c) $y = \frac{1}{2}x + 5$ (d) $y = -\frac{6}{5}x - \frac{1}{2}$

    (e) $y = 4x$ (f) $y = x$

    (g) $y = -x$ (h) $y = 5$.

2. Find the gradient and intercept with the y-axis of each of the lines without drawing the graphs.

    (a) $4x + y = 9$ (b) $x + 2y = 6$

    (c) $3x - 2y = 4$ (d) $4y - 2x + 6 = 0$

3. A line passes through the point (5, 1) and has a gradient of 3. Find the equation of the line.

4. A line with gradient $-\frac{1}{3}$ passes through the point (4, 6). Find the equation of the line, leaving fractions in your answer.

5. Find the equation of the straight line which passes through the points (2, -1) and (6, 7).

6. A line which is parallel to $y = 2x$ passes through the point (3, -2). Find the equation of this line.

*7. A line which is parallel to $4x + 3y - 6 = 0$ passes through the origin. Find its equation

7.3   Linear laws through experiment

Many experiments are set up to validate mathematical 'laws' which are then used to predict future behaviour or consequences - straight line graphs can help find these laws, although often ingenuity will have to be used to make sure that the line should be straight, as you will see below.

Lens equation

A convex lens produces an image of an object on a flat screen; such lenses are used in cameras, some spectacles, contact lenses and telescopes.

The distance, v metres, between the lens and the sharp, inverted image on the screen, depends on the distance between the lens and the object , u metres. In general terms, the further the object is from the lens, the nearer the image is to the lens.

An experiment was carried out to find the relationship between u and v, and the data collected is shown opposite on a graph of v against u.

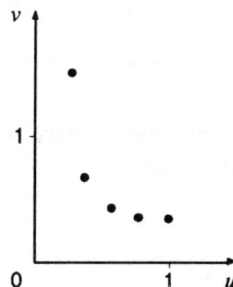

u(m)	1	0.8	0.6	0.4	0.3
v(m)	0.35	0.36	0.43	0.66	1.51

Plotting the data on a graph of v against u doesn't help you to find the law connecting u and v. However, if you are trying to verify a law of the form

$$\frac{1}{v} = m\left(\frac{1}{u}\right) + c \qquad (m, c \text{ constant})$$

you should obtain a straight line if you plot $y = \frac{1}{v}$ against $x = \frac{1}{u}$

Activity 4

Using the giving data, plot $y\left(=\frac{1}{v}\right)$ against $x\left(=\frac{1}{u}\right)$ and obtain

estimates for the constants m and c by drawing the line of best fit through the data points. 'Line of best fit' means the line which is as close as possible to all the data points - it will balance points each side of the line.

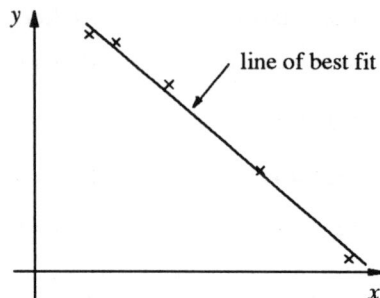

Activity 5   Fishing competitions

To win a fishing competition usually means catching the heaviest fish, and this usually means greatest length (it is much easier to measure the length rather than weigh each fish); but in some competitions the winner is the person who has the 'Greatest total weight of fish caught during the day'.

To save each competitor from having weighing scales, it is proposed to model the weight / length relationship by an equation of the form

$$w = \rho \ell^3$$

where w is the weight in grams and ℓ the length of the fish in cms and ρ a constant.

Using data given below, draw an appropriate straight line graph to see if the model is adequate, and if so, to estimate the value of ρ.

length (cm)	36.3	31.3	43.1	36.3	31.6	44.4	35.3	31.6
weight (g)	675	425	1025	650	425	1225	575	400

Criticize the model constructed in Activity 5. What factors have been neglected?

7.4 Linear inequalities

Often real life problems can have more than one solution, but some solutions may be better than others.

Activity 6

A club is organising a trip, and needs to transport at least 225 people using mini buses, coaches or some of each. A coach holds 45 people and costs £120 to hire, whilst a minibus holds 15 and costs £60. The club only has £720 to spend on the transport. Find a possible solution to this problem. Is it unique?

Find the most economic way to arrange the transport.

To solve the problem in Activity 6 in a logical and precise way requires the use of inequalities. Another similar problem is given below.

A farmer has 100 hectares available on which to sow 2 crops, wheat and sugar beet. Each hectare of wheat is expected to produce a profit of £20, whilst each hectatre of sugar beet should produce a profit of £30. However EEC quota regulations will not allow the farmers to grow more than 50 acres of sugar beet and 70 hectares of wheat. The time taken in soil preparation, seeding and tendering is estimated to be 5 man hours per hectare of wheat and 10 man hours per hectare of sugar beet. The available man power is up to 700 hours.

Can you find the optimum solution - that is a solution which satifies all the conditions and maximises the profit?

You can use intelligent trial and error methods, but an algebraic approach is ideal.

Let x = no of hectares sown with wheat

y = no of hectares sown with sugar beet,

The land restriction can be written as

$$x + y \le 100$$

It should also be noted that

$$x \ge 0,$$

and $y \ge 0;$

these three inequalities can be illustrated on a graph.

Note that the actual lines shown are $x+y=100$, $x=0$, $y=0$ and that, for example, all the region to the left of the line $x+y=100$ satisfies the inequality

$$x+y \leq 100$$

(for example, the point $x=y=0$ satisfies this inequality).

So to find the 'solution' of an inequality, you first draw the equality, and then identify the allowable region.

The region to be excluded, is shown partially shaded.

There are further inequalities to be satisfied, because of the regulations; namely

$$x \leq 70$$
$$y \leq 50$$

These can be shown on the graph so that the allowable (or feasible) region is further restricted.

There is yet one more inequality to consider; that is the available number of man-hours. This gives

$$5x+10y \leq 700$$

This is added to the graph as shown opposite.

The feasible region, which contains all possible solutions, is a convex polygon with vertices at O, A, B, C, D, and E.

Any point inside this polygon is a possible solution, but the optimum solution is the one which gives rise to a maximum value of

$$P = 20x + 30y$$

Why should P obtain its maximum value at a vertex of the polygon?

In this case the optimum solution will occur at one of the vertices, and the table below evaluates P at each vertex.

Point	Coordinates	Profit
0	$x=0$, $y=0$	$P=0$
A	$x=70$, $y=0$	$P = 20 \times 70 + 30 \times 0 = 1400$
B	$x=70$, $y=30$	$P = 20 \times 70 + 30 \times 30 = 2300$
C	$x=60$, $y=40$	$P = 20 \times 60 + 30 \times 40 = 2400$ ← maximum value
D	$x=40$, $y=50$	$P = 20 \times 40 + 30 \times 50 = 2300$
E	$x=0$, $y=50$	$P = 20 \times 0 + 30 \times 50 = 1500$

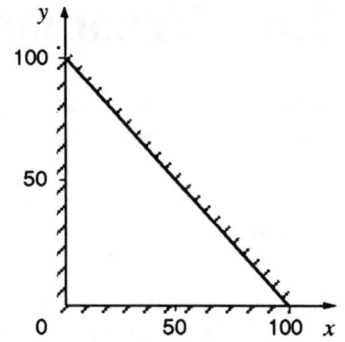

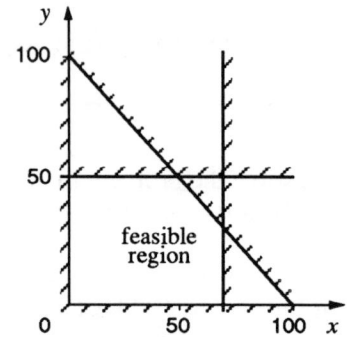

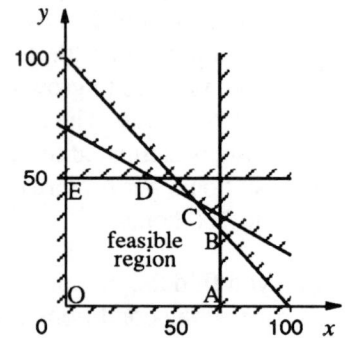

So the optimum solution occurs at $x = 60$, $y = 40$ showing that the farmer should sow 60 hectares of wheat and 40 hectares of sugar beet.

This example illustrates the technique of **linear programming**, which has been used extensively in business and commerce.

The complete theory is beyond the scope of this text (it is, in fact, dealt with in detail in the Decision Mathematics text), but it does indicate the importance of being able to illustrate inequalities.

Activity 7

Return to the problem in Activity 6 and use graphical analysis of the type shown above to find the most economic way of organising the transport.

Exercise 7C

1. Show graphically the region defined by

   $x + y \leq 2$, $y - 2x \leq 0$, $y \geq -1$

2. Show that the region defined by

   $x + y \leq 1$, $x - y \leq 1$, $y \leq 1$

   is finite.

3. Determine the region defined by

   $x + 2y \leq 4$, $y \geq x$, $y \leq 2x$

   If, in addition, $x \leq \frac{1}{2}$, are there any values of x and y which satisfy all the inequalities?

7.5   Further problems

There are a number of practical problems which require a solution to two equations at the same time. A pair of equations like this are called **simultaneous** ('at the same time'). They can be solved using graphs or algebra, as you have seen earlier. Here are some further problems of this type.

Example

A business woman frequently has to fly to Edinburgh to see clients. On her arrival, she hires a car from one of two firms, which charge different rates. One company hires cars at a rate of 20p per mile, whilst the other charges £24 per day, plus 8p per mile. Find the length of journey for which both companies charge the same amount, assuming that the business woman never requires a car for more than one day. Find the corresponding cost.

Solution

If x is the mileage travelled and y is the charge made by a company in pence, then the charge made by the first company is given by the equation :

$$y = 20x \qquad (1)$$

Similarly, the charge for the second company is given by :

$$y = 8x + 2400 \qquad (2)$$

These two lines can be drawn on a graph, as shown opposite.

The distance for which both companies charge the same is 200 miles, shown in the graph. This is the value of x at which the two lines cross; that is, the point $x = 200$ and $y = 4000$ lies on both lines. So $x = 200$, $y = 4000$ make equations (1) and (2) true at the same time. The mileage required is 200 and the cost 4000p or £40.

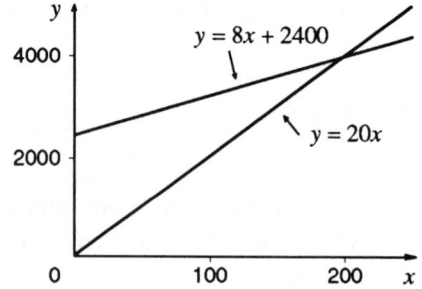

The two equations above could easily be solved by drawing their graphs and reading the x and y values of the point where they cross.

This method of drawing graphs will work for any problem which requires two linear equations to be solved simultaneously, but there are times when the accuracy of the method is not satisfactory. On these occasions, algebraic methods are used.

For example, for the above problem, the lines intersect when

$$20x = 8x + 2400$$
$$\Rightarrow \quad 12x = 2400$$
$$\Rightarrow \quad x = 200$$

Example

A business man is travelling from Britain to Germany and needs to change pounds to Deutschmarks for the trip. If he does this in Britain, he is charged a £2 administration fee, with an exchange rate of £1 to 2.86 DM. In Germany, the administration fee is £3.50, with an exchange rate of £1 to 2.9 DM. He needs to know the amount of money in pounds when it becomes more economical to exchange the money in Germany.

Solution

Let x be the amount of money being changed in pounds, and y the number of Deutschmarks received. Then in Britain bureau, x and y are related by the equation

$$y = (x - 2) \times 2.86$$

(£2 is deducted from the number of pounds to be changed, x, before multiplying by the exchange rate of 2.86 DM/£). Thus

$$y = 2.86x - 5.72$$

Similarly, in Germany , x and y are related by the equation

$$y = (x-3) \times 2.9$$
$$\Rightarrow \quad y = 2.9x - 8.7$$

The graphs of these equations could be drawn on graph paper, and the value of x at the point where they cross will solve the problem. However, it is unlikely that the solution will be easy to read off the graph, so an algebraic method gives better accuracy.

The two lines cross when

$$2.86x - 5.72 = 2.9x - 8.7$$
$$\Rightarrow \quad 8.7 - 5.72 = 2.9x - 2.86x$$
$$\Rightarrow \quad 2.98 = 0.04x$$

giving $x = \dfrac{2.98}{0.04} = 74.50$

So, if the business man requires to change more than £74.50, it will be better to change the money in Germany.

Exercise 7D

1. By drawing the graphs of both equations on the same axes, solve them simultaneously. Both x and y values are required at the point of intersection.

   (a) $x + y = 5$ and $x + 2y + 2 = 0$

   (b) $2y = -x$ and $y - 3x - 7 = 0$

   (c) $y = 2x + 1$ and $3y - x + 2 = 0$

2. By considering the gradients of

   $y = \frac{1}{2}x + 4$ and $2y - x + 7 = 0$, explain why there is no simultaneous solution for these two equations.

3. A farmer has enough land to keep 300 animals in all. He can buy sheep at £23 each, or goats at £36 and has £10,000 to spend altogether. Let x be the number of sheep bought, and y the number of goats.

   (a) Form an equation for x and y which states that the total number of sheep and goats is 300.

   (b) Form another equation stating that the total cost of x sheep and y goats is £10 000.

   (c) Solve simultaneously the equations you have formed using a graph.

*4. A triangle is formed by three lines with the equations given below.

   $$y = 2x, \quad y = -3x + 5 \quad \text{and} \quad 2y = x - 6.$$

   It has a corner (vertex) when any pair of the lines meet. Find the co-ordinates of all three vertices.

7.6 Cartesian coordinates

The x and y axes frequently used in mathematics form part of a system called Cartesian coordinates, named after a French mathematician and philosopher, *René Descartes*. It has been found to be one of the most convenient ways of describing how things are related in space.

As an example of the usefulness of such a system, consider the grid reference system which in affect is a set of Cartesian coordinates, each 100 m standing for a unit on a typical OS (Ordnance Survey) map.

Example

A helicopter pilot is told to fly from grid reference (115, 208) to (205, 088) (Luton Airport to Hatfield Aerodrome). The figure opposite illustrates the journey.

The pilot needs to know the distance in order to estimate his time of his arrival. Using Pythagoras' theorem, calculate the distance to be flown.

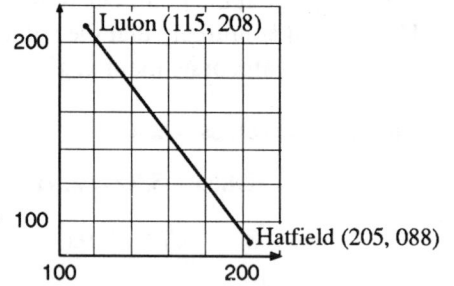

Solution

The figure shows the right angled triangle to be used. Note that the coordinates are given in 100 m units. By Pythagoras' theorem, the distance d in metres can be calculated from

$$d^2 = (20800 - 8800)^2 + (20500 - 11500)^2$$
$$\Rightarrow \quad d^2 = 12000^2 + 9000^2$$
$$\Rightarrow \quad d^2 = 225000000$$

Hence

$$d \quad = \sqrt{225000000}$$
$$\Rightarrow \qquad = 15000 \text{ m}$$
$$\Rightarrow \qquad = 15 \text{ km}$$

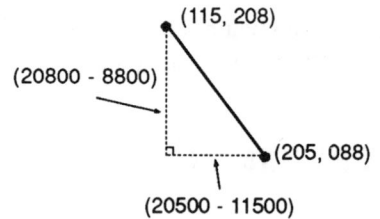

This example illustrates the method for calculating the distance between any two points on a Cartesian co-ordinate grid.

If a point A has coordinates (x_A, y_A), and point B is (x_B, y_B), then using Pythagoras' theorem the distance AB given by

$$AB = \sqrt{(y_B - y_A)^2 + (x_B - x_A)^2}$$

The formula applies even when coordinates are negative.

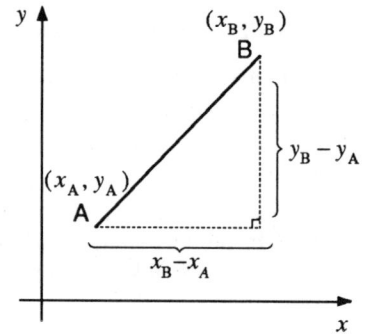

Example

The points A(1, 1), B(–3, 3), C(–1, 7) and D(3, 5) form a square. Find the area of the square and the point P at which its diagonals cross.

Solution

The figure opposite shows the square ABCD. The area of ABCD is simply the square of the length of one of its sides. The length of any of the four sides can be found using Pythagoras' thoerem, as above.

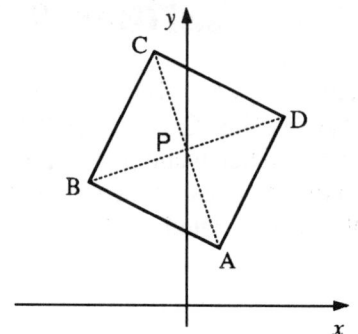

For example

$$AB = \sqrt{\left(y_A - y_B\right)^2 + \left(x_A - x_B\right)^2}$$
$$= \sqrt{\left(1 - 3\right)^2 + \left(1 - (-3)\right)^2}$$
$$= \sqrt{\left(-2\right)^2 + \left(4\right)^2}$$
$$= \sqrt{4 + 16}$$
$$= \sqrt{20}$$

So the area of $ABCD = (AB)^2 = \left(\sqrt{20}\right)^2 = 20$ units^2.

Now P is the midpoint of either diagonal. One of the diagonals, BD, is shown opposite. Since P is the midpoint, its coordinates will be the average of those of B and D.

$$x: \quad \frac{1}{2}(-3+3) = \frac{1}{2} \times 0 = 0$$

$$y: \quad \frac{1}{2}(3+5) = \frac{1}{2} \times 8 = 4$$

So P has coordinates $(0, 4)$, which can be verified from the figure.

In general, the coordinates of the midpoint of a line AB are found by averaging the coordinates of A and B, to give are the midpoint coordinates as

$$\left(\frac{x_A + x_B}{2}, \frac{y_A + y_B}{2} \right)$$

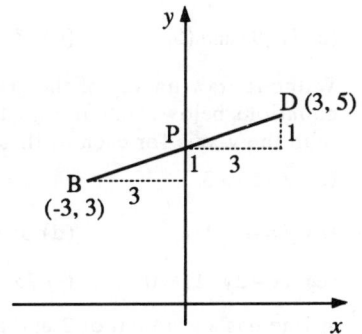

Exercise 7E

1. Find the distances between these pairs of points using Pythagoras theorem

   (a) $(0, 0)$ and $(3, 4)$ (b) $(-1, 3)$ and $(11, 2)$

   (c) $(4, 1)$ and $(-2, -5)$ (d) $(0, 6)$ and $(3, -8)$

   Give answers to three significant figures where necessary.

2. Find the midpoints of the lines joining the pairs of points in Question 1.

3. Calculate the area of the right angled triangle formed by the points $A(1, 3)$, $B(2, 5)$ and $C(5, 1)$. The right angle is formed at the point A.

4. Show that the quadrilateral PQRS is a square, where P is the point $(-2, 5)$, Q is $(-5, 1)$, R is $(-1, -2)$ and S is $(2, 2)$.

7.7 Miscellaneous Exercises

1. A bank charges £10 for 850Fr and £20 for 1900Fr. Plot these data on a graph, with pounds on the horizontal axis, and draw a straight line graph through the points. Find the gradient of the graph, stating its units and its meaning.

2. A car entering a town decelerates from 26.8 ms^{-1} to 13.4 ms^{-1} s in 8 seconds. Draw a graph showing the velocity, v, on the vertical axis and time t, on the horizontal. Find the gradient of the line joining the points, stating the units and the meaning of the gradient.

3. Find the gradient of the lines joining these pairs of points :

   (a) $(2, 1)$ and $(8, 7)$ (b) $(-3, 6)$ and $(1, 12)$

   (c) $(1, 9)$ and $(5, 3)$ (d) $(3, -6)$ and $(12, -20)$

4. Without drawing any of the graphs for the equations below, state the gradient and intercept with the y-axis for each of these lines :

   (a) $y = 5x + 3$ (b) $y = -x + 1$

   (c) $y = \frac{1}{2}x$ (d) $3y = x + 6$

   (e) $5x - 2y - 11 = 0$ (f) $3x + 4y + 1 = 0$

5. A line has a gradient of 2 and passes through the point $(5, -1)$ Find its equation.

6. Find the equation of the line parallel to the line $y = 4x - 1$ which passes through the point $(-3, 9)$

7. Find the equation of the straight line which passes through $(0, 5)$ and $(3, 1)$.

8. An experiment was carried out to see the extension, e cm, produced when different weights were hung from a spring. The results are shown in this table.

Weight, w (grams)	0	100	200	300	400
Extension, e (cm)	0	5	11	14.5	21

Plot these data on a graph, with e on the vertical axis, and draw a line of best fit. Find the gradient of the line and the intercept with the vertical axis, and hence find an equation relating e to w. Use your equation to estimate the extension produced when the weight on the spring is 1 kg.

9. Find the equation of the line which crosses the line $y = 4x + 2$ at the point $(3, 14)$ at right angles.

10. Find the equation of the line which is perpendicular to the line $3y - 2x + 5 = 0$ and which passes through the point $(1, 1)$.

11. Find the equation of the line parallel to $y = 5x - 1$ which passes through the point $(4, 0)$.

12. Two sides of a square are formed by the lines $y = 3x$ and $3y + x - 6 = 0$. Find the coordinates of the corner of the square at which these sides meet using an algebraic method.

13. Find the area of the rectangle formed by the points $P(-3, 5), Q(-8, -7)$ $R(0, -13)$ and $S(5, 1)$. Find also the midpoints of the sides PQ and RS, and so find the equation of the line joining these midpoints.

8 RATES OF CHANGE

Objectives

After studying this chapter you should

- appreciate the connection between gradients of curves and rates of change;
- know how to find the gradient at any point on a curve;
- be able to find the maximum and minimum points;
- understand and know how to find equations of tangents and normals to curves.

8.0   Introduction

Most things change: the thickness of the ozone layer is changing with time; the diameter of a metal ring changes with temperature; the air pressure up a mountain changes with altitude. In many cases, however, what is important is not whether things change, but how fast they change.

The study of rates of change has an important application, namely the process of **optimisation**. An example of an optimisation problem that you have already met is deciding what proportions a metal can should have in order to use the least material to enclose a given volume.

You may have seen signs like the one in the photograph before. They are often put by the roadside to discourage drivers on main roads from driving too fast through small towns or villages; it is in such places that the police often set up 'speed traps' to catch drivers who are exceeding the speed limit.

Activity 1

The town of Dorchester in Dorset is 2 km from end to end, and a 30 mph speed limit is in force throughout. Although the A35 road now by-passes the town, many drivers consider it quicker, late at night when traffic is light, to drive through the centre.

A driver takes 2 minutes 40 seconds to drive through the town. Was the speed limit broken?

What is the shortest time a driver can take to drive through Dorchester and not break the speed limit?

In reality cars do not travel at a constant speed. Suppose the driver's progress through the town was described by the distance-time graph in Activity 2.

Activity 2   When was the driver speeding?

(a)   Travelling through Dorchester one encounters a major roundabout. Where do you think it is, and how can you tell?

(b)   What was the driver's average speed between

      (i)   Grey's Bridge and the Night Club;

      (ii)   the Night Club and the Hospital;

      (iii)   Cornhill and Glyde Path Road;

      (iv)   the Military Museum and St. Thomas Road

      (v)   Wessex Road and Damers Road?

(c)   Cut out a right-angled triangle as shown in the diagram below.

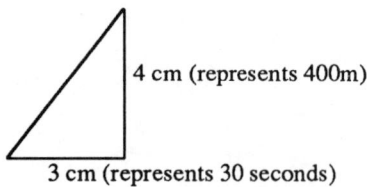

4 cm (represents 400m)

3 cm (represents 30 seconds)

It represents a speed of 30 mph, or $13\frac{1}{3}$ ms^{-1}. Use this to determine when the instantaneous speed of the car was greater than 30 mph.

(d)   Suppose the police set up a speed-trap somewhere in Dorchester. They decide to stop any vehicle going faster than 40 mph. Use a similar method to (c) to determine where along the route the car's speed exceeds 40 mph.

(e)   Assuming it was working accurately, what would the car's speedometer have shown as the car passed

    (i)   the Night Club;

    (ii)   Wessex Road?

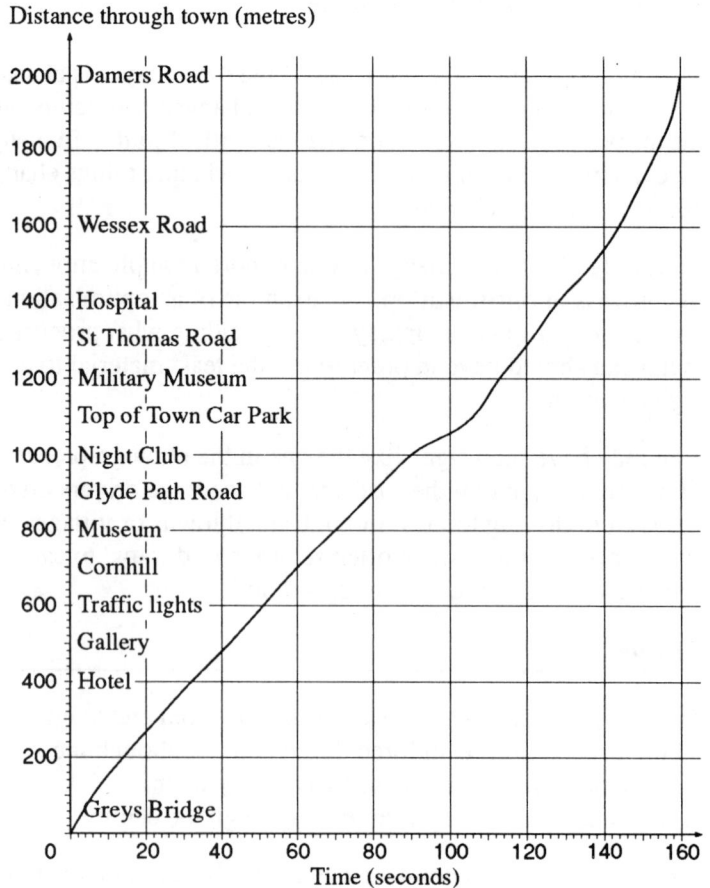

8.1 Instantaneous speed

One method the police use to discover whether or not a car is speeding is to use video cameras to time it between two fixed points. In the above example, suppose the car was timed between Glyde Path Road and the Hospital; then the speed would have been calculated like this

$$\frac{\text{distance travelled}}{\text{time taken}} = \frac{500}{50} = 10 \text{ ms}^{-1} \left(\text{about } 22\tfrac{1}{2} \text{ mph}\right)$$

This figure is only an average speed however. However the car's actual speed varied between these two points, and it may have gone faster than 30 mph and then slowed down.

Another method of finding the speed is to use a 'radar gun', which is focussed on the car as it passes. This gives the **instantaneous** speed of the vehicle, as shown on the speedometer.

On a distance-time graph, the instantaneous speed is indicated by the steepness, or gradient, but when the graph is a complicated curve the gradient is difficult to pin down accurately. One way is to draw a tangent to the curve and to work out it gradient as you saw in Activity 2.

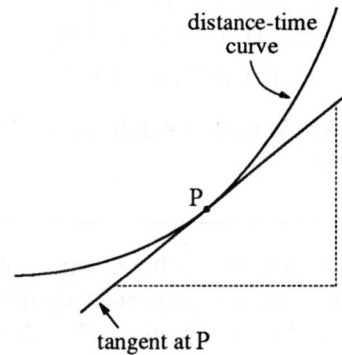

distance-time curve

P

tangent at P

Activity 3   Unemployment Statistics

The table below shows the number of people unemployed in the two years between August 1989 and July 1991. Draw a graph to represent these figures.

Month	Thousands	Month	Thousand
Aug 1989	1741.1	Aug 1990	1657.8
Sept	1702.9	Sept	1673.9
Oct	1635.8	Oct	1670.9
Nov	1612.4	Nov	1728.1
Dec	1639.0	Dec	1850.4
Jan 1990	1687.0	Jan 1991	1959.7
Feb	1675.7	Feb	2045.4
Mar	1646.6	Mar	2142.1
Apr	1626.3	Apr	2198.5
May	1578.5	May	2213.8
June	1555.6	June	2241.0
July	1623.6	July	2367.5

(a) What figures are missing from these political press releases?

Labour Party
In the 12 months
following Jan. 1990
unemployment rose
at an average rate of
?
per month.

Conservative Party
In the 3 months
following Jan. 1990
unemployment fell
at an average rate of
?
per month.

(b) Which of these statements gives a truer impression of :
   (i) the rate of change of unemployment at the start of 1990;

   (ii) the unemployment trend during 1990 as a whole?

(c) Use your graph to find the rate of change of unemployment
   (i) at January 1991

   (ii) at October 1989

Discuss the meaning, relevance and accuracy of your answers.

Activity 3 gave another example of a graph where the **gradient**, or steepness, had an important significance. It gave the rate of change of unemployment. This property of the gradient has important applications to all kinds of graphs.

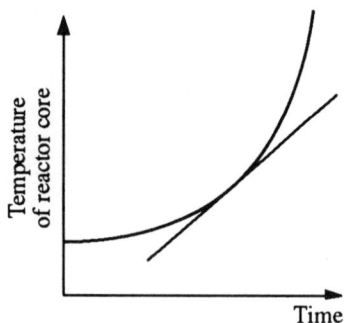

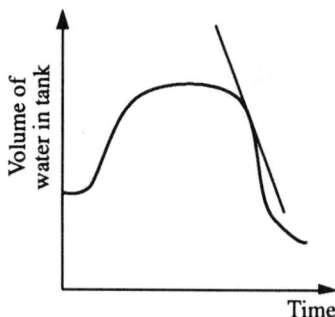

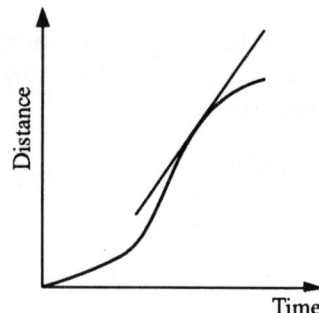

Rates of change have meaning even when neither of the variables is time. For example, the graph opposite shows how the temperature changes with height above sea level. The gradient of the tangent at P is $3°C/km$. This indicates that, 1km above P, the temperature will be approximately $3°C$ higher. The rate of change of temperature with distance is sometimes called the **temperature gradient**.

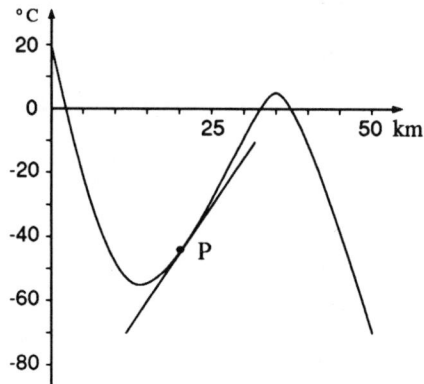

In **rugby union** a try, worth 4 points, is scored by touching the ball down behind the line of the goal-posts. An extra 2 points can be scored by subsequently kicking the ball between the posts, over the cross-bar. This is known as a 'conversion': 4 points is converted into 6 points. The conversion kick must be taken as shown in the diagram, from somewhere on the line perpendicular to the goal line through the point where the ball was touched down.

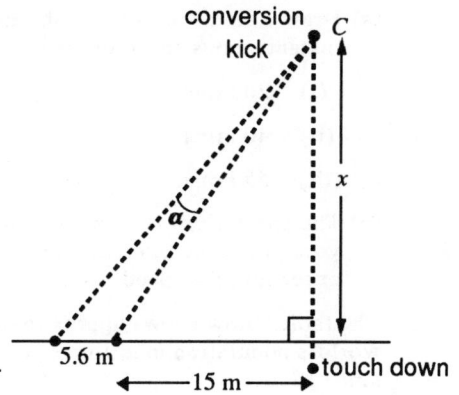

Activity 4   Optimum kicking point

Imagine that a try has been scored 15 m to one side of the goal. The angle in which the ball must be propelled is marked α; the smaller the angle, the trickier the kick. The angle depends on how far back the kick is taken. The way it changes is shown in the table.

Distance (x)	0	5	10	15	20	25	30	35	40	45	50
Angle (α)	0	4.8	7.8	8.9	9.0	8.5	7.9	7.3	6.7	6.2	5.7

(a)   Draw a graph of angle against distance.

(b)   Estimate the gradient of the graph at $x = 5$ and $x = 15$.

(c)   What connection do the figures in (b) have with rates of change? What do these figures tell you?

(d)   Estimate the gradient of the curve when $x = 40$. Interpret your answer.

(e)   What is the best point from which to take the kick? What is the gradient at this point?

What assumptions have been made about taking the 'conversion' in this activity?

Exercise 8A

1.   Harriet is a passenger in a car being driven along a motorway. She monitors the progress of the journey by counting the distance markers by the roadside. She writes down the distance travelled from the start every 5 minutes.

Time (minutes)	0	5	10	15	20	25	30	35	40	45	50	55	60
Distance travelled (miles)	0	6.4	13.1	20.0	26.2	31.3	35.5	39.2	43.4	47.9	53.1	59.8	67.0

(a) Draw a distance-time graph. Estimate the instantaneous speed of the car in mph after

  (i)   20 mins.

  (ii)   40 mins.

  (iii)   55 mins.

(b) The speed limit is 70 mph. Estimate from your graph the times at which the car was exceeding the speed limit.

2. The table below shows approximately how the world's population in millionshas increased since 1700.

Year	1700	1720	1740	1760	1780	1800	1820	1840
Pop.	560	610	670	730	790	850	940	1050

Year	1860	1880	1900	1920	1940	1960	1980
Pop.	1170	1330	1550	1870	2270	3040	4480

(a) Draw a graph to represent these data. Draw tangents at the years 1750, 1800, 1850, 1900 and 1950 and measure their gradients.

(b) Explain what meaning can be attached to the gradients in (a) and write a brief account of what they show.

(c) Find the gradient at the year 1880 and use your answer to estimate the population in 1881.

3. The height h of a stone above the ground is given by the formula $h = 2 + 21t - 5t^2$, where h was measured in metres and t in seconds.

(a) Draw a graph of h against t, for values of t between 0 and 5.

(b) Estimate the velocity of the stone after 1, 2, 3 and 4 seconds. Make sure your method is clear.

(c) Use one of your answers to (b) to estimate the value of h when $t = 1.1$. Check the accuracy by substituting $t = 1.1$ into the original formula.

4. The time taken to travel 120 miles depends on the average velocity v, according to the formula $t = 120 / v$,

where t is in hours and v is in miles per hour. This relationship is shown in the graph below.

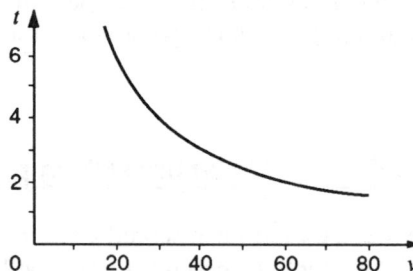

(a) The gradient of the graph when $v = 20$ is -0.3. What does this figure mean?

(b) The gradient when $v = 50$ is -0.048. Given that $t = 2.4$ at this point, estimate t when $v = 51$. How accurate is this estimate?

(c) Describe briefly how the gradient changes as v increases. Explain in everyday terms what this means.

8.2   Finding the gradient

Question 3 of the last exercise required you to draw the graph of the function $h = 2 + 21t - 5t^2$ and to find the gradient at certain points. If you compare your answers to someone else's you may well find that they do not agree precisely; this is because the process of drawing tangents is not a precise art - different people's tangents will have slightly different slopes. Moreover, the process of drawing and measuring tangents can become tiresome if repeated too often - you may well agree!

If the function had been $2 + 21t$ then finding the gradient would have been easy. The function $2 + 21t - 5t^2$ is not linear but there is more than one way of getting an accurate value for the gradient at any point, as you will see in the next activity.

Activity 5   Finding the gradient

(a) Plot the graph of $y = 2 + 21x - 5x^2$ using a graph-plotting facility. Zoom in on the curve in the region of $x = 1$. The further in you go, the more the curve will resemble a straight line.

Use the calculator or computer to give the coordinates of two points very close to $(1, 18)$. Use these coordinates to give an estimate of the gradient.

What do you think the exact gradient is?

Repeat this process for the point $(2, 24)$.

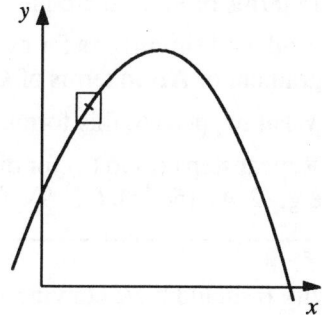

(b) Consider the point A $(1, 18)$. A nearby point on the curve, B, has coordinates $(1.1, 19.05)$. What is the gradient of the line AB?

Find the y-coordinates when $x = 1.01$ and $x = 1.001$.

Label these points B_1 and B_2. Find the gradients of AB_1 and AB_2. Can you infer the exact value of the gradient of the curve at A?

Repeat this process for the point $(3, 21)$.

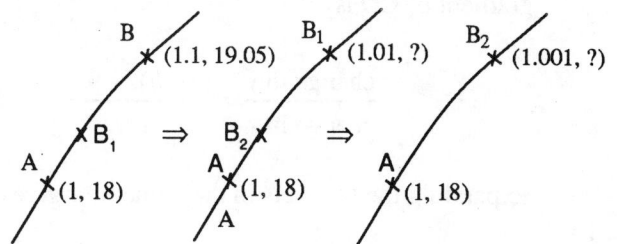

The methods developed in Activity 5 have the advantage of accuracy, but they still take time. A more efficient method is desirable, and the next two Activities examine the simplest non-linear function of all with a view to finding one.

Activity 6   Gradient of x^2

Use the methods above to find the gradient of $y = x^2$ at different points. (Do not forget negative values of x). Make a table of your results and describe anything you notice.

Activity 7   General approach

The diagram shows $y = x^2$ near the point (1, 1), labelled A. The point B is a horizontal distance h along from A.

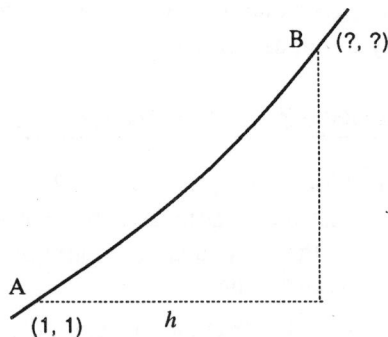

(a) In terms of h, what are the coordinates of B?

(b) Find, and simplify as far as possible, a formula for the gradient of AB in terms of h.

(c) What happens to this formula as B gets closer to A?

(d) Repeat steps (a) to (c) for different positions of the point A, e.g. (2, 4), (5, 25), (–3, 9). Generalise as far as you can.

Activity 6 should have convinced you that at any point of $y = x^2$ the gradient is double the x–coordinate. For example, at the point (3, 9) the gradient is 6.

Activity 7 gives an algebraic way of getting the same answer. In the second diagram, D has coordinates $(3 + h, (3 + h)^2)$. Thus the gradient of CD is

$$\frac{\text{change in } y}{\text{change in } x} = \frac{(3 + h)^2 - 9}{h}$$

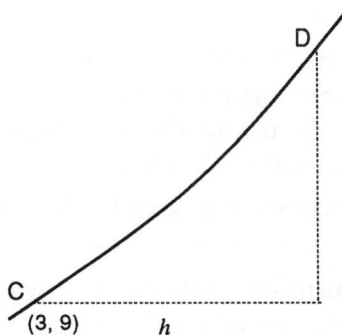

Expanding the brackets in the numerator gives

$$\frac{9 + 6h + h^2 - 9}{h} = \frac{6h + h}{h}$$

$$= \frac{h(6 + h)}{h}$$

$$= 6 + h$$

Hence, as D gets closer to C (i.e. as $h \to 0$), the gradient of CD gets closer to 6.

This procedure can be generalised. Suppose the point (3, 9) is replaced by the general point (x, x^2). In the diagram this point is denoted P, and Q has coordinates $(x + h, (x + h)^2)$.

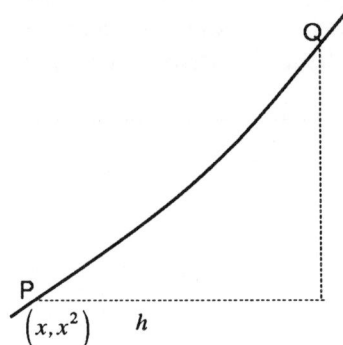

The gradient of PQ is

$$\frac{(x+h)^2 - x^2}{h} = \frac{x^2 + 2hx + h^2 - x^2}{h}$$

$$= \frac{2hx + h^2}{h}$$

$$= \frac{h(2x+h)}{h}$$

$$= 2x + h \text{ (dividing by } h)$$

This suggests that the gradient at (x, x^2) is $2x$, that is, double the x-coordinate.

Activity 8   Alternative derivations

The above approach is based on considering a point B 'further along' the curve from A.

(a) Suppose A is the point $(3, 9)$ and B is a distance h along the x-axis in the negative direction. Find a formula for the gradient of AB and see whether it gets closer to 6 as B gets closer to A.

(b) Now, suppose B is one side of A and C is the other. Find a formula for the gradient of BC and comment on anything of interest.

(c) Generalise (a) and (b) to the point (x, x^2).

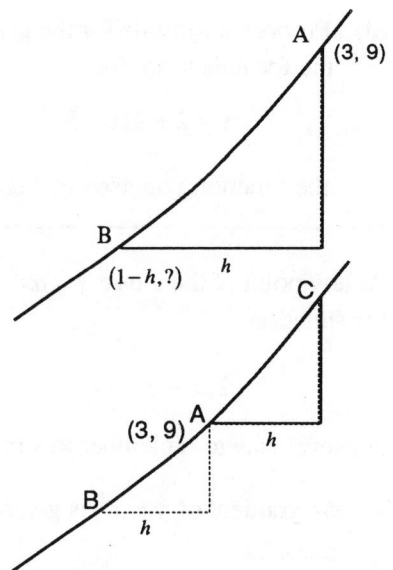

8.3   Gradient of quadratics

You should now be familiar with three methods of finding gradients of curves :

- using a graphic calculator to zoom in;
- calculating gradients of chords;
- using algebra.

The next activity may take some time, according to the method you choose. Its purpose is to establish a formula for the gradient of any quadratic curve, that is, any curve with an equation of the form $y = ax^2 + bx + c$. The method you use, out of the three above, is up to you.

Activity 9   Gradient of any quadratic

(a)   Find formulas for the gradients of

$$y = 2x^2, \ y = \tfrac{1}{4}x^2, \ y = 3x^2, \ y = \tfrac{1}{2}x^2.$$

What is the formula for the gradient of $y = ax^2$?

(b)   Find formulas for the gradients of

$$y = x^2 + 3, \ y = x^2 - 8, \ y = x^2 + 1.$$

What is the formula for the gradient of $y = x^2 + c$? Explain why this is true.

(c)   Repeat for equations of the form $y = x^2 + bx$, where b is any number.

(d)   Propose a formula for the gradient of $y = ax^2 + bx + c$. Does the formula work for

$$y = 2 + 21x - 5x^2,$$

the function you used in Activity 5?

At any point of the curve $y = ax^2 + bx + c$ the gradient is given by the function

$$2ax + b.$$

A useful way to remember this rule is as follows :

-   the gradient of $y = x^2$ is given by $2x$, so the gradient of $y = ax^2$ is $a \times 2x$

-   the gradient of $y = bx$, a straight line, is just b;

-   adding a constant, c, merely moves the curve up or down and does **not** alter the gradient;

-   the formula $2ax + b$ comes from combining these properties.

Example

Find formulas for the gradients of these curves :

(a) $y = 5x^2 - 7x + 10$

(b) $s = \tfrac{5}{8}t - \tfrac{1}{6}t^2$

(c) $q = \dfrac{2p^2 - 7p + 8}{3}.$

Solution

(a) gradient $= 5 \times (2x) - 7 = 10x - 7$

(b) gradient $= \frac{5}{8} - \frac{1}{6} \times (2t) = \frac{5}{8} - \frac{1}{3}t$

(c) gradient $= \dfrac{2 \times (2p) - 7}{3}$

$\qquad\qquad = \frac{4}{3}p - \frac{7}{3}$

Can you see why the denominator of 3 is 'untouched' in part (c)?

Example

If $y = \dfrac{x^2 - 7x}{10} + 17$ find the gradient when $x = 15$.

Solution

To answer this question, find the gradient function and then substitute the value 15 for x. Now

$$\text{gradient} = \frac{2x - 7}{10}.$$

When $x = 15$, gradient $= \dfrac{2 \times 15 - 7}{10} = \dfrac{23}{10} = 2.3$

Exercise 8B

1. Find formulas that give the gradients of these curves

   (a) $y = 2x^2$

   (b) $y = x^2 + x$

   (c) $s = t^2 + 4t - 8$

   (d) $y = x^2 - x + 10$

   (e) $h = 6l^2 - 7$

   (f) $y = 10x - \dfrac{x^2}{5}$

   (g) $T = \dfrac{1}{9}Y^2 + 3Y - 1$

   (h) $A = \dfrac{n^2 - 5n + 10}{2}$

   (i) $u = v - 6v^2 + \dfrac{1}{15}$

   (j) $y = \dfrac{3}{4}x^2 + \dfrac{x}{5} + 2$

2. Find the gradients of

   (a) $y = 10 + 5x - 3x^2$ when $x = 2$

   (b) $p = \dfrac{T^2}{2} + 8T - 16$ when $T = -3$

   (c) $9 = 3u^2 - \dfrac{u}{6}$ when $u = 4$

   (d) $y = \dfrac{x^2 + 7x - 3}{12}$ when $x = 10$

   (e) $m = 100 + 65N - \dfrac{N^2}{5}$ when $N = -15$.

3. The gradient of the graph of $h = 2 + 21t - 5t^2$ gives the speed of a stone where h is the height in metres and t the time in seconds. Find the speed

    (a) when $t = 0.5$

    (b) when $t = 2.8$.

8.4 Differentiation

The process of finding 'gradient functions' is called **differentiation**. The proper name for the gradient function is the derivative or derived function. Hence the function

$3x^2 - 12x + 5$ is **differentiated** to give $6x - 12$

$6x - 12$ is the **derivative** of $3x^2 - 12x + 5$.

The inventor of this technique is generally thought to have been *Sir Isaac Newton*, who developed it in order to explain the movement of stars and planets. However, the German mathematician *Gottfried Wilhelm Leibniz* ran him close, and it was Leibniz who was the first actually to publish the idea, in the year 1684. Much vigorous and acrimonious discussion ensued as to who discovered the technique first. Today both are saluted for their genius.

The notation used by Leibniz is still used today. The gradient of a straight line is

$$\frac{\text{change in } y}{\text{change in } x}$$

which he shortened to $\dfrac{dy}{dx}$ (read as 'dy by dx').

The above example could be written thus :

$$y = 3x^2 - 12x + 5 \Rightarrow \frac{dy}{dx} = 6x - 12$$

or alternatively

$$\frac{d}{dx}(3x^2 - 12x + 5) = 6x - 12,$$

the symbol $\dfrac{'d'}{dx}$ standing for the derivative with respect to x.

Another way of denoting a derived function is to use the symbol f', as follows:

$$f(x) = 3x^2 - 12x + 5 \qquad \text{(function)}$$

$$f'(x) = 6x - 12 \qquad \text{(derived function)}$$

Activity 10   Differentiating $y = x^3$

The aim of this Activity is to find the derivative of the function $y = x^3$. There is more than one way to accomplish this; it can be done numerically by finding gradients at different points; or it can be done algebraically. You should attempt at least one of (a) or (b) in the Activity.

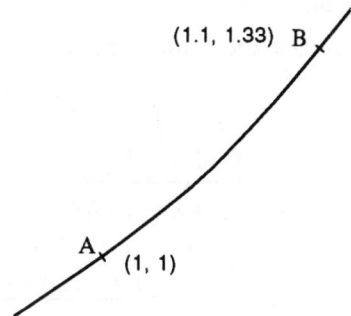

(a) Either by using a graphic calculator or by considering a nearby point, find the gradient of the curve $y = x^3$ at the point $(1, 1)$. Repeat for the points $(2, 8)$, $(3, 27)$, $(4, 64)$, and for negative values of x. Can you establish a formula for the gradient?

(b) In the diagram, C is the point with x-coordinate $1 + h$ and A is the point $(1, 1)$. Explain why the gradient of AC is

$$\frac{(1+h)^3 - 1}{h}$$

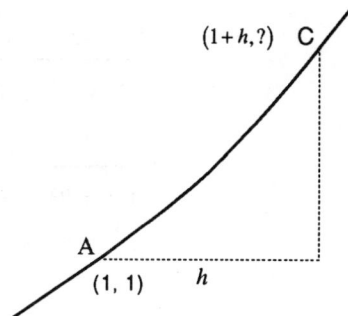

Expand $(1 + h)^3$ by treating it as

$$(1+h)(1+h)^2 = (1+h)(1+2h+h^2).$$

Hence simplify the formula for the gradient and deduce the gradient of the graph at $(1, 1)$.

Repeat this process for the points $(2, 8)$, $(3, 27)$, $(4, 64)$. Generalise to any point on the curve.

You may have established the derivative of the function x^3. In case you didn't the details are given below.

With reference to the diagram opposite, the gradient of PQ is given by

$$\frac{(x+h)^3 - x^3}{h}$$

$$= \frac{x^3 + 3x^2 + 3xh^2 + h^3 - x^3}{h}$$

$$= 3x^2 + 3xh + h^2$$

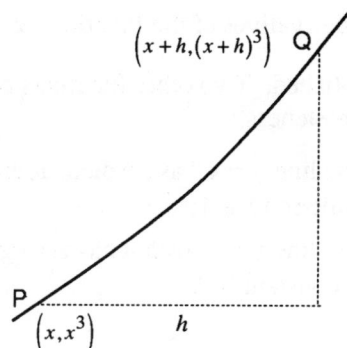

Whatever the value of x, this gradient gets closer and closer to $3x^2$ as $h \to 0$, so

$$\frac{dy}{dx} = 3x^2$$

Example

Find the derivative of $y = \frac{1}{x}$ $(x \neq 0)$

Solution

In this case the gradient of AB is

$$\left(\frac{1}{x+h} - \frac{1}{x} \right) \frac{1}{h}$$

$$= \left(\frac{x - (x+h)}{(x+h)x} \right) \frac{1}{h}$$

$$= \left(\frac{-h}{x(x+h)} \right) \frac{1}{h}$$

$$= \frac{-1}{x(x+h)}$$

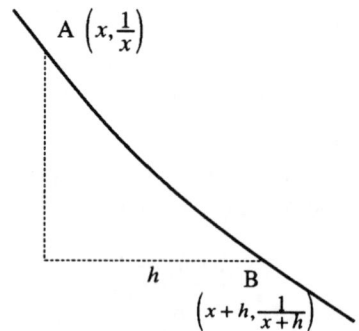

As h gets closer to zero, this formula gets closer to $\frac{-1}{x^2}$. Hence

$$\frac{dy}{dx} = -\frac{1}{x^2} \ (x \neq 0)$$

The derivatives of the functions x^2, x^3 and $\frac{1}{x}$ have now been established. Two other functions can be added to those, for completeness :

- the line $y = x$ has gradient 1, and so the derivative of the function x is 1;

- the line $y = $ constant has zero gradient, and so the derivative of a constant is 0.

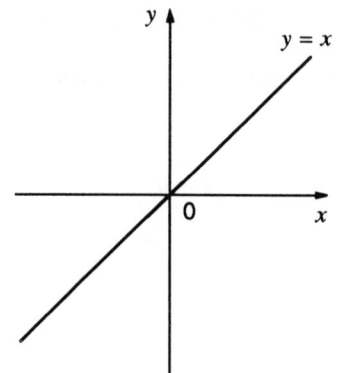

A summary of the results obtained so far is as follows :

Function	Derivative
constant	0
x	1
x^2	$2x$
x^3	$3x^2$
$\dfrac{1}{x}$	$-\dfrac{1}{x^2}$

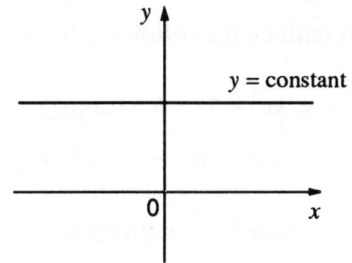

You may be able to guess the derivatives of higher powers of x; this and other matters will be covered in the last section of this chapter.

In Section 8.2, it was observed that the derivative of, for example, $5x^2$ was 5 times the derivative of x^2.

Similarly, the derivative of $5x^2$ is 5 times the derivative of x^3 :

$$\frac{d}{dx}(5x^3) = 5 \times \frac{d}{dx}\left(x^3\right)$$
$$= 5 \times 3x^2$$
$$= 15x^2$$

Another assumption to make is that functions such as $x^2 + \dfrac{1}{x}$ can be differentiated by adding together the derivatives of x^2 and $\dfrac{1}{x}$:

e.g. $\dfrac{d}{dx}\left(x^2 + \dfrac{1}{x}\right) = \dfrac{d}{dx}(x^2) + \dfrac{d}{dx}\left(\dfrac{1}{x}\right) = 2x - \dfrac{1}{x^2}$

Justification for this assumption will be given later on.

Example

Differentiate the following functions :

(a) $y = 3x^3 - 5x + 6$ with respect to x

(b) $y = x(x-3)(x+4)$ with respect to x

(c) $A = 10q^3 - \dfrac{5}{q}$ with respect to q

(d) $P = \dfrac{(h^3 + 3)}{2h}$ with respect to h

Solution

(a) $\dfrac{dy}{dx} = 3\dfrac{d}{dx}(x^3) - 5\dfrac{d}{dx}(x) + \dfrac{d}{dx}(6) = 3(3x^2) - 5 = 9x^2 - 5$

(b) $y = x^3 + x^2 - 12x$ (brackets must first be multiplied out)

   $\dfrac{dy}{dx} = 3x^2 + 2x - 12$

(c) $\dfrac{dA}{dq} = 10(3q^2) - 5(-\dfrac{1}{q^2})$ (note $\dfrac{dA}{dq}$ instead of $\dfrac{dy}{dx}$ as the

derivation is of A with respect to q)

   $= 30q^2 + \dfrac{5}{q^2}$

(d) $P = \dfrac{h^2}{2} + \dfrac{3}{2h}$ (function must be divided out)

   $\dfrac{dP}{dh} = \dfrac{2h}{2} + \dfrac{3}{2}(-\dfrac{1}{h^2}) = h - \dfrac{3}{2h^2}$

Exercise 8C

1. Find the derivative of the following functions :

   (a) $y = x^3 + 5x^2 + 3x$ with respect to x

   (b) $r = 6t^3 - 10t^2 + 2t$ with respect to t

   (c) $f(x) = 5x^2 + \dfrac{1}{x}$ with respect to x

   (d) $g(x) = x^2(x - \dfrac{1}{x})$ with respect to x

   (e) $f(t) = \dfrac{t^3 + 3t}{5}$ with respect to t

2. Differentiate these functions :

   (a) $(x+2)^2$

   (b) $x(x+1)(x-1)$

   (c) $s(s+\tfrac{1}{3})^2$

   (d) $\dfrac{8y^2 + 3y^2}{9} + 3$

   (e) $\dfrac{x^4 - 5x^2 - 1}{x}$

3. (a) What is the gradient of the curve

    $y = x^3 - 3x^2 + 6$ at the point (3, 6)?

   (b) What is the gradient of the curve

    $$y = 2x - \frac{5}{x}$$

    at the point (2, 1)?

   (c) At what point is the gradient of

    $y = x^2 + 6x + 3$ equal to 10?

   (d) When is the tangent to the curve

    $y = 3x^2 - 5x + 10$ parallel to the line
    $y = 20 - 11x$?

   (e) At what two points is the gradient of

    $y = 2x^3 - 9x^2 + 36x - 11$ equal to 24?

4. A student suggests that the height of the average male (beyond the age of 3) can be modelled according to the formula

    $$h = 6 - \frac{12}{y}$$

   where h is the height in feet and y is the age in years.

   Use this model to find the rate of growth of the average male (in inches per year) at the ages of

   (a) 6   (b) 8

8.5 Optimisation

Here is a problem similar to that at the start of Chapter 6. A piece of card 20 cm by 20 cm has four identical square pieces removed from the corners so that it forms a net for an open-topped box. The problem this time is not to make a specific volume but to find the dimension of a box with the largest volume.

Activity 11 Maximising the volume

(a) Write down a formula for the volume V in terms of x.

(b) Sketch a graph of V against x for all the allowable values of x.

(c) Find the gradient of the graph when $x = 1, 2$ and 3. Interpret these figures, in terms of rates of change.

(d) What is the gradient when $x = 4$? Interpret your answer.

(e) Find the co-ordinates (x, V) where the gradient is zero. What is the significance of this?

Activity 12   Stationary points

The graph opposite shows a function $f(x)$. Copy the graph and underneath sketch a graph of the derivative $f'(x)$.

(The graph of $f(x)$ should not attempt to be accurate. It should be made clear where the gradient is positive, where it is negative, and where it is zero.)

The graph in the last Activity contained three examples of stationary points. This is the general term used to describe maximum and minimum points. At a stationary point the gradient of the graph is zero; the tangent is exactly horizontal.

Activity 11 showed how useful this fact is. The **maximum** and **minimum** points of any function can be found by working out where the gradient is zero. The process of finding maximum and minimum points is sometimes called **optimisation**. It should also be noted that stationary points can also turn out to be points of inflection, as illustrated opposite.

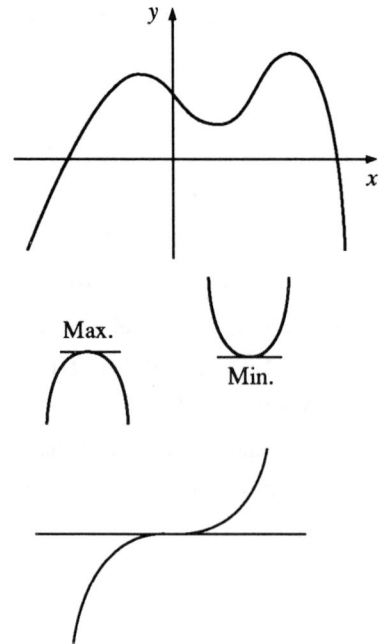

Max.

Min.

Example

Find the largest volume of an open top box that can be made from a piece of A4 paper (20.9 cm by 29.6 cm).

Solution

Suppose squares of side x are cut from each corner. Then the volume is given by

$$V = x(20.9 - 2x)(29.6 - 2x)$$

$$= 618.64x - 101x^2 + 4x^3$$

(Remember: brackets must be multiplied out before differentiation).

The volume is a maximum when the gradient is zero.

$$\frac{dV}{dx} = -202x + 12x^2$$

The required value of x can be obtained by solving the quadratic equation

$$12x^2 - 202x + 618.64 = 0$$

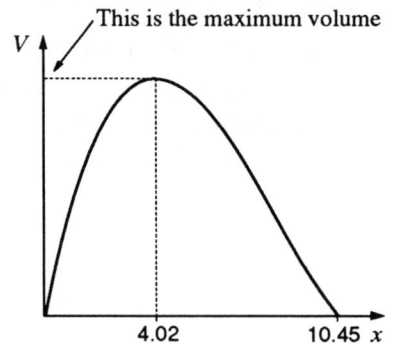

This is the maximum volume

4.02   10.45 x

Hence
$$x = \frac{202 \pm \sqrt{202^2 - 4 \times 12 \times 618.64}}{24}$$

$$= 4.02 \text{ cm or } 12.8 \text{ cm}$$

12.8 cm is clearly inappropriate to this problem. Hence $x = 4.02$ cm is the size of square that maximises the volume. The largest volume is therefore the value of V when $x = 4.02$:

$$V_{\text{max}} = 4.02(20.9 - 2 \times 4.02)(29.6 - 2 \times 4.02)$$

$$= 1115 \text{ cm}^3 \text{ to nearest whole number}$$

A potential snag with this method is that it only tells you where the stationary points are, but does not distinguish between maxima and minima. There are two simple ways round this problem.

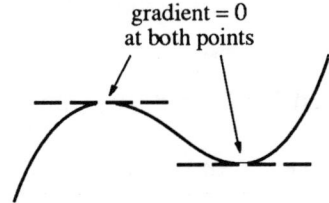

gradient = 0
at both points

Activity 13   Maximum or minimum?

(a) Show that the graph of $y = 2x^3 + 3x^2 - 72x + 15$ has stationary points at (–4, 223) and (3, –120).

(b) Copy and complete these tables :

x	-4.1	-4	-3.9
y		223	

x	2.9	3	3.1
y		-120	

Use these answers to infer which point is a maximum and which is a minimum.

(c) Here is another possible way. Copy and complete these tables.

x	-4.1	-4	-3.9
gradient		0	

x	2.9	3	3.1
gradient		0	

Do these answers support your conclusions in part (b)?

(d) Spot the flaw in this argument :

"(– 4, 223) is higher than (+3, – 120). Therefore (– 4, 223) must be the maximum and (3, –120) the minimum."

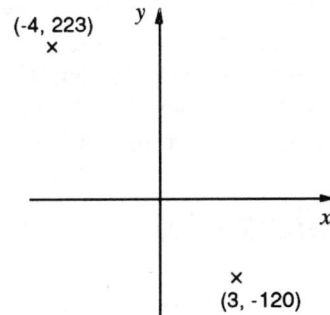

Example

Find the two stationary points of the function $T = 2k + \frac{8}{k}$, and determine which is a maximum and which is a minimum.

Solution

Now $\dfrac{dT}{dk} = 0$ for stationary point, and $\dfrac{dT}{dh} = 2 - \dfrac{8}{k^2}$

$\Rightarrow \quad 2 - \dfrac{8}{k^2} = 0$

$\Rightarrow \quad 2 = \dfrac{8}{k^2}$

$\Rightarrow \quad k^2 = 4$

$\Rightarrow \quad k = 2 \text{ or } -2$

When $k = 2, T = 8,$ and when $k = -2, T = -8$

k	-2.1	-2	-1.9
T	-8.01	-8	-8.01

maximum

k	1.9	2	2.1
T	8.01	8	8.01

minimum

Hence the function has a

maximum at $(-2, -8)$

minimum at $(2, 8)$

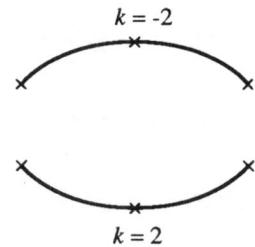

Two important points arise from the last worked example :

1. Note that the maximum point is lower than the minimum.
2. The word 'maximum' is always taken to mean 'local maximum'. In the diagram, P is higher than any neighbouring point, but there are other points on the curve that are higher. Similarly, the word 'minimum' is taken to mean 'local minimum'.

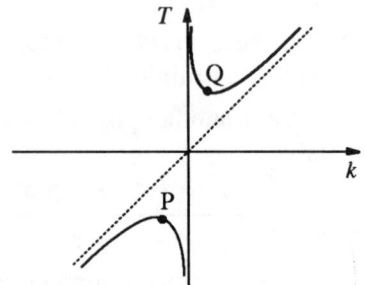

Exercise 8D

1. A function $f(x)$ is defined as follows :

    $f(x) = x^3 - 6x^2 - 36x + 15$

    Show that $f'(-2) = f'(6) = 0$, and hence find the co-ordinates of the maximum and minimum points.

2. Find the maximum and minimum points of these curves :

    (a) $y = 2x^2 - 6x + 7$

    (b) $y = 3x + \dfrac{27}{x}$

    (c) $y = 70 + 105x - 3x^2 - x^3$

    (d) $y = x^2 + \dfrac{16}{x}$.

3. A manufacturing company has a total cost function

    $$C = 5Q^2 + 180Q + 12500$$

    This gives the total cost of producing Q units.

    (a) Find a formula for the unit cost U, in terms of Q, where $U = C/Q$.

    (b) Find the value of Q that minimises the unit cost. Find this minimum unit cost.

4. The makers of a car use the following polynomial model to express the petrol consumption M miles per gallon in terms of the speed v miles per hour,

    $$M = \frac{v^3 - 230v^2 + 15100v - 145000}{4000}$$

    (a) Find the speed that maximises the petrol consumption, M.

    (b) The manufacturers only use this model for $30 < v < 90$. Give two reasons why this restriction is sensible.

8.6 Real problems

Activity 14   Maximising subject to a constraint

You have 120 m of fencing and want to make two enclosures as shown in the diagram. The problem is to maximise the area enclosed.

Let A be the area in square metres. Clearly $A = xy$.

(a) To find the maximum area, differentiate the expression for A and put it equal to zero. What is the problem with doing this?

(b) Use the fact that the total length of fencing is 120 m to write an equation connecting x and y.

(c) Make y the subject of this equation. Hence write a formula for A in terms only of x. Now differentiate with respect to x to solve the original problem.

(d) Try doing (c) the other way round. That is, make x the subject, express A in terms of y alone, and see if you get the same answer.

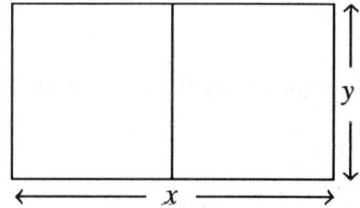

The problem posed in the Activity above was different to those earlier in this section. The quantity that needed maximising was first expressed in terms of two quantities, x and y. However, x and y were connected by the condition that the total length of fencing had to be 120 m. This sort of condition is known as a constraint. It allowed A to be expressed in terms of one quantity only, and thus the problem could be solved.

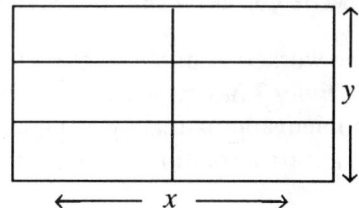

Example

Find the maximum area that can be enclosed by 120 m of fencing arranged in the configuration on the right.

Solution

Let the overall dimensions be x metres and y metres and the area be A square metres.

$$A = xy \quad \text{(the quantity to be minimised)}$$

$$4x + 3y = 120 \quad \text{(constraint from total length of fencing)}$$

$$y = \frac{120 - 4x}{3} \quad \text{(make } y \text{ the subject)}$$

$$A = \frac{x(120 - 4x)}{3}$$

$$= 40x - \frac{4}{3}x^2$$

$$\frac{dA}{dx} = 40 - \frac{8}{3}x$$

At a stationary point $\frac{dA}{dx}$ must be zero; this gives

$$40 - \frac{8}{2}x = 0$$

$$\Rightarrow \quad x = 15$$

The question asked for the maximum area. From the equation for y

$$y = \frac{120 - 4 \times 15}{3} = 20$$

So maximum area $= 20 \times 15 = 300 \text{ m}^2$.

How do you know that the area is actually a maximum?

The worked example below is identical to the problem in Activity 7 of Chapter 1. However, whereas before you solved the problem approximately, using a graph, it is now possible to obtain an accurate solution.

Example

A cylindrical can has a volume of 350 cm³. Find the dimensions of the can that minimise the surface area.

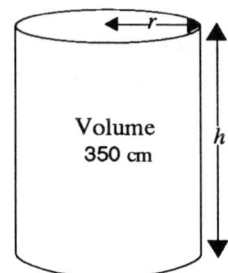

Volume
350 cm

Solution

Let the radius be r cm and the height h cm. Let the surface area be S cm^2; then

$$S = 2\pi r^2 + 2\pi rh \ (\text{ the quantity to be minimised})$$

At present, S involves two variables, r and h. The fact that the volume has to be 350 cm^3 gives a connection between r and h; namely

$$\pi r^2 h = 350 \ (\text{constraint})$$

So $\qquad h = \dfrac{350}{\pi r^2}$ (make h the subject)

and $\qquad S \quad = 2\pi r^2 + 2\pi r\left(\dfrac{350}{\pi r^2}\right)$ (substitute for h in the S formula)

$$= 2\pi r^2 + \dfrac{700}{r}$$

giving $\qquad \dfrac{dS}{dr} = 4\pi r - \dfrac{700}{r^2}.$

At a stationary point, $\dfrac{dS}{dr} = 0$,

giving

$$4\pi r - \dfrac{700}{r^2} = 0$$

$$\Rightarrow \quad 4\pi r = \dfrac{700}{r^2}$$

$$\Rightarrow \quad r^3 = \dfrac{700}{4\pi} \approx 55.7$$

$$\Rightarrow \quad r = 3.82 \text{ cm to 3 s.f.}$$

$$\Rightarrow \quad h = \dfrac{350}{\pi r^2} = 7.64 \text{ cm to 3 s.f. (from equation above)}$$

Could the problem have been solved by making r the subject of the constraint instead of h?

Exercise 8E

1. The rectangular window frame in the diagram uses 20 m of window frame altogether. What is the maximum area the window can have?

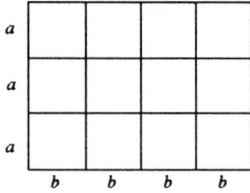

2. Repeat Question 1 for these window designs.

(a)

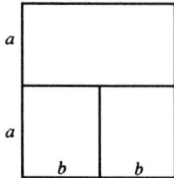

(b)

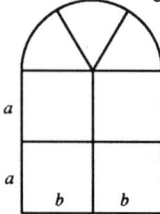

3. A rectangular paddock is to have an area of 50 m². One side of the rectangle is a straight wall; the remaining three sides are to be made from wire fencing.

   What is the least amount of fencing required?

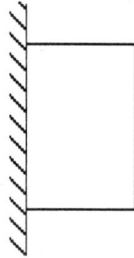

4. The enclosure shown has a total area of 300 m². Find the minimum amount of fencing required.

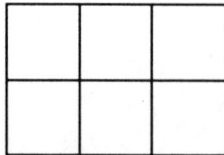

5. A small closed water tank is in the shape of a cuboid with a square base. The total surface area is 15000 cm². The problem here is to maximise the volume.

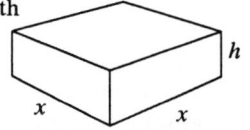

   (a) Let x cm be the side of the square and h cm be the height. Write down an expression for the volume V.

   (b) Show that $h = \dfrac{3750}{x} - \dfrac{x}{2}$

   (c) Show that the maximum volume is 125 litres.

6. An emergency petrol tank is designed to carry 1 gallon of petrol (4546 cm³). Its shape can be considered to be a cuboid.

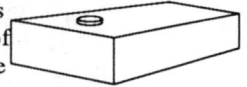

   The base of the cuboid is a rectangle with the length double the width.

   Find the dimensions of the tank that minimise the surface area required. Give the answers to the nearest millimetre.

7. The solution to the last worked example was such that the diameter and the height were equal. Show that this is true for any fixed volume when the surface area is to be minimised.

8.7   More complicated problems

The principle behind the questions in the next exercise is the same as that used in the exercise you have just done. The algebra is more complicated to set up, however. Working through the next Activity will show you what is required.

Activity 15   Fitting a cylinder into a sphere

What is the volume of the largest cylinder that can be fitted into a sphere of radius 10 cm?

To solve this problem, start off in the usual way :

Let the cylinder have base radius r and height h. Then the volume of the cylinder $V = \pi r^2 h$.

Finding the constraint is not so easy.

(a) By considering the diagram oppopsite, explain why

$$r^2 + \frac{h^2}{4} = 100.$$

Use this to express V in terms of only h. Hence solve the problem.

(b) Suppose the sphere had radius R cm. Show that the volume

of the largest cylinder that can be cut out is $\dfrac{4\pi R^3}{3\sqrt{3}}$

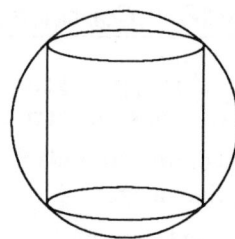

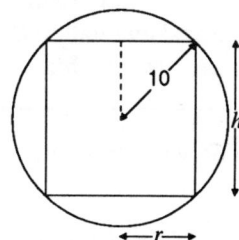

Exercise 8F

1. The interior of a space capsule is a cone with height 8 m and base radius 3 m.
    A cylinder is placed inside so that the base of the cylinder rests firmly on the base of the capsule and the top of the cylinder touches the sloping sides. The cylinder has height h m and base radius r m.

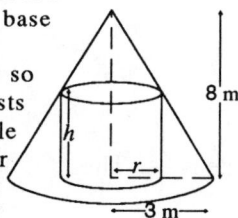

(a) Similar triangles are shown in the diagram. Use these to prove that $8r + 3h = 24$.

(b) Show that the volume of the largest such cylinder is 33.5 m³ to 3 s.f.

2. A sphere has a radius of 10 cm. A cone is placed inside so that it fits exactly.

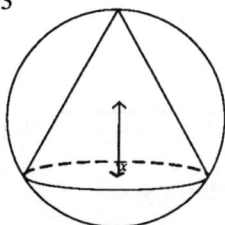

The base of the cone is a distance x below the centre of the sphere.

(a) Find the value of x that maximises the volume of the cone. Hence find the volume of the largest cone that can be cut from the sphere.

(b) Suppose the sphere has radius R cm. Show that the maximum volume of the cone is

$$\frac{32\pi}{81}R^3.$$

3. This shape consists of a rectangle topped by an equilateral triangle. Find the minimum perimeter if the area enclosed is 100 cm². (The dotted line does not count as part of the perimeter).

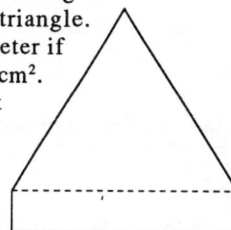

8.8   Differentiating other functions

From the work done so far, you should be able to differentiate any function involving sums and differences of $x^3, x^2, x, \dfrac{1}{x}$ and constants. This section extends this to other powers of x.

Activity 16   Continuing the pattern.

Function	Derivative
x	1
x^2	$2x$
x^3	$3x^2$
x^4	
x^5	
.	
.	
.	
x^n	

(a) The table opposite shows some of the derivatives you already know. Guess the derivatives lower down the table and conjecture a formula for the derivative of x^n, where n is any positive integer.

(b) Use the techniques of earlier sections to see whether your guess for x^4 is correct. You may wish to restrict yourself to numerical evaluation of gradients at particular points but, if you can, use algebra.

You can now differentiate any polynomial function. For example :

$$y = x^6 - 3x^5 + 8x^3 + 2x - 6$$

$$\Rightarrow \quad \frac{dy}{dx} = 6x^5 - 3 \times (5x^4) + 8 \times (3x^2) + 2$$

$$= 6x^5 - 15x^4 + 24x^2 + 2$$

Another function which you know how to differentiate is $\dfrac{1}{x}$. The next Activity suggests how function such as $\dfrac{1}{x^2}$, can be differentiated.

Activity 17   Differentiation of $1/x^n$

(a) Another way of writing $\frac{1}{x}$ is x^{-1}. In the activity above you found that the derivative of x^n is nx^{n-1}. What happens if you put $n = -1$ in this formula? Does it give the right answer?

(b) Extend this to find the derivatives of $\dfrac{1}{x^2}, \dfrac{1}{x^3}$, and $\dfrac{1}{x^{10}}$.

*Activity 18   Differentiating $1/x^2$ using algebra.

In the Activity above you found that the derivative of $\dfrac{1}{x^2}$ was $-\dfrac{2}{x^3}$. The objective here is to prove this result formally.

(a)   Show that the gradient of the chord AB is

$$\frac{1}{h}\left\{\frac{1}{(1+h)^2}-1\right\}$$

and show that this simplifies to

$$-\frac{(2+h)}{(1+h)^2}$$

What is the gradient of the tangent at $(1,1)$?

(b)   Now consider finding the tangent at $\left(x,\frac{1}{x^2}\right)$

The overall summary of these results is as follows :

> If $y = x^n$, then $\dfrac{dy}{dx} = nx^{n-1}$ for n any integer.

Is this result true for $n = 0$?

Example

If $y = \dfrac{7}{x^5}$, find $\dfrac{dy}{dx}$

Solution

$\dfrac{7}{x^5}$ can be written $7x^{-5}$, so

$$\frac{dy}{dx} = 7 \times (-5x^{-6}) = -35x^{-6} = -\frac{35}{x^6}.$$

Example

If $A = 5t^3 + 2t^{-3}$, find $\dfrac{dA}{dt}$.

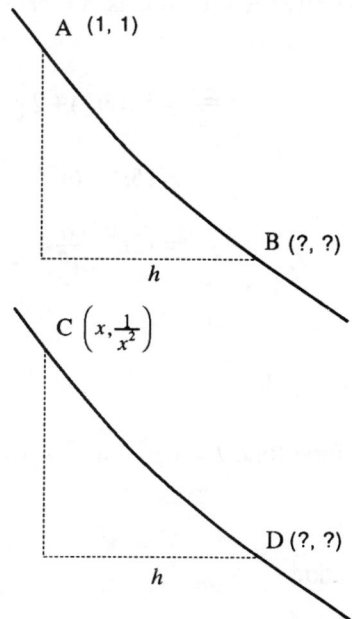

Solution

Re-write the function as $A = 5t^3 + 2t^{-3}$

So $\qquad \dfrac{dA}{dt} = 5x(3t^2) + 2x(-3t^{-4})$

$\qquad\qquad = 15t^2 - 6t^{-6}$

$\qquad\qquad = 15t^2 - \dfrac{6}{t^6}$

Example

Differentiate $T = 6p^5 - 8p^4 + 10p - \dfrac{4}{p^2}$

Solution

Since $\qquad T = 6p^5 - 8p^4 + 10p - 4p^{-2}$

$\Rightarrow \qquad \dfrac{dT}{dp} = 6\times(5p^4) - 8\times(4p^3) + 10 - 4\times(-2p^{-3})$

$\qquad\qquad = 30p^4 - 32p^3 + 10 + \dfrac{8}{p^3}$

Example

If $f(x) = (x^2 - 2)^2$ find $f'(2)$

Solution

$\qquad\qquad f(x) = x^4 - 4x^2 + 4$

$\Rightarrow \qquad f'(x) = 4x^3 - 8x$

$\Rightarrow \qquad f'(2) = 4\times2^3 - 8\times2 = 16$

Exercise 8G

1. Differentiate

    (a) $y = \dfrac{1}{x^6}$

    (b) $y = 3x^3 + \dfrac{2}{x^2}$

    (c) $C = 5q^4 + 6q^2 + 15 - \dfrac{3}{q^3}$

    (d) $G = t^8 - \dfrac{3}{t^6}$

    (e) $y = \dfrac{1}{2x^4}$

    (f) $L = \dfrac{3}{5x^2}$

    (g) $S = \dfrac{2}{t} - \dfrac{7}{2t^4}$

    (h) $y = \dfrac{x^5 - 3}{4x^3}$

2. (a) $f(x) = 6 - \dfrac{10}{x^2}$ find $f'(2)$

    (b) $g(t) = 15t + \dfrac{4}{t}$ find $g'(-1)$

    (c) If $h(w) = w^7 - \dfrac{8}{w^3}$ find $h'(-2)$

3. Find the gradients of :

    (a) $y = x^2 - \dfrac{1}{x^2}$ at the point $(1, 0)$;

    (b) $y = 4x^5 + 3x^2$ at the point $(-2, -116)$;

    (c) $y = \dfrac{54}{x^2} - \dfrac{81}{x^3}$ at the point $(3, 3)$.

8.9   Linearity

In this chapter the assumption has been made, that differentiation is a **linear** process. This means for example that the function $x^3 + x^5$ can be differentiated as follows :

$$\frac{d}{dx}(x^3 + x^5) = \frac{d}{dx}(x^3) + \frac{d}{dx}(x^5)$$

(Differentiate x^3 and x^5 separately, then add).

Similarly, to differentiate $6x^3$:

$$\frac{d}{dx}\left(6x^3\right) = 6\frac{d}{dx}(x^3)$$

(Differentiate x^3 and multiply by 6).

In general, given two functions f and g and two constants a and b, linearity means that

$$\frac{d}{dx}(af(x) + bg(x))$$

$$= a\frac{d}{dx}(f(x)) + b\frac{d}{dx}(g(x))$$

or $\qquad (af(x) + bg(x))' = af'(x) + bg'(x)$

Is this assumption valid?

The only evidence in favour of it, is that obtained in Section 8.2 when quadratic functions were being investigated. To prove that differentiation is indeed linear, first of all a more formal definition of the derivative is needed. This is found as follows :

In the diagram, A and B are nearby points on the general curve $y = f(x)$. A is the general point $(x, f(x))$. B is at a horizontal distance h further along and has co-ordinates $(x + h, f(x + h))$.

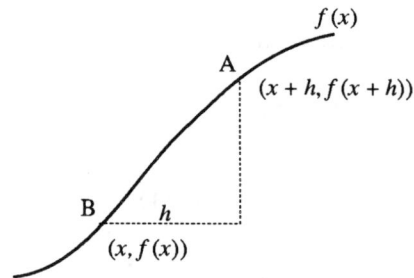

(Compare this to the way derivatives were established for functions like x^2, x^3 etc.)

The gradient of AB is given by

$$\frac{f(x+h) - f(x)}{h}$$

The gradient at A is defined as the tangent at A to the curve, which is the limit of the gradient of AB as $h \to 0$. This is written

$$f'(x) = \lim_{h \to 0} \left\{ \frac{f(x+h) - f(x)}{h} \right\}$$

Before tackling the final Activity, make sure you clearly understand the above definition and how it was formulated.

Activity 19   Proving linearity

(a) Suppose $p(x) = f(x) + g(x)$. Then

$$p'(x) = \lim_{h \to 0} \left\{ \frac{p(x+h) - p(x)}{h} \right\}$$

Show that

$$p'(x) = \lim_{h \to 0} \left\{ \frac{f(x+h) - f(x)}{h} + \frac{g(x+h) - g(x)}{h} \right\}$$

What conclusion can be inferred about $p'(x)$?

(b) Suppose $q(x) = kf(x)$, where k is a constant. Find an expression for $q'(x)$ in terms of $k, h, f(x + h)$ and $f(x)$ and explain why $q'(x) = kf'(x)$

(c) Suppose $r(x) = f(x) g(x)$. Is it true that $r'(x) = f'(x) g'(x)$? Justify your answer.

8.10 Using the results

This chapter ends with some practice in some traditional problems involving differentiation. Follow through these worked examples and then attempt Exercise 8H.

Example

Find the equation of the tangent to the curve

$$y = x^2 - \frac{1}{x^2}$$

at the point $(1, 0)$

Solution

The gradient is first found when $x = 1$.

$$\frac{dy}{dx} = 2x + \frac{2}{x^3} \text{ and when } x = 1, \frac{dy}{dx} = 4.$$

The tangent is thus a straight line with gradient 4 passing through $(1, 0)$.

Equation must be of the form $y = 4x + c$, where the constant c can be found by substituting $(1, 0)$ for (x, y) :

$$0 = 4 + c \Rightarrow c = -4$$

So the equation is $y = 4x - 4$.

Example

Find the equation of the normal to the curve $y = x^3 - 3x + 2$ when $x = 2$.

(The normal is the line perpendicular to the tangent.)

Solution

$$\frac{dy}{dx} = 3x^2 - 3 \text{ and when } x = 2, \frac{dy}{dx} = 9.$$

When $x = 2$ the gradient of the tangent is 9. The gradient of the normal is therefore $-\frac{1}{9}$. Also when $x = 2$,

$$y = 2^3 - 3 \times 2 + 2 = 4.$$

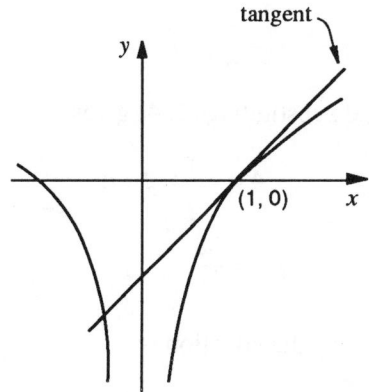

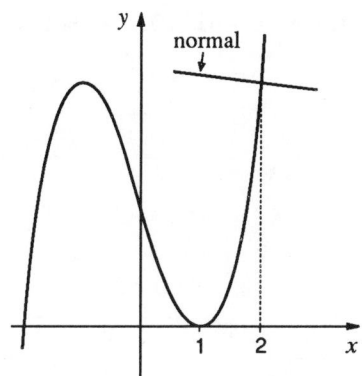

To find the equation of the normal, one point that lies on the normal needs to be found. The one point known so to lie is (2, 4), the point on the curve through which the normal passes.

So the equation is given by

$$y = -\frac{1}{9}x + c$$

and substituting (2, 4) gives

$$4 = -\frac{1}{9} \times 2 + c$$

$$\Rightarrow c = \frac{38}{9}$$

Hence the equation is

$$y = -\frac{1}{9}x + \frac{38}{9}$$

or $9y + x = 38$

Exercise 8H

1. Find the equation of the tangent to :

   (a) $y = x^2 + 4x - 3$ at $(3, 18)$

   (b) $y = 5x^3 - 7x^2 + x$ at $(1, -1)$

   (c) $y = 2x - \frac{16}{x^2}$ when $x = -2$

2. Find the equation of the normal to :

   (a) $y = 3x^2 - 5x + 10$ at $(1, 8)$

   (b) $y = 2(1 - \frac{1}{x^2})$ when $x = -4$

   (c) $y = x^4 - 4x^3 + 7x + 9$ when $x = 3$

3. You are given that

   $$f'(3) = 2 \qquad f'(5) = -1$$
   $$g'(3) = -6 \qquad g'(5) = 8$$

   Find the following, where possible :

   (a) $p'(3)$ where $p(x) = 2f(x)$

   (b) $q'(5)$ where $q(x) = f(x)g(x)$

   (c) $r'(5)$ where $r(x) = 5f(x) + g(x)$

   (d) $f'(8)$

8.11 Miscellaneous Exercise

1. Differentiate with respect to x

   (a) $x^2 + 4x - 3$ (b) $x^3 - 4x^2 + 17x + 10$

   (c) $x(x+1)^2$ (d) $x^2 + \dfrac{2}{x}$

   (e) $x(x - \dfrac{1}{x})^2$.

2. Find the derived function for :

   (a) $f(x) = 5x^4 + \dfrac{3}{x^2}$ (b) $f(t) = \dfrac{9}{2t^4}$

   (c) $f(y) = y^2(y^4 - \dfrac{5}{y^4})$ (d) $f(p) = \dfrac{(p+1)^2}{p^2}$

   (e) $f(x) = \left(\dfrac{3+2x^2}{5x}\right)^2$.

3. (a) Find the equation of the tangent to the curve

   $y = 2x^2 - 7$ at the point $(2, 1)$.

   (b) Find the equation of the normal to the curve

   $y = 2x - \dfrac{1}{x}$ when $x = -1$.

4. The definition of the derived function is :

   $$f'(x) = \lim_{h \to 0} \left\{ \frac{f(x+h) - f(x)}{h} \right\}$$

   (a) Use this definition to show that the derivative of $x^2 + 3x$ is $2x + 3$. (This is known as differentiating from first principles.)

   (b) Differentiate the function $5x^2 - 2x + 4$ from first principles.

5. A lampshade is shaped in the form of an open cylinder of radius 6 cm. The base of the lampshade is 220 cm above a flat horizontal floor.

   The bulb is a distance x above the bottom of the lampshade and lights up a circular area of floor.

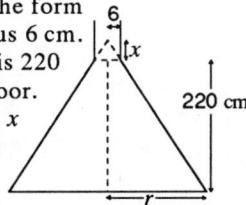

   (a) Let the radius of this circle be r. Find a formula for r in terms of x.

   (b) If A cm^2 is the area of floor illuminated, show that

   $$A = \frac{36\pi(220+x)^2}{x^2}$$

   (c) Sketch a graph of A against x.

   (d) The bulb is positioned so that $x = 10$. Find the area of floor that is illuminated.

   (e) Find the gradient of the graph when $x = 10$.

   (f) Use your answers to (d) and (e) to estimate A when $x = 10.01$ cm.

6. Wire from a construction kit is used to make a skeleton for a cuboid, which has a square base. The total length of wire is 300 cm. What is the maximum volume enclosed?

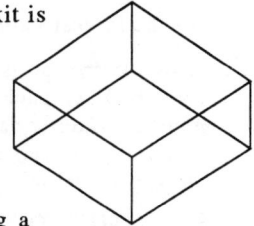

7. A particle is moving along a straight line. At time t seconds its distance s metres from a fixed point F is given by

   $$s = t^3 - 12t^2 + 45t + 10$$

   (a) Its velocity v in ms can be obtained by differentiating s with respect to t. Find v in terms of t.

   (b) Find the two values of t for which the particle is stationary.

   (c) The acceleration of the particle can be obtained by differentiating v with respect to t. When is the particle's acceleration zero?

8. Another particle moves along the line so that its distance from F is given by :

   $$s = 25 + 40t - 8t^2.$$

   (a) Find s when the particle is stationary.

   (b) Show that the particle's acceleration is constant.

9. (a) If $y = x^4 - 8x^3 - 62x^2 + 144x + 300$, show that

   $$\frac{dy}{dx} = 4(x^3 - 6x^2 - 31x + 36)$$

   (b) Show that there is a stationary point where $x = 1$ and find the two other points where $\dfrac{dy}{dx} = 0$.

   (c) Sketch the curve, showing clearly the co-ordinates of the three stationary points.

10. If $A = \dfrac{1}{p^3} + \dfrac{1}{p^2} - \dfrac{1}{p}$ show that

$$\frac{dA}{dp} = \frac{p^2 - 2p - 3}{p^4}$$

and hence find

(a) the smallest value taken by A when $p > 0$;

(b) the largest value taken by A when $p < 0$.

11. A shopkeeper sells packets of home-made sweets. Each packet costs 50p to make. The number sold per day depends on the price at which the packets are sold. Some typical figures are shown below.

Selling price (x)	60p	65p	70p
Number sold per day (N)	60	48	36

Assuming N to be a linear function of x.

(a) Find a formula for the profit made on these sweets per day in terms of x;

(b) Find the value of x that maximises the daily profit, and calculate this profit.

*12. Differentiating $y = \sqrt{x}$ from first principles

(a) Show that

$$(\sqrt{x+h} - \sqrt{x})(\sqrt{x+h} + \sqrt{x}) = h$$

(b) Hence simplify $\dfrac{\sqrt{x+h} - \sqrt{x}}{h}$

(c) If $y = \sqrt{x}$, what is $\dfrac{dy}{dx}$

*13. Differentiating $y = x\sqrt{x}$.

(a) Show that

$$\frac{(x+h)\sqrt{x+h} - x\sqrt{x}}{h} = x\left(\frac{\sqrt{x+h} - \sqrt{x}}{h} \right) + \sqrt{x+h}$$

(b) Find $\dfrac{dy}{dx}$ when $y = x\sqrt{x}$.

*14. If $y = \dfrac{1}{\sqrt{x}}$, find $\dfrac{dy}{dx}$.

9 POWERS

Objectives

After studying this chapter you should

*   understand fractional indices;
*   know how to use the binomial theorem for any rational index;
*   be able to answer simple combinational problems.

9.0 Introduction

You are already familiar with expressions like $3^2, 4^{10}$ and 10^{-6}, all of which involve **powers** (or **indices**). But can any meaning be attached to an expression like $2^{0.6}$? If so, does it have any relevance? This chapter starts off by answering these questions. The rest of the chapter is concerned with a famous and important piece of mathematics known as the **binomial theorem**.

The topic is introduced through a case study on bacterial growth. Bacteria perform the roles of friend and foe at the same time. They are micro-organisms that perform a crucial function in nature by causing plant and animal debris to decay in the soil, but at the same time they can cause disease. Under favourable conditions they reproduce freely. Lone bacterium will first split into two bacteria, then both of these bacteria will themselves split into two and so on.

This growth can be observed by placing a lone bacterium onto a petri dish and positioning the dish in a warm environment. The splitting process under these sorts of conditions will take place twice per hour; hence after one hour there will be 4 bacteria, after two hours 16 bacteria and so on.

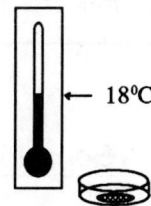

← 18°C

Petri dish with jelly

Activity 1 Bacterial growth

(a) Copy and complete the table shown. Write a formula for the number of bacteria after t hours.

Time (hours)	0	1	2	3	4	...
Number of bacteria	1	4	16	...	..	

(b) Interpret this formula when $t = \frac{1}{2}, 1\frac{1}{2}, 2\frac{1}{2}, 3\frac{1}{2}$.

9.1 Fractional indices

Before continuing, you will find it useful to revise your knowledge of how integer indices work.

Activity 2   Revision

Complete these general statements :

For any non-zero number x, and any integers m and n :

(a) $x^m x^n =$ (b) $\dfrac{x^m}{x^n} =$ (c) $\left(x^m\right)^n =$

(d) $x^0 =$ (e) $x^{-n} =$

Activity 3

(a) You have already seen that $2^0 = 1, 2^1 = 2, 2^2 = 4$ and $2^3 = 8$. Draw an accurate graph of $y = 2^x$ by joining together these points with as smooth a curve as you can.

(b) The value of $2^{1\frac{1}{2}}$ must be double that of $2^{\frac{1}{2}}$. Why? Make a similar statement about $2^{2\frac{1}{2}}$. Are these statements supported by your graph?

(c) What is your interpretation of $2^{\frac{1}{2}}$?

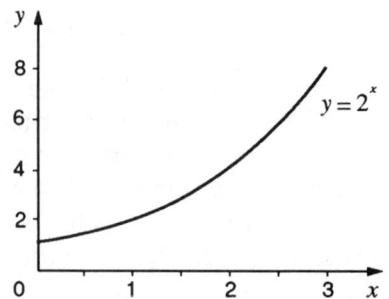

Activity 4   Some logical deductions

(a) Let p stand for the value of $2^{\frac{1}{2}}$. What can you say about p^2? What rule of indices did you use?

(b) What is the meaning of $x^{\frac{1}{2}}$?

(c) Suppose q stands for the value of $2^{\frac{1}{3}}$. What is q^3?

(d) What does $x^{\frac{1}{3}}$ mean?

In general :

$$x^{\frac{1}{2}} \text{ is another way of writing } \sqrt{x}$$

$$x^{\frac{1}{3}} \text{ is another way of writing } \sqrt[3]{x}$$

$$x^{\frac{1}{n}} \text{ is another way of writing } \sqrt[n]{x}.$$

This alternative notation is often used on calculators. Your calculator may have a function $x^{\frac{1}{y}}$. If so, it can be used to find nth roots as follows.

To find $\sqrt[10]{2}$ or $2^{\frac{1}{10}}$ press $\boxed{2}$ $\boxed{x^{\frac{1}{y}}}$ $\boxed{10}$ $\boxed{=}$ The answer should be 1.071773463.

Activity 5   nth root

Find $\sqrt[10]{100}$ to 3 s.f. without using the $x^{\frac{1}{y}}$ or $\sqrt[x]{}$ function on your calculator. You MAY use the x^y function.

(Try simpler examples like $\sqrt{100}$ or $\sqrt[3]{100}$ first if you're not sure how to do this.)

Repeat for $\sqrt[100]{1\ 000\ 000\ 000}$ and $\sqrt[10]{0.1}$.

The problem at the beginning of this chapter is an example of 'exponential' growth. This will be dealt with in more detail in Chapter 12 but for the present it can be defined as a growth whereby the number of bacteria present is multiplied by 4 every hour.

Fractional indices are useful in many situations. A good example concerns credit card accounts where any unpaid debt grows exponentially by a certain percentage each month. Credit cards quote a monthly interest rate and the equivalent APR (annual percentage rate).

A typical APR is 29.84%. This means that a debt of £100 at the start of the year becomes one of £129.84 by the end. However, the interest is worked out monthly at a monthly rate of 2.2%. This figure arises because

$$(1.022)^{12} = 1.2984$$

giving an APR of $(1.2984 - 1)\%$.

Another way of writing this is to say

$$1.2984^{\frac{1}{12}} = 1.022.$$

Example

What is the monthly interest rate if the APR is 34.45%?

Solution

Since $(1.3445)^{\frac{1}{12}} = 1.02498...,$

monthly interest rate = 2.5%.

Example

Joel has an outstanding credit card bill of £162. The APR is 30.23%. He leaves it two months before paying. How much does he have to pay?

Solution

2 months is $\frac{1}{6}$ of a year.

But $(1.3023)^{\frac{1}{6}} = 1.045$

and $1.045 \times 162 = £169.29$

Exercise 9A

1. Calculate these without a calculator:

    (a) $16^{\frac{1}{2}}$ (b) $8^{\frac{1}{3}}$ (c) $81^{\frac{1}{4}}$

    (d) $\left(\frac{1}{4}\right)^{\frac{1}{2}}$ (e) $\left(\frac{16}{625}\right)^{\frac{1}{4}}$ (f) $(-1)^{\frac{1}{3}}$

2. Use a calculator to work these out to 3 s.f.

    (a) $6^{\frac{1}{2}}$ (b) $10^{\frac{1}{3}}$ (c) $56^{\frac{1}{3}}$

    (d) $(0.03)^{\frac{1}{20}}$ (e) $(0.5)^{\frac{1}{2}}$

3. Solve these equations:

    (a) $4^x = 2$ (b) $125^x = 5$

    (c) $100\,000^x = 10$ (d) $81^x = \frac{1}{3}$

4. The APR on a credit card is 26.08%.

    (a) Jan has an outstanding bill of £365. To how much does this grow in 6 months, assuming no other transactions take place?

    (b) Mark has a bill of £218. What is his bill a month later?

5. Water lilies on a pond grow exponentially so that the area they cover doubles every week. On Sunday they cover 13% of the surface. What percentage do they cover on Monday?

6. An investment policy boasts exponential growth and guarantees to treble your investment, at least, over 20 years. Work out the minimum guaranteed value, to the nearest pound, of

    (a) a £1000 investment after 10 years;

    (b) a £600 investment after 5 years;

    (c) a £2400 investment after 4 years.

9.2 Further problems

Piano tuners also deal with exponential growth. When tuning a piano to 'concert pitch' the first thing to do is to make sure the note A in the middle of the piano is in tune. A properly tuned 'middle A' has a vibrating frequency of 440 hz (hz is short for Hertz and means cycles per second).

There are lots of notes called A on the piano. The distance between consecutive As is called an octave. Every octave up, the frequency doubles, as follows :

two As above	1760
one A above	880
middle A	440
one A below	220
two As below	110

Activity 6   Finding the frequency

(a)   Write down a formula for the frequency of the note p octaves above middle A.

(b)   Use this to find the frequency of the note half an octave above A.

(c)   An octave actually consists of 12 notes, so 'half an octave higher' means '6 notes higher'. Find the frequency of the note

   (i)   3 notes above middle A

   (ii)   5 notes above middle A

   (iii) 6 notes below middle A

   (iv) 11 notes below middle A

(d)   Adapt your formula in (a) to find the frequency of the note n above middle A. Does this formula apply to notes below middle A?

Activity 7

Before reading on, discuss the meaning of these numbers and hence find their numerical value to 3 s.f., explaining your method clearly. Can you find more than one method for some of them?

$$2^{\frac{5}{12}} \quad 5^{\frac{3}{2}} \quad 10^{1.2} \quad 2^{-\frac{11}{12}} \quad 3^{-0.4}$$

There are always at least two ways of thinking about expressions like $3^{\frac{2}{5}}$.

Remembering that $\left(x^m\right)^n = x^{mn}$

$$3^{\frac{2}{5}} = \left(3^2\right)^{\frac{1}{5}} \quad \text{or} \quad 3^{\frac{2}{5}} = \left(3^{\frac{1}{5}}\right)^2$$

$$= \left(\sqrt[5]{3^2}\right) \qquad\qquad = \left(\sqrt[5]{3}\right)^2$$

$$= \left(\sqrt[5]{9}\right)$$

In general

$$x^{\frac{p}{q}} = \sqrt[q]{x^p} \text{ or } \left(\sqrt[q]{x}\right)^p$$

Since $\frac{2}{5} = 2 \times \frac{1}{5}$ or $\frac{1}{5} \times 2$, the order of the root and the power does not matter.

Does this extend to negative indices?

Examples are

$$3^{-2} = \frac{1}{3^2} = \frac{1}{9}$$

and

$$3^{-\frac{2}{5}} = \frac{1}{3^{\frac{2}{5}}}.$$

Example

Calculate the numerical values of

(a) $2^{\frac{4}{7}}$ (b) $10^{\frac{3}{2}}$ (c) $6^{\frac{3}{8}}$ (d) $(0.6)^{\frac{-7}{3}}$

Solution

There is always more than one method of working these out. Only one method is shown for each one here. Pay particular regard to the methods in (b) and (d).

(a) $\quad 2^{\frac{4}{7}} = \left(2^4\right)^{\frac{1}{7}} = \sqrt[7]{16} = 1.49$ to 3 s.f.

(b) $\quad 10^{\frac{3}{2}} = 10^1 \times 10^{\frac{1}{2}} = 10\sqrt{10} = 31.6$ to 3 s.f.

(c) $\quad 6^{\frac{3}{8}} = \left(6^{\frac{1}{8}}\right)^3 = (1.251033...)^3 = 1.96$ to 3 s.f.

$$6^{-\frac{3}{8}} = \frac{1}{6^{\frac{3}{8}}} = \frac{1}{1.9579731...} = 0.511 \text{ to 3 s.f.}$$

(d) $(0.6)^{\frac{7}{3}} = (0.6)^{2\frac{1}{3}} = (0.6)^2 \times (0.6)^{\frac{1}{3}}$

$$= 0.36 \times \sqrt[3]{0.6} = 0.3036358... = 0.304 \text{ to 3 s.f.}$$

$$(0.6)^{-\frac{7}{3}} = \frac{1}{0.3036358...} = 3.29 \text{ to 3 s.f.}$$

(Calculator note : always carry through as many figures as you can until the end of the calculation.)

Exercise 9B

1. Work these out without a calculator.

   (a) $4^{1\frac{1}{2}}$ (b) $27^{\frac{2}{3}}$ (c) $100^{\frac{5}{2}}$

   (d) $1000^{1\frac{1}{3}}$ (e) $16^{\frac{5}{4}}$ (f) $32^{0.4}$

2. Use a calculator to work these out to 3 s.f.

   (a) $120^{\frac{3}{2}}$ (b) $(0.7)^{\frac{5}{3}}$ (c) $5^{3.25}$

   (d) $1000^{\frac{2}{9}}$ (e) $\left(\frac{1}{4}\right)^{\frac{3}{7}}$ (f) $(0.36)^{3.1}$

3. Write these as fractions.

   (a) $16^{-\frac{1}{2}}$ (b) $4^{-1\frac{1}{2}}$ (c) $32^{-0.4}$ (d) $125^{-\frac{4}{3}}$

4. Calculate these to 3 s.f.

   (a) $10^{-\frac{1}{2}}$ (b) $15^{-\frac{1}{3}}$ (c) $(0.2)^{-\frac{5}{2}}$ (d) $(3.5)^{-\frac{4}{3}}$

5. Solve these equations. A calculator is not required.

   (a) $100^x = 100$ (b) $4^x = 32$

   (c) $9^x = \frac{1}{3}$ (d) $8^x = \frac{1}{2}$

   (e) $64^x = 16$ (f) $16^x = \frac{1}{8}$

6. Rewrite these formulas without using fractional or negative indices.

   $\left(\text{e.g. } x^{-1} = \dfrac{1}{x}; \ x^{\frac{3}{2}} = x\sqrt{x} \text{ or } \sqrt{x^3} \ .\right)$

   (a) $5p^{\frac{1}{2}}$ (b) $6q^{-1}$

   (c) $10x^{-\frac{1}{2}}$ (d) $\frac{3}{4}y^{-\frac{1}{2}}$

   (e) $\frac{1}{2}m^{\frac{3}{2}}$ (f) $12t^{-\frac{5}{2}}$

7. The population of a city is expected to grow exponentially and to double in 15 years. By what percentage would you expect the population to have risen after

   (a) 4 years;   (b) 10 years.

9.3   Binomial expansions

'Binomial' is a word meaning 'two terms', and is used in algebra to mean expressions such as $a + 2$ and $2x - y$. (Compare with the word 'polynomial').

Binomial expressions were used extensively in Chapter 8; for example the gradient of the chord in this diagram is

$$\frac{(2+h)^2 - 4}{h}$$

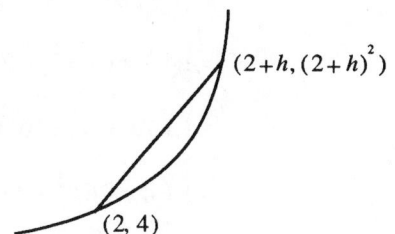

$(2+h, (2+h)^2)$

$(2, 4)$

At the heart of this is the binomial $2 + h$, raised to a power. For a more complicated curve the gradient might be

$$\frac{(2+h)^7 - 128}{h},$$

which is not so easily simplified because the binomial is raised to a high power. Dealing with $(2+h)^7$ and the like is the focus of the next three sections.

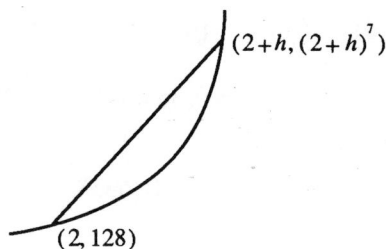

The diagram shows a curve with points labelled $(2+h, (2+h)^7)$ and $(2, 128)$.

Activity 8   True or False?

Here are six statements about binomials. Which of these are true, and which false?

(a) $2(a+b) = 2a + 2b$ (b) $(a+b) \div 2 = (a \div 2) + (b \div 2)$

(c) $(a+b)^2 = a^2 + b^2$ (d) $(a+b)^3 = a^3 + b^3$

(e) $\sqrt{(a+b)} = \sqrt{a} + \sqrt{b}$ (f) $2^{(a+b)} = 2^a + 2^b$

Writing $(a+b)^2$ for $a^2 + b^2$ is a common mistake (though hopefully one you no longer make!). During Chapter 8 you should have got used to handling expansions like

$$(a+b)^2 = a^2 + 2ab + b^2$$

$$(a+b)^3 = a^3 + 3a^2b + 3ab^2 + b^3$$

These are simple examples of **binominal expansions**.

Example

Expand $(1+x)^4$

Solution

$$(1+x)^4 = (1+x)^2(1+x)^2$$

$$= (1 + 2x + x^2)(1 + 2x + x^2)$$

$$= 1(1 + 2x + x^2) + 2x(1 + 2x + x^2) + x^2(1 + 2x + x^2)$$

$$= 1 + 2x + x^2 + 2x + 4x^2 + 2x^3 + x^2 + 2x^3 + x^4$$

$$= 1 + 4x + 6x^2 + 4x^3 + x^4$$

Exercise 9C

1. Write out the expansion of these :

    (a) $(1+x)^2$ (b) $(1+x)^3$

    (c) $(1+x)^4$ (d) $(1+x)^5$

2. Repeat for these :

    (a) $(a+x)^2$ (b) $(a+x)^3$

    (c) $(a+x)^4$ (d) $(a+x)^5$

3. Write out the expansions of

    (a) $(3-2x)^2$ (b) $(2+5p)^3$ (c) $(\tfrac{1}{2}m-5)^4$

You will be thankful that Exercise 9C was neither lengthy nor involved powers higher than the 5th power. As the index rises, so the expansions become more tedious. Expanding $(1+x)^{50}$ by this method, for example, would be somewhat daunting.

9.4 Binomial coefficients

Activity 9

(a) Using your answers to Question 1 of Exercise 9C, complete this table :

Coefficients of

	1	x	x^2	x^3	x^4	x^5	x^6	x^7
$(1+x)^1$	1	1						
$(1+x)^2$	1	2	1					
$(1+x)^3$	1	3	3	1				
$(1+x)^4$	1	4	6	4	1			
$(1+x)^5$								

(b) What are the coefficients in the 6th and 7th rows?

(c) Write out the expansions of $(1+x)^6$ and $(1+x)^7$.

(d) How does the table above correspond to the expansions of $(a+x)^2, (a+x)^3$ etc.? What do all the terms in the expansion of, say $(a+x)^4$ have in common?

(e) Write out the expansions of $(a+x)^6$ and $(a+x)^7$.

(f) What is the **sum** of the numbers in the nth row?

The numbers in the table are called the **binomial coefficients**. It is evident that they obey a pattern, but finding the numbers in a particular row depends on knowing the numbers in the row before.

To expand $(1+x)^{50}$ you would need to know all the coefficients up to and including the 49th row first.

To get a formula for the binomial coefficients it is important to see how they arise. Look more closely at the expression $(a+x)^5$. Remember that it is short for

$$(a+x)(a+x)(a+x)(a+x)(a+x)$$

and that the expansion is

$$a^5 + 5a^4x + 10a^3x^2 + 10a^2x^3 + 5ax^4 + x^5.$$

In each term the two index numbers add up to 5. Moreover, the sum of the coefficients is 32. These two facts reflect what is going on here : each term is the product of a mixture of 'a's and 'x's, one letter from each of the 5 brackets. Each bracket can supply one of two letters, so the number of different combinations of 'a's and 'x's is $2 \times 2 \times 2 \times 2 \times 2$.

Only one such combination is all 'a's, hence the expansion starts with a single a^5 term. Five of the 32 possibilities give a^4x, 10 of them a^3x^2, and so on. It is the general method of finding the binomial coefficients 1, 5, 10, 10, 5, 1 that is being sought. That is the aim of the next section.

Activity 10   Bridge problem

Bridge is one of the best of all card games. In particular, it is one in which the skills of the player can overcome the luck of the deal. Here is a situation often faced by bridge players.

Suppose you are sitting at S (south). The opposing players are sitting either side of you at W (west) and E (east). You know that between them they have four cards in the heart suit. It is important to know the relative likelihood of various distributions of these four cards between W and E.

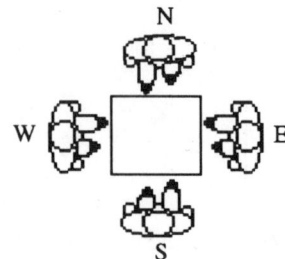

Each of the four cards can be in one of two places. Hence there are $2^4 = 16$ different arrangements.

(a) How many of these arrangements involve all four hearts being with W and none with E?

(b) How many of them involve E having only **one** heart?

(c) Complete the table below.

W	E	No. of ways
4	0	1
3	1	4
2	2	6
1	3	4
0	4	1
	Total	16

W W W W
W W W E
W W E W
W E W W
E W W W

W E W E
E W E W
W W E E
E W W E
W E E W
E E W W

(d) Bridge players sometimes claim that in these circumstances a '1-3 split' (either way) is more likely than an even split. Is this true? Explain.

(e) Extend this to the case where S knows that W and E have 5 hearts between them.

You should by now have realised that it could be dangerous to play bridge with a mathematician!

Activity 11 Streets of Manhattan

Many U.S. cities have a street layout based on a rectangular pattern. On Manhattan Island, for instance, more than 200 streets run across the island and ten avenues run at right angles to them.

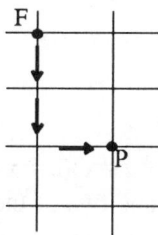

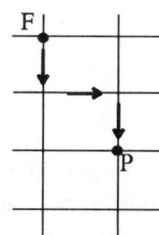

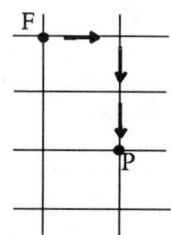

F represents a fire station. A fire is reported at junction P and the fire services need to take the shortest route from F to P. In fact, there are three equally short routes, as the three diagrams show.

Copy the street plan shown opposite and, by each crossroad, write the number of shortest routes there are from F. Some have been written in to help you.

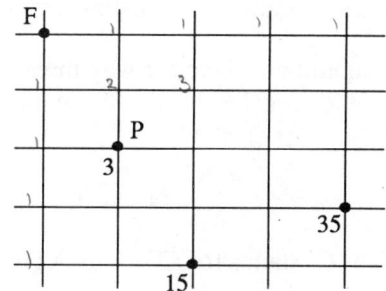

See if you can discover the link between this problem and the binomial coefficients.

Activities 10 and 11 give the same pattern as that of the binomial coefficients. This is because all three situations can be reduced essentially to the same thing. Here are specific questions and solutions from all three.

Bridge problem

How many ways can 5 cards be divided so that 3 are with E and 2 with W?

Solution

Imagine all five cards in a row. Each can be marked either E or W. There must be 3 Es and 2 Ws in any order. The answer to the question is, therefore, the number of different combinations possible with 3 Es and 2 Ws. As the list opposite shows, that number is 10.

EEEWW	EWWEE
EEWEW	WEEEW
EEWWE	WEEWE
EWEWE	WEWEE
EWEEW	WWEEE

Streets problem

In this plan, how many shortest routes are there from F to Q?

Solution

F and Q are 5 blocks away. To get to Q you must go across (A) 3 and down (D) 2, in any order. The number of such routes is therefore the number of different combinations of 3 As and 2 Ds. The answer is 10.

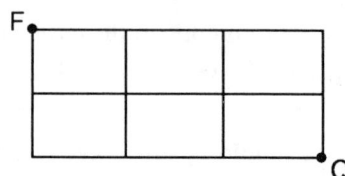

F and Q plan diagram with F at top-left and Q at bottom-right.

AAADD	ADDAA
AADAD	DAAAD
AADDA	DADAA
ADADA	DAADA
ADAAD	DDAAA

Binomial coefficients

In the expansion of $(a+x)^5$, what is the coefficient of the term a^3x^2?

Solution

$(a+x)^5$ is short for $(a+x)(a+x)(a+x)(a+x)(a+x)$. To get a^3x^2 three brackets must supply an 'a' and two of them an 'x'. The coefficient will be the number of ways this can be done. A list of the possible ways is shown on the right.

It should now be clear why three apparently unrelated problems give identical answers. The number of ways of arranging 3 As and 2 Ds is denoted

$$\binom{5}{3} \quad \text{or} \quad {}^5C_3.$$

The 'C' stands for COMBINATION.

To take another example.

$$\binom{7}{2} \quad \text{or} \quad {}^7C_2$$

means the number of different combinations of 7 objects, 2 of one type and 5 of another.

aaaxx	*axxaa*
aaxax	*xaaax*
aaxxa	*xaaxa*
axaax	*xaxaa*
axaxa	*xxaaa*

Exercise 9D

1. From what you have done in this section, write down the values of

   (a) $\binom{4}{2}$ (b) $\binom{5}{1}$ (c) $\binom{5}{2}$ (d) $\binom{6}{3}$ (e) $\binom{3}{1}$

2. How many different ways are there of arranging the letters of the words BOB and ANNA? Express your answers in terms of combinations,

   i.e $\binom{p}{q}$

9.5   Factorials

Activity 12   Combinations of letters

(a) In the 8-letter word DOMINATE all the letters are different. Including the one given, how many arrangements are there of these letters.

(b) The word NOMINATE also has 8 letters, but two of them are the same. How many arrangements are there now?

(c) The word ADDITION has eight letters with TWO pairs of identical ones. How many different arrangements are there?

(d) How many arrangements are there of the word CALCULUS?

(e) The word DIVISION has one letter that appears three times. How many arrangements of these eight letters are there?

(f) How many arrangements are there of the words COCOONED and ASSESSES?

(g) Write down a general rule that finds the number of arrangements of the letters of an 8-letter word.

In general, the number of **arrangements** of an n-letter word with one letter repeated p times is given by the formula

$$\frac{n!}{p!}$$

$n!$ read as 'n factorial', is short for the product of all the natural numbers up to and including n; so for example

$$7! = 7 \times 6 \times 5 \times 4 \times 3 \times 2 \times 1 = 5040.$$

The number of different arrangements of letters in the word **SIMILAR** is therefore

$$\frac{7!}{2!} = \frac{7 \times 6 \times 5 \times 4 \times 3 \times 2 \times 1}{2 \times 1} = 2520.$$

This rule can be extended to cover the repetition of more than one letter. The number of arrangements of letters in the word **SENSES** is

$$\frac{6!}{3!2!} = \frac{6 \times 5 \times 4 \times 3 \times 2 \times 1}{3 \times 2 \times 1 \times 2 \times 1} = 60.$$

This automatically provides a way of finding binomial coefficients.

$\binom{5}{3}$ is the number of arrangements of the letters AAADD

so $\qquad \binom{5}{3} = \dfrac{5!}{2!3!} = \dfrac{5\times4\times3\times2\times1}{2\times1\times3\times2\times1} = \dfrac{5\times4}{2\times1} = 10$

Similarly the binomial coefficient

$$\binom{10}{7} = \dfrac{10!}{3!7!} = \dfrac{10\times9\times8\times7\times6\times5\times4\times3\times2\times1}{(3\times2\times1)\times(7\times6\times5\times4\times3\times2\times1)}$$

$$= \dfrac{10\times9\times8}{3\times2\times1} = 120$$

Thus the general formula for binomial coefficients is given by

$$\binom{n}{r} = \dfrac{n!}{(n-r)!r!}$$

Your calculator probably has a factorial function on it. It may also have a function for working out binomial coefficients; consult your manual if you are not certain. In any case, it is often possible to work them out without a calculator; though the formula looks difficult there is always a good deal of cancelling that can be done, as the examples so far have shown.

Example

How many ways are there of selecting a tennis team of 5 from a squad of 9 players?

Solution

Imagine all 9 players in a line, 5 of them are in the team : label them T; 4 are not : label them N. The answer to the question is the number of arrangements of 5 Ts and 4 Ns, since each different arrangement gives a different team.

$$\binom{9}{5} = \dfrac{9!}{4!5!} = \dfrac{9\times8\times7\times6\times5\times4\times3\times2\times1}{(4\times3\times2\times1)(5\times4\times3\times2\times1)} = 126$$

Activity 13

Compose a question to which the answer is $\binom{5}{5}$. Why does this mean that $0! = 1$?

Exercise 9E

1. How many 'words' can be made from

   (a) 6 As and 3 Xs;

   (b) 7 Us and 5 Ws;

   (c) 10 Ys and 13 Zs?

2. How many different ways are there of arranging the letters in these names?

   SADIA

   WILLIAM

   BARBARA

   CHRISTOPHER

3. A committee of four is to be selected from a club of 25 people. How many different possibilities are there?

4. A florist puts ten pot plants in the window. Six of them are red, four green. How many different arrangements of the colours can be made?

5. Eleven hockey players are to be selected from a squad of 16. How many possible selections are there?

*6. 12 people are to be split up into

   group A - four people;

   group B - six people;

   group C - two people.

   How many ways are there of doing this?

9.6   Binomial theorem

First, there are some results concerning binomial coefficients which will be very useful later.

Activity 14

Show that $\begin{pmatrix} n \\ 0 \end{pmatrix} = 1, \begin{pmatrix} n \\ 1 \end{pmatrix} = n$ and $\begin{pmatrix} n \\ 2 \end{pmatrix} = \dfrac{n(n-1)}{2}$

Write down corresponding formulas for $\begin{pmatrix} n \\ 3 \end{pmatrix}$ and $\begin{pmatrix} n \\ 4 \end{pmatrix}$

What is $\begin{pmatrix} n \\ r \end{pmatrix}$?

Activity 15

$$\begin{pmatrix} 21 \\ 12 \end{pmatrix} = 293\ 930 \text{ and } \begin{pmatrix} 21 \\ 13 \end{pmatrix} = 203\ 490$$

Without using a calculator , write down the values of

$$\begin{pmatrix} 21 \\ 9 \end{pmatrix} \text{ and } \begin{pmatrix} 22 \\ 13 \end{pmatrix}$$

What properties of the binomial coefficients have you used?
Express this symbolically.

The expansion of $(1+x)^5$ can now be written down as follows :

$$1+\binom{5}{1}x+\binom{5}{2}x^2+\binom{5}{3}x^3+\binom{5}{4}x^4+x^5$$

$$=1+5x+10x^2+10x^3+5x^4+x^5$$

Similarly the expansion of $(1+a)^8$, where a is any number, can be found thus :

$$a^8+\binom{8}{1}a^7x+\binom{8}{2}a^6x^2+\binom{8}{3}a^5x^3+\binom{8}{4}a^4x^4+\binom{8}{5}a^3x^5$$
$$+\binom{8}{6}a^2x^6+\binom{8}{7}ax^7+x^8$$

$$=a^8+8a^7x+28a^6x^2+56a^5x^3+70a^4x^4+56a^3x^8$$
$$+28a^2x^6+8ax^7+x^8$$

The general result is known as the **binomial theorem** for a positive integer index.

For $a, x \in \mathbb{R}$ and $n \in \mathbb{N}$

$$(a+x)^n = a^n+\binom{n}{1}a^{n-1}x+\binom{n}{2}a^{n-2}x^2+...$$

$$...+\binom{n}{r}a^{n-r}x^r+...+x^n$$

Note the way this result is set out. Since n is unknown it is impossible to show the whole expansion. So the formula shows how the expansion starts and finishes, and gives a formula for the general term, that is, the rth term

$$\binom{n}{r}a^{n-r}x^r.$$

One way of using the binomial theorem was to give rough approximations to numbers like $(1.01)^8$.

$$(1.01)^8 = 1+\binom{8}{1}(0.01)+\binom{8}{2}(0.01)^2+\binom{8}{3}(0.01)^3+...$$

$$=1+8\times0.01+28\times0.0001+56\times0.000001+...$$

If 3 s.f. accuracy is required, clearly only the first three terms are needed at most :

$$(1.01)^8 = 1.0828...=1.08 \text{ to 3s.f.}$$

The first four terms give at least 6 s.f.

$$(1.01)^8 = 1 + 0.08 + 0.0028 + 0.000056$$
$$= 1.08286 \text{ to 6 s.f.}$$

And all this without a calculator!

Example

Expand $(2x + 3y)^4$.

Solution

$$(2x + 3y)^4 = (2x)^4 + \binom{4}{1}(2x)^3(3y) + \binom{4}{2}(2x)^2(3y)^2$$
$$+ \binom{4}{3}(2x)(3y)^3 + (3y)^4$$
$$= 16x^4 + 96x^3y + 216x^2y^2 + 216xy^3 + 81y^4$$

Example

Expand $\left(3 - \tfrac{1}{2}p\right)^5$

Solution

$$(3 - \tfrac{1}{2}p)^5 = 3^5 + \binom{5}{1}3^4(-\tfrac{1}{2}p) + \binom{5}{2}3^3(-\tfrac{1}{2}p)^2 + \binom{5}{3}3^2(-\tfrac{1}{2}p)^3$$
$$+ \binom{5}{4}3(-\tfrac{1}{2}p)^4 + (-\tfrac{1}{2}p)^5$$
$$= 243 - \frac{405}{2}p + \frac{135}{2}p^2 - \frac{45}{4}p^3 + \frac{15}{16}p^4 - \frac{1}{32}p^5.$$

Example

Find the coefficient of x^5 in the expansion of $(4 - 3x)^9$.

Solution

The x^5 term must be $\binom{9}{5}4^4(-3x)^5$

$$= 126 \times 256 \times (-243)x^5$$
$$= -7\ 838\ 208x^5$$

So the coefficient of x^5 is $-7\ 838\ 208$.

Exercise 9F

1. Expand in full :

   (a) $(1+2x)^4$ (b) $(5-2p)^3$

   (c) $(6-\frac{1}{2}a)^6$ (d) $(2m+3n)^5$

   (e) $(2-\frac{1}{2}r)^4$

2. Find the coefficients of

   (a) x^3 in $(2+x)^{10}$ (b) a^5 in $(5-2a)^7$

   (c) q^6 in $(2p+5q)^9$ (d) k^4 in $(3k-2l)^{12}$

   (e) n^7 in $(5-\frac{5}{6}n)^{10}$

3. Find these to 3 s.f. without using a calculator.

   (a) $(1.01)^5$ (b) $(1.02)^{10}$

   (c) $(0.99)^7$ (d) $(2.01)^3$

   (e) $(99.5)^4$ (f) 11^6

9.7 Binomial expansion

The next task is to extend the theory to expressions such as

$$(1+x)^{\frac{1}{2}} \text{ or } (3+y)^{-\frac{1}{2}}$$

What is the problem in applying the binomial theorem for fractional powers?

A rule has been established for expanding $(a+x)^n$ where n is any **natural** number; but what happens when n is not a natural number?

The answer is that the binomial theorem can be extended to such cases. The next few activities are designed to give you a 'feel' for what happens.

Activity 16 $\sqrt{(1+x)}$

(a) Type into your calculator any number greater than 1. Take successive square roots. You will find that the numbers in the display get closer and closer to 1. But what else do you notice as the numbers get below about 1.2?

(b) Does this also happen for numbers less than 1?

(c) Experiment with cube roots and other fractional indices.

(d) Try to express your results symbolically.

You may have noticed that, for small x,

$$(1+x)^{\frac{1}{2}} \approx 1+\tfrac{1}{2}x.$$

The '≈' sign shows that this is not exact. For example, for $x = 0.01$,

$$\sqrt{1.01} = 1.0049876, \text{ to 7 d.p.'s.}$$

But the approximate formula gives $1 + \frac{1}{2}(0.01) = 1.005$

Now   error = true answer – approximation

$$= -1.25 \times 10^{-5} \text{ to 3 s.f.}$$

Can you find a formula for the 'error' in terms of x? If you can, you should be able to improve the approximate formula.

Activity 17

Square $1 + \frac{1}{2}x$. How close is it to $1 + x$? Square the improved approximate formula suggested above.

Check your improved approximation.

Activity 18 $\quad (1+x)^{-1}$

Use your calculator to work out $1.05^{-5}, 1.01^{-1}$ and other expressions of the form $(1+x)^{-1}$ where x is small. Include negative values of x. Try to obtain an approximate formula for $(1+x)^{-1}$.

(The **exact** value of 1.01^{-1} might give a clue to a precise formula.)

The binomial theorem for a rational index is as follows :

$$(1+x)^n = 1 + nx + \frac{n(n-1)}{2!}x^2 + \frac{n(n-1)(n-2)}{3!}x^3 + \dots$$
$$\dots + \frac{n(n-1)\dots(n-(n-1))}{r!}x^r + \dots$$

This works only for $-1 \le x \le 1$ or $|x| \le 1$.

Note that

1. This is an infinite series : this formula shows how it starts and gives the general coefficient of x^r.

2. The condition $|x| \le 1$ or $-1 \le x \le 1$ is important : it is known as the domain of validity.

3. The formula is for $(1+x)^n$, not $(a+x)^n$.

Compare this with your work on the preceding activities.

Example

Expand $(1+x)^{\frac{2}{3}}$ up to and including the term in x^3.

Solution

The best procedure is to apply the formula first and then tidy up each term.

$$(1+x)^{\frac{2}{3}} = 1 + \frac{2}{3}x + \frac{\left(\frac{2}{3}\right)\left(-\frac{1}{3}\right)}{1 \times 2}x^2 + \frac{\left(\frac{2}{3}\right)\left(-\frac{1}{3}\right)\left(-\frac{4}{3}\right)}{1 \times 2 \times 3}x^3 + \ldots$$

$$= 1 + \frac{2}{3}x - \frac{1}{9}x^2 + \frac{4}{81}x^3 - \ldots \text{ for } |x| \le 1.$$

Example

Expand $(1-x)^{-\frac{1}{2}}$ up to and including the term in x^3.

Solution

The domain of validity will be $|x| < 1$ since $|-x| = |x|$

$$(1-x)^{-\frac{1}{2}} = 1 + \left(-\frac{1}{2}\right)(-x) + \frac{\left(-\frac{1}{2}\right)\left(-\frac{3}{2}\right)}{1 \times 2}(-x)^2 + \frac{\left(-\frac{1}{2}\right)\left(-\frac{3}{2}\right)\left(-\frac{5}{2}\right)}{1 \times 2 \times 3}x^3$$

$$\approx 1 + \frac{1}{2}x + \frac{3}{8}x^2 + \frac{5}{16}x^3 \text{ for } |x| < 1$$

Example

Expand $(1+2x)^{\frac{1}{2}}$ up to and including the term in x^3.

Solution

The 'x' in the general formula must be replaced by '2x'. This necessitates a different domain of validity. It is $|2x| \le 1$, which simplifies to $|x| \le \frac{1}{2}$

$$(1-x)^{-\frac{1}{2}} = 1 + \left(-\frac{1}{2}\right)(-x) + \frac{\left(-\frac{1}{2}\right)\left(-\frac{3}{2}\right)}{1 \times 2}(-x)^2 + \frac{\left(-\frac{1}{2}\right)\left(-\frac{3}{2}\right)\left(-\frac{5}{2}\right)}{1 \times 2 \times 3}x^3$$

$$\approx 1 + x + -\frac{1}{2}x^2 + \frac{1}{2}x^3 \text{ for } |x| \le \frac{1}{2}$$

The binomial expansion can also be used for expressions of the form $(a+x)^n$ where $a \neq 1$.

Example

Expand $(3+x)^{-1}$

Solution

In the binomial expansion, $3+x$ must be re-written as '$1 \pm$ something' for the formula to be used. This is done as follows:

$$3+x = 3\left(1+\tfrac{x}{3}\right)$$

So $\quad (3+x)^{-1} = 3^{-1}\left(1+\tfrac{x}{3}\right)^{-1} = \tfrac{1}{3}\left(1+\tfrac{x}{3}\right)^{-1}$

$$= \tfrac{1}{3}\left\{1+(-1)\left(\tfrac{x}{3}\right)+\frac{(-1)(-2)}{1\times 2}\left(\tfrac{x}{3}\right)^2 + \frac{(-1)(-2)(-3)}{1\times 2\times 3}\left(\tfrac{x}{3}\right)^3 +...\right\}$$

$$= \tfrac{1}{3}\left(1-\tfrac{x}{3}+\tfrac{x^2}{9}-\tfrac{x^3}{27}+....\right) \text{ for } \left|\tfrac{x}{3}\right| \leq 1 \text{ i.e.} |x| \leq 3.$$

Example

Expand $(5-3x)^{\frac{3}{2}}$ up to and including the term in x^3, and find the domain of validity.

Solution

Change to '1 + something', so

$$5-3x = 5\left(1-\tfrac{3x}{5}\right)$$

and $\quad (5-3x)^{\frac{3}{2}} = 5^{\frac{3}{2}}\left(1-\tfrac{3x}{5}\right)^{\frac{3}{2}}$

$$= 5\sqrt{5}\left\{1+\left(\tfrac{3}{2}\right)\left(-\tfrac{3x}{5}\right)+\frac{\left(\tfrac{3}{2}\right)\left(\tfrac{1}{2}\right)}{1\times 2}\left(-\tfrac{3x}{5}\right)^2 + \frac{\left(\tfrac{3}{2}\right)\left(\tfrac{1}{2}\right)\left(-\tfrac{1}{2}\right)}{1\times 2\times 3}\left(-\tfrac{3x}{5}\right)^3 +...\right\}$$

$$= 5\sqrt{5}\left(1-\tfrac{9}{10}x+\tfrac{27}{200}x^2 + \tfrac{27}{2000}x^3+....\right)$$

or $\quad 5\sqrt{5}\left(1-0.9x+0.135x^2 +0.0135x^3+....\right)$

Domain of validity: $\left|\tfrac{-3x}{5}\right| \leq 1 \Rightarrow |x| \leq \tfrac{5}{3}.$

Exercise 9G

1. Simplify these expressions. Write the coefficients as fractions :

   (a) $\dfrac{\left(\frac{3}{4}\right)\left(-\frac{1}{4}\right)}{1\times 2}x^2$ (b) $\dfrac{\left(\frac{2}{3}\right)\left(-\frac{1}{3}\right)}{1\times 2}(3x)^2$

   (c) $\dfrac{(-2)(-3)(-4)}{1\times 2\times 3}\left(\frac{x}{2}\right)^3$ (d) $\dfrac{\left(\frac{5}{4}\right)\left(\frac{1}{4}\right)\left(-\frac{3}{4}\right)}{1\times 2\times 3}x^3$

   (e) $\dfrac{\left(-\frac{5}{2}\right)\left(-\frac{7}{2}\right)\left(-\frac{9}{2}\right)}{1\times 2\times 3}\left(-\frac{x}{3}\right)^3$

2. Write these as series up to and including the term in x^3. State the domain of validity in each case.

   (a) $(1-x)^{\frac{1}{2}}$ (b) $(1-x)^{-2}$ (c) $(1-x)^{\frac{3}{2}}$

   (d) $(1+2x)^{-1}$ (e) $\left(\sqrt[3]{1+\frac{x}{2}}\right)$ (f) $(1-3x)^{\frac{5}{4}}$

   (g) $\dfrac{1}{\sqrt{(1-6x)}}$ (h) $\left(1+\frac{3x}{4}\right)^{0.6}$

3. Repeat Question 2 for these :

   (a) $\dfrac{1}{2+x}$

   (b) $\sqrt{(4-x)}$

   (c) $(8-3x)^{-\frac{1}{3}}$

4. Expand $\sqrt{\left(1-x^2\right)}$ up to and including the term in x^6. State the domain of validity.

9.8   Further examples

The binomial theorem in this form is particularly useful for making approximations, especially to complex formulas involving surds. The two worked examples demonstrate how this is done.

Example

Calculate (a) $\sqrt{4.8}$ (b) $\sqrt[3]{1100}$ to 4 s.f. without a calculator.

Solution

(a)   Now $4.8 = 4+0.8 = 4(1+0.2)$, so

$$\sqrt{4.8} = \sqrt{4}\sqrt{(1+0.2)} = 2\sqrt{1+0.2}$$

$$= 2(1+0.2)^{\frac{1}{2}}$$

$$= 2\left\{1 + \left(\tfrac{1}{2}\right)0.2 + \frac{\left(\tfrac{1}{2}\right)\left(-\tfrac{1}{2}\right)}{1 \times 2}(0.04) + \frac{\left(\tfrac{1}{2}\right)\left(-\tfrac{1}{2}\right)\left(-\tfrac{3}{2}\right)}{1 \times 2 \times 3}(0.008) + \ldots\right\}$$

$$= 2\{1 + 0.1 - 0.005 + 0.0005\ldots\}$$

$$= 2 \times 1.0955$$

$$= 2.191 \text{ to 4 s.f.}$$

(b)   Now $1100 = 1000 + 100 = 1000(1 + 0.1)$, so

$$\sqrt[3]{1100} = \left(\sqrt[3]{1000}\right)\left(\sqrt[3]{1 + 0.1}\right) = 10\left(\sqrt[3]{1 + 0.1}\right)$$

$$= 10(1 + 0.1)^{\frac{1}{3}}$$

$$= 10\left\{1 + \left(\tfrac{1}{3}\right)(0.1) + \frac{\left(\tfrac{1}{3}\right)\left(-\tfrac{2}{3}\right)}{1 \times 2}(0.01) + \frac{\left(\tfrac{1}{3}\right)\left(-\tfrac{2}{3}\right)\left(-\tfrac{5}{3}\right)}{1 \times 2 \times 3}(0.001) + \ldots\right\}$$

$$= 10\{1 + 0.03333\ldots - 0.001111\ldots + \ldots\}$$

$$= 10 \times 1.03222\ldots$$

$$= 10.32 \text{ to 4 s.f.}$$

Example

The time T seconds taken for a simple pendulum of length l cm to swing to and fro once is given approximately by the formula

$$T = 0.2\sqrt{l}$$

Suppose the length increases by $p\%$ and that this causes T to increase by $q\%$. Show that

$$q = 100\left[\sqrt{1 + \frac{p}{100}} - 1\right]$$

and find an approximate linear formula for q in terms of p when p is small.

Solution

% increases: $l \to l\left(1 + \frac{p}{100}\right)$

$$T \to T\left(1 + \frac{q}{100}\right)$$

$$\Rightarrow \quad T\left(1 + \frac{q}{100}\right) = 0.2\sqrt{l\left(1 + \frac{p}{100}\right)}$$

$$\Rightarrow \quad T\left(1+\frac{q}{100}\right)=0.2\sqrt{l}\sqrt{1+\frac{p}{100}}$$

$$\Rightarrow \quad 1+\frac{q}{100}=\sqrt{1+\frac{p}{100}}, \qquad \text{since } T=0.2\sqrt{l}$$

$$\Rightarrow \quad q=100\left[\sqrt{1+\frac{p}{100}}-1\right]$$

To find an approximate formula, expand $\sqrt{1+\frac{p}{100}}$ using the binomial theorem

$$\left(1+\frac{p}{100}\right)^{\frac{1}{2}}\approx1+\left(\frac{1}{2}\right)\left(\frac{p}{100}\right)=1+\frac{p}{200}$$

$$\Rightarrow \quad q\approx100\left[1+\frac{p}{200}-1\right]$$

$$\Rightarrow \quad q\approx\frac{p}{2}.$$

Exercise 9H

1. Work these out to 4 s.f. without using a calculator:

   (a) 1.2^{-1} (b) $\sqrt{9.09}$ (c) $8.24^{\frac{2}{3}}$

2. The APR on a credit card account is $x\%$. The corresponding monthly interest rate is $y\%$.

   (a) Show that $y=100\left[\left(1+\frac{x}{100}\right)^{\frac{1}{12}}-1\right]$.

   (b) Show that this reduces to an approximate formula $y=\frac{x}{12}-kx^2$ under certain conditions. State these conditions and find the value of k (as a fraction).

9.9 Miscellaneous Exercises

1. Write these as fractions
   (a) $81^{-\frac{1}{4}}$ (b) $\left(\frac{9}{4}\right)^{\frac{1}{2}}$ (c) $\left(\frac{1}{8}\right)^{\frac{1}{3}}$ (d) $16^{-\frac{3}{4}}$

2. Solve these equations. A calculator is not needed.
   (a) $16^x=2$ (b) $100^x=0.1$ (c) $8^x=\frac{1}{4}$
   (d) $25^{2x+1}=0.04$

3. Given that $3^{2.524}=16$, solve these equations to 2 s.f.
   (a) $3^x=4$ (b) $3^x=\frac{1}{2}$ (c)$3^x=\frac{1}{256}$ (d) $9^x=32$

4. A class of 23 students selects a Year Council representative, four duty team members and two quiz team members. How many ways are there of doing this if

   (a) no-one is allowed to fill more than one of these roles;

   (b) there is no such restriction?

5. Use the binomial theorem to expand:

   (a) $(1+x)^5$ (b) $\left(3+\dfrac{x}{2}\right)^4$ (c) $(2-3p)^6$

6. Expand these up to and including the term in x^3

   (a) $(1+x)^{-1}$ (b) $(1+3x)^{\frac{1}{2}}$ (c) $\left(1-\dfrac{x}{3}\right)^{0.7}$

   (d) $\dfrac{1}{\sqrt[3]{(1+2x)}}$

7. Expand these up to and including the term in x^2:

   (a) $(100+x)^{\frac{-3}{2}}$ (b) $(a-x)^{-4}$

8. Find the coefficient of

   (a) x^6 in $\left(x+x^2\right)^4$ (b) p^3 in $\left(p^2-\dfrac{3}{p}\right)^6$

   (c) n^4 in $(2+n)^3(1+2n)^4$

   (d) y^6 in $(5-2y)^4(y-1)^5$

9. The 'gravitational potential energy' of an object is given by the formula $mgr^2\left(\dfrac{1}{r}-\dfrac{1}{r+h}\right)$ where m is the mass of the object, g is the acceleration due to gravity, r is the earth's radius and h is the height of the object above the ground. In practice this formula is frequently too complicated to use. Show that, when $\dfrac{h}{r}$ is small, the formula is approximately equal to mgh.

10. The speed v of waves travelling along a string depends on the tension T in the string, according to the formula $v=k\sqrt{T}$ where k is constant.

    In a particular experiment, when the tension is 50 newtons the speed of the wave is 200 ms^{-1}.

    The tension is increased by a small amount h. Show that the speed of the move is increased approximately by $2h$.

    (You can check this is true by putting $T=51$ in the exact formula).

11. (a) Which gives the better annual percentage rate (APR) of interest: 2% compound interest per month or 6% compound interest per quarter? Your answer should be fully explained.

    (b) What is the APR if

       (i)   1% is compounded 24 times a year;

       (ii) 0.5% is compounded 48 times a year;

       (iii) 0.25% is compounded 96 times a year?

    (c) What happens as you continue this pattern? (i.e. as you halve the % interest and double the number of payments.)

10 CIRCULAR MEASURE

Objectives

After studying this chapter you should

- know that radians are another unit for measuring angles;
- be able to convert from degrees to radians and vice versa;
- know and be able to use formulae for arc length and sector area in terms of radians;
- know and be able to use small angle approximations for $\sin x$, $\cos x$ and $\tan x$.

10.0 Introduction

You are familiar with using degrees to measure angles, but this is not the only way to do it. In fact, as you see in some areas of mathematics, it is not even the most convenient way.

Activity 1 History of circular measure

(a) Your calculator will have settings for 'degrees', 'radians' and 'gradians'. Look up gradians in an encyclopaedia, and find out when these were first introduced. When did they last get used in Europe? How many gradians are there in a full turn, or in a right angle?

(b) Find out why 360 degrees make a full turn, and which civilizations invented them. What were the reasons for adopting 360? Why are 360 degrees still used?

Activity 2

The figure opposite shows a circle of radius r and arc PQ which is subtended by an angle of θ degrees.

(a) What is the length of the complete circumference?

(b) What is the arc length PQ in terms of θ and r?

(c) If PQ $= r$, what is the value of θ?

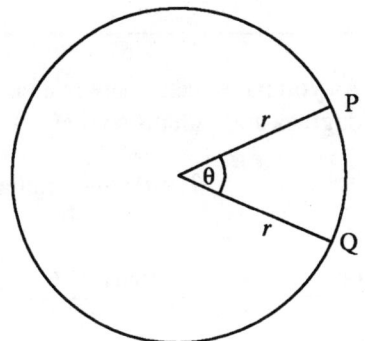

10.1 Radian measure

The second activity gives the clue as to how radians are defined. One radian corresponds to the angle which gives the same arc length as the radius. So if θ is the angle POQ in degrees.

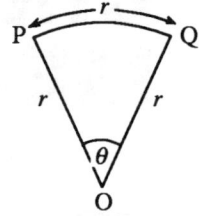

$$\frac{\theta}{360} = \frac{r}{2\pi r} = \frac{1}{2\pi}$$

Thus $\theta = \dfrac{360}{2\pi} = \dfrac{180}{\pi}$ showing that

$$1 \text{ radian} \equiv \frac{180}{\pi} \text{ degrees}$$

This is not a very convenient definition, so it is usual to note that

$$\boxed{\pi \text{ radians} \equiv 180 \text{ degrees}}$$

Example

Convert the following angles in degrees to radians.

(a) 90°   (b) 360°   (c) 720°   (d) 60°

Solution

Since $180° = \pi$ radians,

(a) $90° = \dfrac{\pi}{2}$ radians   (b) $360° = 2\pi$ radians

(c) $720° = 4\pi$ radians   (d) $60° = \dfrac{\pi}{3}$ radians

Activity 3   Conversion between degrees and radians

Copy the diagram opposite, putting in the equivalent radian measure for each angle given in degrees.

As you have seen, there is a one to one correspondence between degrees and radians so that

$$\theta° \equiv \theta \times \frac{\pi}{180} \text{ radians}$$

or $\qquad \theta \text{ radians} \equiv \left(\theta \times \dfrac{180}{\pi} \right)°$

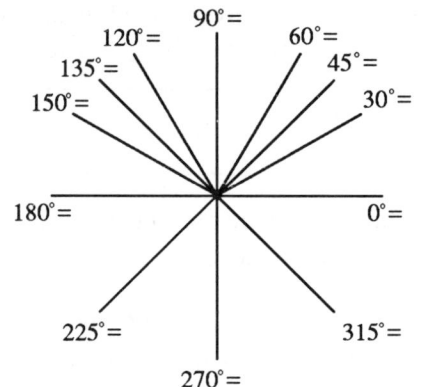

Example

Use the formulae above to convert

(a)   to radians   (i) 45°   (ii) 30°   (iii) 150°

(b)   to degrees (i) $\frac{2\pi}{3}$ radians (ii) $\frac{\pi}{12}$ radians

Solution

(a)   (i) $45° = 45 \times \frac{\pi}{180}$ radians $= \frac{\pi}{4}$ radians

(ii) $30° = 30 \times \frac{\pi}{180}$ radians $= \frac{\pi}{6}$ radians

(iii) $150° = 150 \times \frac{\pi}{180}$ radians $= \frac{5\pi}{6}$ radians

(b)   (i) $\frac{2\pi}{3}$ radians $= \left(\frac{2\pi}{3} \times \frac{180}{\pi}\right)° = 120°$

(ii) $\frac{\pi}{12}$ radians $= \left(\frac{\pi}{12} \times \frac{180}{\pi}\right)° = 15°$

Exercise 10A

1. Convert these angles in radians to degrees.

   (a) $\frac{7\pi}{6}$ radians   (b) 3π radians

   (c) 2 radians   (d) $\frac{11\pi}{12}$ radians

2. Convert these angles in degrees to radians.

   (a) $12\frac{1}{2}^{°}$ (b) $72\frac{1}{2}^{°}$

   (c) 210°   (d) 20°

10.2 Arc length and sector area

You may already have seen how to find the arc length. If θ is measured in degrees then

$$\text{arc length PQ} = \frac{\theta}{360} \times 2\pi r = \frac{\pi r \theta}{180}$$

but if θ is measured in radians

$$\text{arc length PQ} = \frac{\theta}{2\pi} \times 2\pi r = r\theta$$

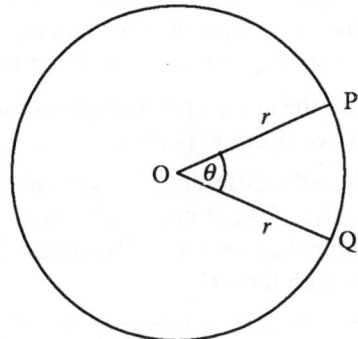

Why is 2π used instead of 360° in this formula?

There is a simple formula for the area of a sector of a circle subtended by an angle.

Activity 4   Sector area

(a)   What is the area of a circle, of radius r?

(b)   The sector subtends an angle of θ radians at the centre of a circle.   What proportion is the area of the sector to the area of the whole circle?

(c)   Deduce the formula for the sector area?

So, using radian measure, the formula for both arc length and sector area take a simple form, namely

$$\text{arc length} = r\theta$$

$$\text{sector area} = \frac{1}{2}r^2\theta$$

You will see how the first of these formulae is used in a practical problem solving activity.

When sheets of tinplate are produced, the edges become 'wavy'. This needs to be corrected, which can be done if the centre portion is 'stretched'.   The manufacturers need to be able to calculate the factor by which the material must be stretched.

Activity 5   A tinplate problem

Suppose the edge is wavy, as shown opposite.   You can assume that the wave is part of a circular arc, with radius r.

(a)   Express OX in terms of r.

(b)   Use Pythagoras' theorem on the right angled triangle OPX to form an equation for the radius, r.

(c)   Solve this equation to find r, correct to one decimal place.

(d)   Find the angle θ, using the triangle OPX.   If necessary, convert θ, once you have found it, to radians, giving your answer to three decimal places.   Double this to find the angle the arc PQ 'subtends' at the point O, namely POQ.

(e)   Use the arc length formula to find the length of the arc PQ, to two decimal places.

(f)   Find the difference between this length and the length of the centre of the tinplate, 250 mm.   Express this difference as a percentage of the 'flat' distance, 250 mm. (This is called the **stretch factor**).

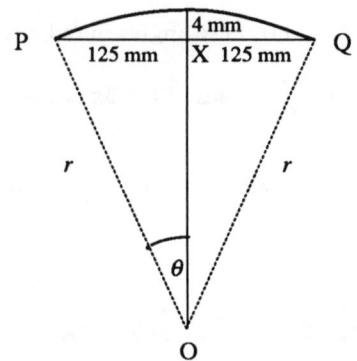

Exercise 10B

1. An oil drum of diameter 60 cm is floating as shown in the diagram below :

    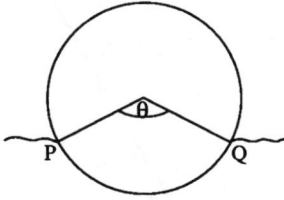

    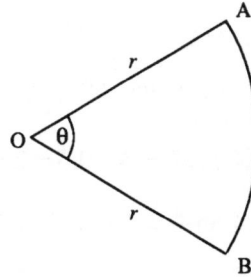

    (a) Given that the arc length PQ is 50 cm, find θ in radians.

    (b) The drum is 1 metre long, find the volume of the drum which lies below the surface level.

2. The length of the arc AB is 20 cm, and the area of the sector AOB is 100cm^2. Form two equations involving r and θ when θ is measured in radians. Solve these equations simultaneously to find r and θ.

3. A stone, swung round on the end of a 200 cm string, completes an arc subtending an angle at the centre of 2 radians every second. Find the speed at which the stone is moving.

10.3 Small angle approximations

From your work at GCSE, you will be familar with the use of sine, cosine and tangent formulae for angles. For example, for this figure opposite, writing 'opp' for opposite, 'hyp' for hypotenuse, and 'adj' for adjacent,

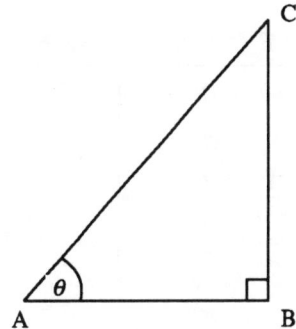

$$\sin \theta = \frac{\text{opp}}{\text{hyp}} = \frac{BC}{AC}$$

similarly

$$\cos \theta = \frac{\text{adj}}{\text{hyp}} = \frac{AB}{AC}$$

and

$$\tan \theta = \frac{\text{opp}}{\text{adj}} = \frac{BC}{AC}$$

What is the relationship between sin θ, cos θ, and tan θ?

You can use these formulae to find lengths and angles in triangles using the θ in either degrees or radians, remembering though to use the correct mode (degrees or radians) on your calculator.

Example

Find the area of the right angled triangle opposite.

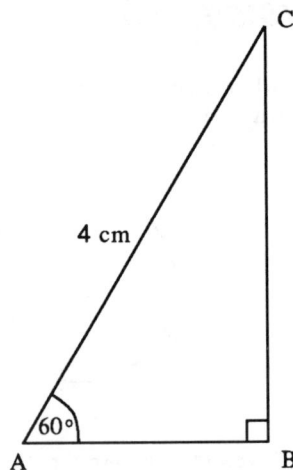

Solution

Working in radians, and since $60° = \frac{\pi}{3}$ radians,

$$AB = 4\cos\frac{\pi}{3} = 4 \times 0.5 = 2 \text{ cm}$$

$$BC = 4\sin\frac{\pi}{3} = 4 \times 0.707 = 2.823 \text{ cm}$$

giving

$$\text{area} = \frac{1}{2} \times AB \times BC = \frac{1}{2} \times 2 \times 2.823$$

$$= 2.823 \text{ cm}^2.$$

Activity 5

Use your calculator to find the following values (remember that you are working in **radians** not degrees)

(a)

θ (radians)	$\sin\theta$	$\cos\theta$	$\tan\theta$
0			
$\frac{\pi}{6}$			
$\frac{\pi}{4}$			
$\frac{\pi}{3}$			
$\frac{\pi}{2}$			
$\frac{3\pi}{4}$			
π			

(b) Sketch the functions

$$f(\theta) = \sin\theta \qquad g(\theta) = \cos\theta \qquad h(\theta) = \tan\theta$$

on the same axis for $0 \le \theta \le \pi$.

(Check your sketches using a graphic calculator)

You may have noticed that the graphs of $\sin\theta$ and $\tan\theta$ for small values of θ look similar.

What is meant by small values of θ?

In fact, for small θ in radians, both $\sin\theta$ and $\tan\theta$ can be approximated by

$$\sin\theta \approx \theta \qquad (1)$$

$$\tan\theta \approx \theta \qquad (2)$$

You will see how accurate this approximation is in the next activity.

Activity 6   Small angle approximations

Use a graphic calculator to plot $\sin\theta$, $\tan\theta$ and θ for θ between 0 and 0.1 radians. (Note that you will need to plot $y = \sin x$, $y = \tan x$ and $y = x$)

What form will the approximation to $\cos\theta$ take for small θ?

Since $\cos\theta = 1$ when $\theta = 0$, the approximation might take the form

$$\cos\theta = 1 + a\theta + b\theta^2 + \ldots..$$

for suitable constants, a and b - remember again that θ here is measured in radians. There are several ways to obtain values for a and b, one of which depends on the relationship

$$\sin^2\theta + \cos^2\theta = 1.$$

Why is this result true for any angle θ?

This can be used to find $\cos\theta$, since

$$\cos^2\theta = 1 - \sin^2\theta$$

$$\Rightarrow \quad \cos\theta = \left(1 - \sin^2\theta\right)^{\frac{1}{2}}$$

But for **small** θ in radians

$$\sin\theta \approx \theta,$$

and $\qquad \cos\theta \approx \left(1 - \theta^2\right)^{\frac{1}{2}}$

In Chapter 9 you met the binomial expansion which can be used here.

Since
$$(1-\theta^2)^{\frac{1}{2}} = 1 + \frac{1}{2}(-\theta^2) + \frac{1}{2}\left(-\frac{1}{2}\right)\frac{(-\theta^2)^2}{2!} + \ldots$$

$$= 1 - \frac{1}{2}\theta^2 + \frac{1}{8}\theta^4 + \ldots$$

for small θ, ignoring θ^4 terms and higher,

$$(1-\theta^2)^{\frac{1}{2}} \approx 1 - \frac{1}{2}\theta^2$$

and so
$$\boxed{\cos\theta \approx 1 - \frac{1}{2}\theta^2}$$ (3)

Activity 7

Use a graphic calculator to plot $\cos\theta$ and $1 - \frac{1}{2}\theta^2$ for θ between 0 and 0.1 radians. Comment on the approximations.

Example

Show that when x is small and in radians,

$$\cos 2x \approx 1 - \frac{1}{2}(\sin 2x)^2$$

Solution

For small x, using equations (1) and (3),

$$\text{L.H.S.} = \cos 2x \approx 1 - \frac{1}{2}(2x)^2 = 1 - 2x^2$$

$$\text{R.H.S.} = 1 - \frac{1}{2}(\sin 2x)^2 = 1 - \frac{1}{2}(2x)^2 = 1 - 2x^2$$

and so the result holds.

Exercise 10C

1. Find small angle approximations when x is measured in radians, for
   (a) $\sin(3x)$ (b) $\sin(x^2)$ (c) $\cos\left(\frac{1}{2}x\right)$
   (d) $\tan(-2x)$

2. Using the approximations given, find the approximate values of
   (a) $\sin(0.0001)$ (b) $\cos(0.003)$ (c) $\cos(0.00001)$
   (d) $\tan(0.01)$ (e) $\cos(-0.004)$

3. Use your calculator for each part of Question 2 in order to find the values as accurately as possible. Compare the two sets of values.

4. What is the value of α to 2 d.p. such that the approximation
   $$\sin\theta \approx \theta$$
   is correct to 3 d.p. for $0 \le \theta \le \alpha$?

10.4 Miscellaneous Exercises

1. Find in radians the angle subtended at the centre of a circle of circumference 36 cm by an arc of length 7.5 cm.

2. Find the length of the arc of a circle of radius 2 cm which subtends an angle of $\frac{\pi}{6}$ at the centre.

3. Find the radius of a circle if a chord of length 10 cm which subtends an angle of 85° at the centre.

4. Change the following angles into radians.

   (a) 45°   (b) 73.78°   (c) 178.83°

5. By using suitable approximations for $\sin\theta$ and $\cos\theta$, obtain an approximation in radians to the positive solution of the equation

$$\cos\theta - \theta\sin\theta = 0.9976$$

   when θ is small.

6. Find the area of a sector of a circle of radius 2 cm which subtends an angle of $\frac{\pi}{4}$ radians at the centre.

7. Determine the angle in radians subtended by the sector of a circle of radius 5 cm such that the area of the sector is $10\,\text{cm}^2$.

11 MODELLING NATURAL CYCLES

Objectives

After studying this chapter you should

- be familiar with basic properties of sine, cosine and tangent functions;

- be able to solve simple trigonometric equations;

- be able to model periodic systems using trigonometric functions;

- understand inverse trigonometric functions including the restrictions of their domains.

11.0 Introduction

There are many systems in nature which have a repeating pattern, like the height and times of high tides, the length of daylight during a year or the pattern of sound waves from a musical instrument. This chapter shows how the trigonometric functions can be used to describe these systems.

You must first though review the uses of sine, cosine and tangents in right angled triangles. Some of these were noted in Chapter 10, but the following activities below will further revise the concepts.

Activity 1

(a) Write down the sine, cosine and tangent of the angles marked in the triangles A and B.

(b) Find the lengths marked x to 2 d.p. in these triangles C and D.

(c) The sketch labelled E shows the frame of a roof. Find the width x.

(d) Find the angle marked, to 1 d.p., in each of the triangles, F and G.

(e) Find the angle δ in the rhombus shown opposite, as H.

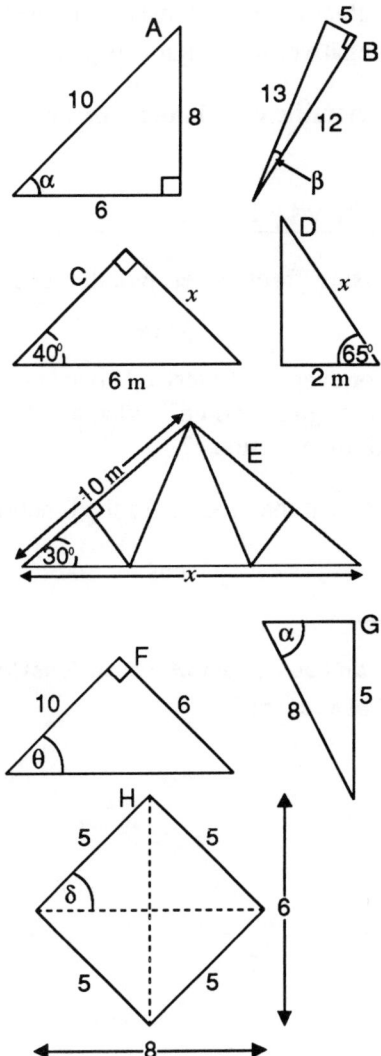

11.1 Sine, cosine and tangent functions

You have seen how sine, cosine and tangent are defined for angles between 0° to 90° but this can be extended to other angles.

Angles of more than 90° can be defined as the angle θ made between a rotating 'arm' OP and the positive x axis, as shown opposite. It is possible to define angles of more than 360° in this way, or even negative angles, as shown opposite.

If the length of OP is 1 unit, then the sine, cosine and tangent of any angle is defined in terms of the x and y co-ordinates of the point P as follows;

$$\sin\theta = y \qquad \cos\theta = x \qquad \tan\theta = \frac{y}{x}$$

Note that $\sin\theta$, $\cos\theta$ and $\tan\theta$ may be negative for certain values of θ. For instance in the third figure opposite, y and x are negative, whilst $\tan\theta$ is positive.

Scientific calculators give sine, cosine and tangent of any angle.

Activity 2

Using graph paper, plot the function

$$y = \sin x$$

for every 10° interval between 0° and 360°. Draw a smooth curve through the points. What are the maximum and minimum values of the function y?

Repeat the process for the function

$$y = \cos x$$

Can you explain why both functions, $\sin x$ and $\cos x$, lie between ± 1?

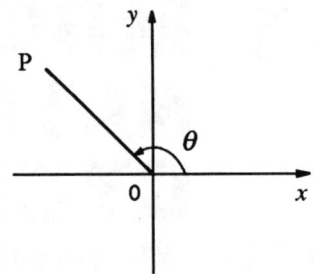

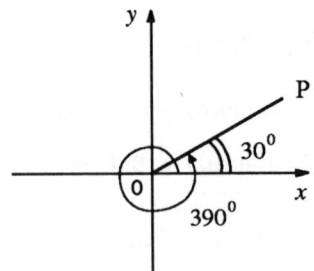

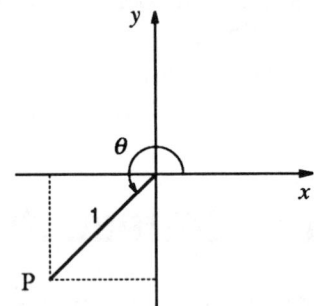

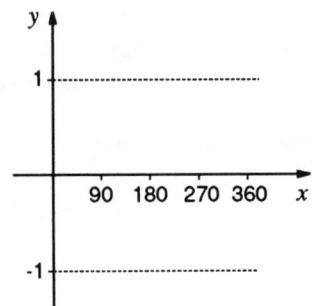

Whilst you have only plotted the functions for $0 \le x \le 360°$, it is easy to see how it can be extended.

What is the period of each function?

For the range $360° \le x \le 720°$, the pattern will repeat itself, and similarly for negative values. The graphs are illustrated below

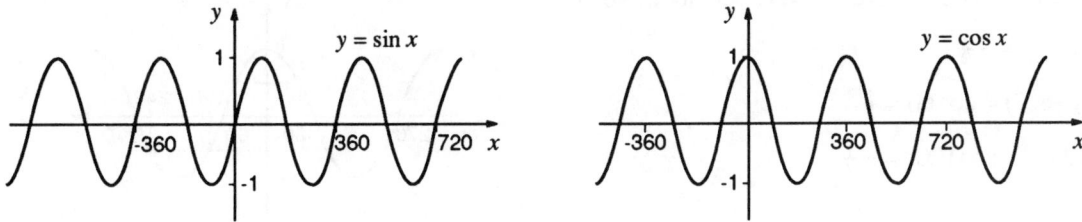

You can see from these graphs that $y = \sin x$ is an ODD function $(y(-x) = -y(x))$, whilst $y = \cos x$ is an EVEN function $(y(-x) = y(x))$.

You can always use your calculator to find values of sine and cosine of an angle, although sometimes it is easier to use the properties of the functions.

Example

Show that $\sin 30° = \frac{1}{2}$, and find the following without using tables:

(a) $\sin 150°$ (b) $\sin(-30°)$ (c) $\sin 210°$

(d) $\sin 390°$ (e) $\cos 60°$ (f) $\cos 120°$

(g) $\cos(-60°)$ (h) $\cos 240°$

Solution

From the sketch opposite of an equilateral triangle $\sin 30° = \dfrac{\frac{1}{2}}{1} = \dfrac{1}{2}$

(a) $\sin 150° = \sin 30° = \frac{1}{2}$

(b) $\sin(-30°) = -\sin 30°$ (function is odd)

      $= -\frac{1}{2}$

(c) $\sin 210° = -\frac{1}{2}$

(d) $\sin 390° = \sin 30° = \frac{1}{2}$

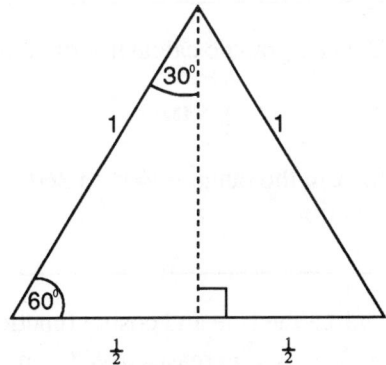

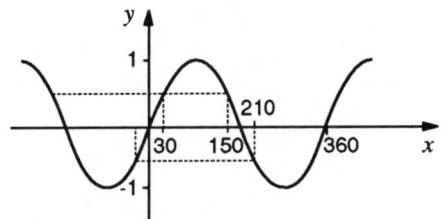

(e)   Since $\cos\theta = \sin(90 - \theta°)$

$$\cos 60° = \sin(90 - 60°)$$
$$= \sin 30°$$
$$= \tfrac{1}{2}$$

(f) $\cos 120° = -\cos 60°$ (function is odd about 90°)

$$= -\tfrac{1}{2}$$

(g) $\cos(-60°) = \cos 60 = \tfrac{1}{2}$

(h) $\cos 240° = \cos 120° = -\tfrac{1}{2}$

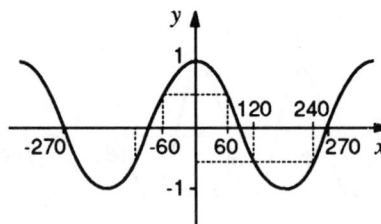

In the above example, the relationship

$$\cos\theta = \sin(90 - \theta°)$$

was used.

Why is this result true?

Similarly, it should be noted that

$$\sin\theta = \cos(90 - \theta°)$$

The last function to consider here is   , which has rather different properties from $\sin x$ and $\cos x$.

Activity 3   tan x

Using a graphic calculator or computer, sketch the function

$$y = \tan x$$

for x in the range $-360°$ to $360°$. What is the period of the function?

Unlike the sine and cosine functions, $\tan x$ is not bounded by ± 1. In fact, as x increases to $90°$, $\tan x$ increases without limit.

Example

Using the value $\tan 45° = 1$ and $\tan 30° = \frac{1}{\sqrt{3}}$ find, without using a calculator

(a) $\tan 135°$ (b) $\tan(-45°)$ (c) $\tan 315°$

(d) $\tan 135° = -1$ (e) $\tan 150°$ (f) $\tan 60°$

Solution

(a) $\tan 135° = -1$

(b) $\tan(-45°) = -\tan 45°$

$\qquad\qquad\quad = -1$

(c) $\tan 315° = -\tan 45°$

$\qquad\qquad\quad = -1$

(d) $\tan(-30) = -\tan 30° = -\dfrac{1}{\sqrt{3}}$

(e) $\tan 150° = -\tan 30° = -\dfrac{1}{\sqrt{3}}$

(f) $\tan 60° = \dfrac{\sin 60°}{\cos 60°} = \dfrac{\cos 30°}{\sin 30°} = \dfrac{1}{\tan 30°}$

$\qquad\quad = \dfrac{1}{\left(1/\sqrt{3}\right)} = \sqrt{3}$

In (f), the result is equivalent to using

$$\tan \theta = \frac{1}{\tan(90 - \theta)}$$

which is set in the exercise below.

The full results are summerised below

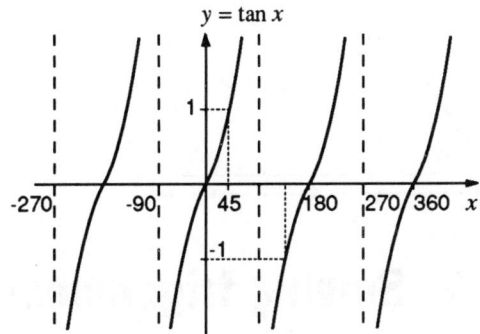

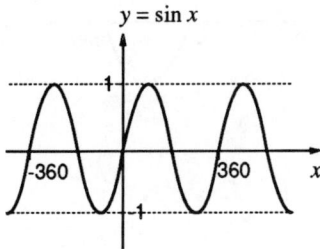

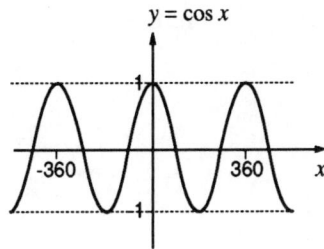

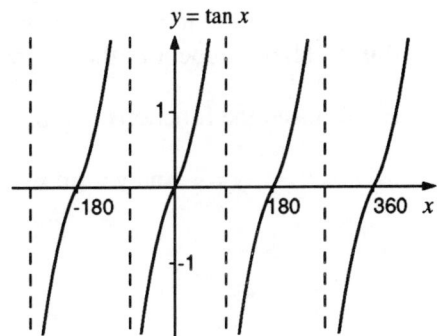

Exercise 11A

1. Find the length marked x to 2 d.p. for each of the figures below.

(a)

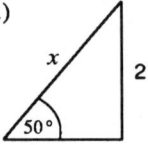

(b)

(c)

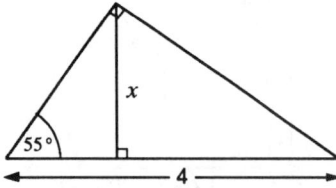

2. Given that $\sin 45° = \dfrac{1}{\sqrt{2}}$, find, without using tables

   (a) $\sin 135°$ (b) $\sin(-45°)$ (c) $\sin 315°$

   (d) $\cos 45°$ (e) $\cos(225°)$ (f) $\tan 45°$

   (g) $\tan 135°$ (h) $\tan(-45°)$

3. Prove that $\sin\theta = \cos(90 - \theta)$

11.2 Solving trigonometrical equations

Later on in this chapter you will see how trigonometric functions can be used to model physical phenomena. Applying these functions to problems in the real world will often result in a trigonometric function to solve. In this section, you will consider some simple equations to solve. These will highlight the difficulties that can occur.

Example

Solve $\sin x = \frac{1}{2}$ for $0° \le x \le 360°$

Solution

In solving equations of this sort it is **vital** to be aware that there may be **more than one** possible solution in the allowable domain - this possibility results from the periodic nature of this function. It is usually helpful to make a sketch of the relevant function, and this will help to identify the number of possible solutions.

In this case, the functions to plot are

$$y = \tfrac{1}{2} \text{ and } y = \sin x.$$

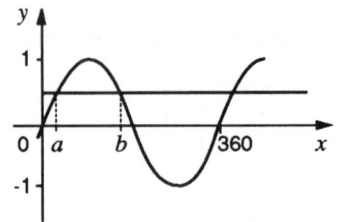

The points of intersection in the range $0° \leq x \leq 360°$ are solutions of the equation.

The figure shows that there are just **two** solutions, denoted by a and b, in the range. You can find the value of a by entering 0.5 in your calculators and using the 'inverse' sin or 'arc sin' buttons.

This gives $a = 30°$ (or $\dfrac{\pi}{6} \approx 0.524$ radians, if you are using radians).

Your calculator will only give you this single value, but it is easy to find b provided you have used a sketch. This shows that the function $y = \sin x$ is **symmetrical** about $x = 90°$ and so $b = 90° + 60° = 150°$ (≈ 2.618 radians). So the equation has two solutions, namely $30°$ or $150°$.

Note: Even though your calculator cannot directly give you both solutions, it can be usefully used to check the answers. To do this enter $30°$ and press the $\boxed{\text{sin}}$ button; similarly for $150°$.

If in the example above, the range had been $-360°$ to $360°$, how many solutions will there be?

Example

Solve $3\cos x = -0.6$ for $0° \leq x \leq 360°$, giving your answer to one d.p.

Solution

Rearranging the equation

$$\cos x = -\frac{0.6}{3} = -0.2$$

As before, you plot the functions

$$y = \cos x \text{ and } y = -0.2$$

There are two solutions, a and b, with a between $90°$ and $180°$, and b between $180°$ and $270°$. Enter -0.2 into your calculator and use the 'inverse' cosine button - this will give you the answer.

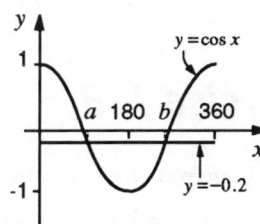

From the graph the second solution b is given by

$$b - 180° = 180° - a \approx 76.5°$$

$$\Rightarrow b = 258.5°$$

and to 1 d.p. the solutions are $101.5°$ or $258.5°$

(Remember to check the solutions on your calculator)

What are the solutions to the previous problem if the range had been given as $-360° \le x \le 0°$?

Example

Solve the equation $\tan x = -2$ where x is measured in radians and $-2\pi \le x \le 2\pi$. Give your answers to 2 d.p. (Set your calculator to radian mode).

Solution

Note that x is given in radians rather than degrees here and remember that π radians $\equiv 180°$.

The graph of $y = \tan x$ and $y = -2$ are shown opposite.

You can see that there are four solutions denoted by a, b, c, and d.

To find one of these solutions, enter -2 in your calculator and use the 'inverse tan' (or 'arctan')button. This should give -1.107 radians, so that $b = -1.107$. Since the function $\tan x$ has period π (or $180°$).

$$a = b - \pi = -4.249$$

$$c = b + \pi = 2.035$$

$$d = b + 2\pi = 5.177$$

To 2 d.p.'s, the solutions in radians are

$$-4.25, \; -1.11, \; 2.04 \text{ and } 5.18$$

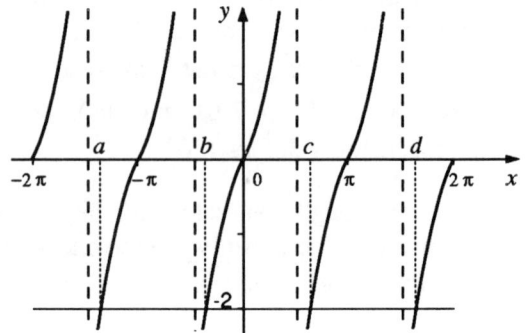

Exercise 11B

Give all answers to 3 significant figures

1. Solve $\cos x = 0.5, \; 0° \le x \le 360°$

2. Solve $\tan x = 1, 0° \le x \le 360°$

3. Solve $\sin x = \frac{1}{4}, 0° \le x \le 180°$

4. Solve $\sin x = -0.5, 0° \le x \le 360°$

5. Solve $4\tan x = 1, 0° \le x \le 720°$

6. Solve $\sin x = -\frac{1}{3}, \; -\pi \le x \le \pi$

7. Solve $3\cos x = 1 - \pi \le x \le \pi$

8. Which of these equations have no solutions?

   (a) $\sin x = 1, \; 0° \le x \le 360°$

   (b) $\cos x = -\frac{1}{4}, \; 0° \le x \le 90°$

   (c) $\cos x = 2, \; 0° \le x \le 360°$

   (d) $\tan x = 2, 0° \le x \le 90°$

   (e) $4\sin x = -5, \; 0° \le x \le 360°$

11.3 Transformations

So far you have met the graphs of $y = \sin x, \cos x$ and $\tan x$, however, in most practical cases you will need more complicated functions. A general sine function, for example takes the form

$$y = a\sin(\alpha x + \beta) + b$$

for constraints a, b, α and β. The next activity will give you a feel for the effect of the various constants before a more systematic approach is made.

Activity 4

Using a graphic calculator with range $0 \le x \le 10$ and $-4 \le y \le 4$ to illustrate the following functions (remember that x is measured in radians). In each case, try and predict the shape before entering the function in your calculator.

(a) $y = \sin 2x$ (b) $y = \sin \frac{1}{2} x$ (c) $y = \sin x + 1$

(d) $y = 2 \sin x$ (e) $y = \sin\left(x + \frac{\pi}{2}\right)$ (f) $y = 2\sin(3x + \pi) - 1$

The effect of each of the constants a, b, α and β will be studied separately before combining them.

(i) $y = a \sin x$

This has the effect of changing the amplitude from 1 to a as shown opposite.

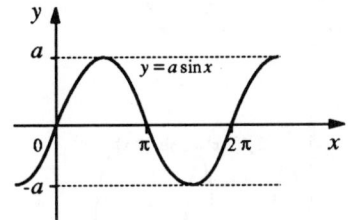

(ii) $y = \sin x + b$

This has the effect of moving the curve up b units as shown below.

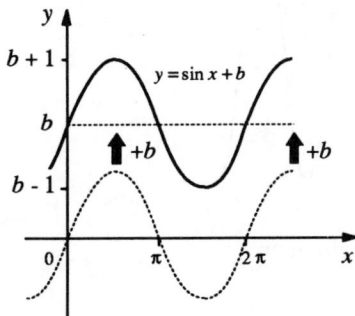

(iii) $y = \sin(x + \beta)$

This has the effect of moving the curve to the left a distance β.

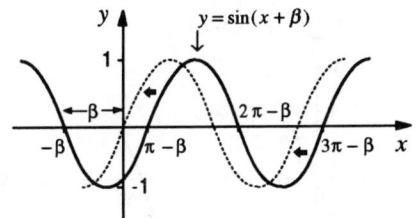

(iv) $y = \sin \alpha x$

This changes the period of the function. Since $y = 0$ when

$$\alpha x = 0, \ \pm \pi, \ \pm 2\pi,...$$

$$\Rightarrow \quad x = 0, \ \pm \frac{\pi}{\alpha}, \ \pm \frac{2\pi}{\alpha}$$

and the function will have period $\frac{2\pi}{\alpha}$ rather than 2π (note again that x is measured in radians rather than degrees).
This curve is sketched opposite.

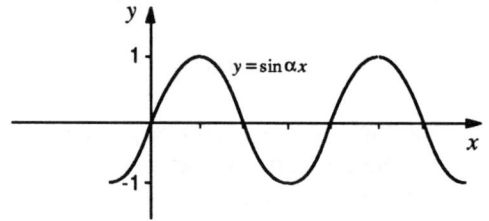

Example

Without using a calculator or computer, sketch the following functions

    (a) $y = 2\sin 2x - 1$ (b) $y = \sin\left(3x - \frac{\pi}{2}\right)$

Solution

(a) Writing $f(x) = \sin x$, then $y = 2f(2x) - 1$; this shows how to transform $f(x)$ to obtain y. Each stage is shown below

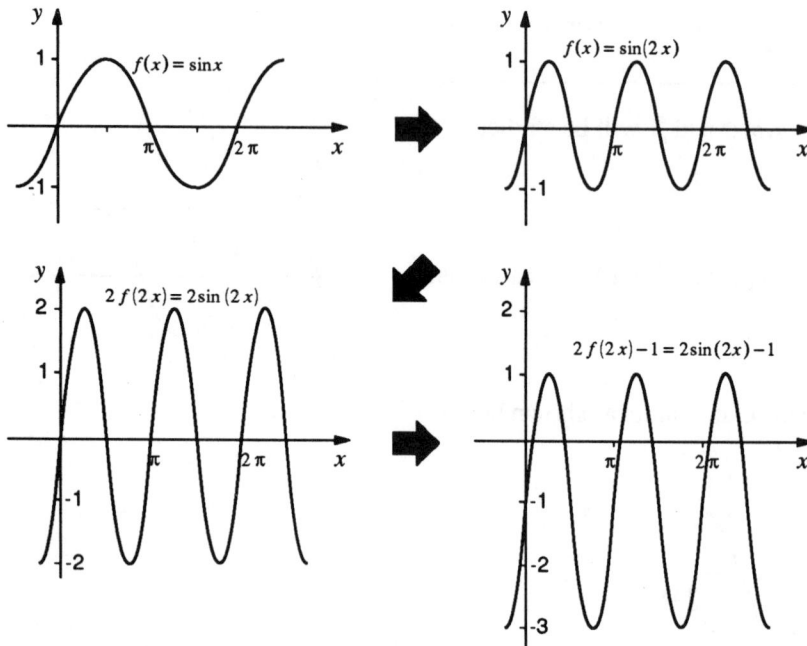

(b) Writing $f(x) = \sin x$, then $y = \sin\left(3x - \frac{\pi}{2}\right)$ and again each stage is shown

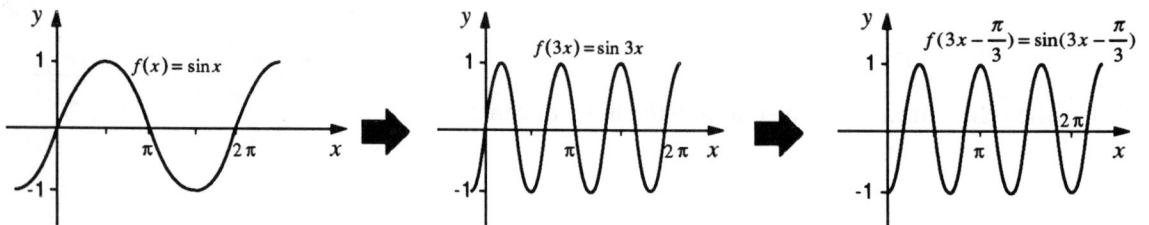

Transformation of $y = \cos x$ to $y = a\cos(\alpha x + \beta) + b$ are dealt with similarly. The same is true for $y = \tan x$ although the discontinuity of the function make it rather more involved as you will see in the next example.

Example

Sketch the curve $y = \tan\left(3x - \frac{\pi}{6}\right)$ for $-\frac{\pi}{2} \leq x \leq \frac{\pi}{2}$

Solution

Defining $f(x) = \tan x$ then $y = f\left(3x - \frac{\pi}{6}\right)$

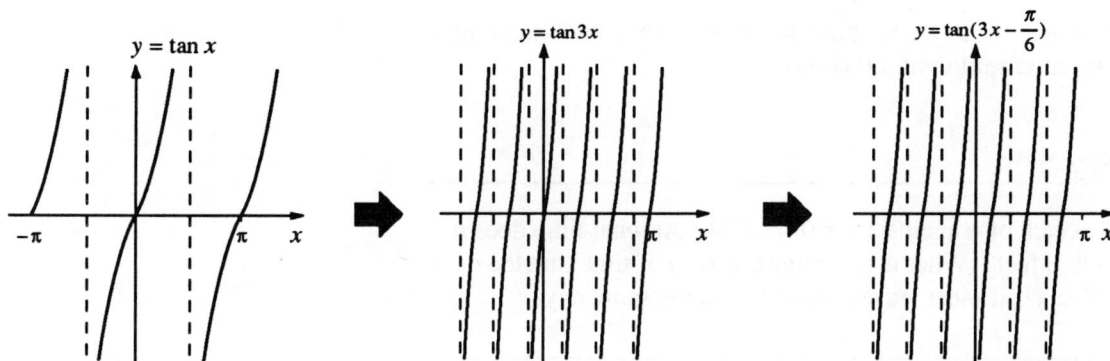

Exercise 11C

Do not use a calculator in this exercise. Sketch the following graphs for the domains stated. You should draw sketches of other graphs to help you on the same axes. Show clearly the points at which the graphs cross the x-axes, and state the period in terms of degrees or radians as indicated. Where appropriate, state the amplitude of the graph.

1. $y = \sin(x + \pi)$ $0° \leq x \leq 2\pi$

2. $y = 3\cos x$ $0° \leq x \leq 180°$

3. $y = -2\sin x$ $0° \leq x \leq 360°$

4. $y = \tan\left(x - \frac{\pi}{2}\right)$ $-\pi \leq x \leq \pi$

5. $y = \frac{1}{2}\sin x - 1$ $0° \leq x \leq 360°$

6. $y = \cos(3x - \pi)$ $0 \leq x \leq 2\pi$

7. $y = 5\sin\left(\frac{1}{2}x\right)$ $0° \leq x \leq 360°$

8. $y = -\tan(x + \pi) - 1$ $0 \leq x \leq 2\pi$

9. $y = 2\cos(2x - \pi)$ $0 \leq x \leq \pi$

10. $y = -3\sin(x + 60°)$ $0° \leq x \leq 360°$

*11. Sketch the graph of
$y = -2\sin(x - \pi) + 1,\ 0 \leq x \leq 2\pi$.

By drawing the graph $y = \frac{1}{2}x$ on the same axes, state the number of solutions to the equation

$$x = 4\sin(3x - \pi) + 4.$$

11.4 Applications

You will now see how trig functions are fundamental in describing physical phenomenon, such as tides, and how mathematics can be used to predict.

Catching the tide

A fishing boat needs the water level in its harbour to be at least three feet above mean level if it is to sail out. You may assume that the height of the water above this mean level can be calculated at any time during a particular day using this formula:

$$h = 5\sin(30t) \qquad 0 \le t \le 24$$

h is the height of the water above the mean levels, and t the time (in hours) after midnight (00:00 hrs).

Activity 5

Draw a sketch of h against t for $0 \le t \le 24$. At what time does h first reach 3 feet? What is the longest period of time that the boat can be at sea if it must leave and return on the same day.

Currents

Alternating currents are generated in power stations which give rise to oscillating voltages. For example if the voltage is given by

$$V = 5\sin(2\pi t)$$

then V will oscillate between ± 5 volts with period $\frac{2\pi}{2\pi} = 1$ second (remember that $\sin(\alpha t)$ has period $\frac{2\pi}{\alpha}$).

Activity 6

Sketch the function V for $0 \le t \le 1$. If it is dangerous for V to be above 4 volts or below - 4 volts for what proportion of time is V in the dangerous zone?

Hours of daylight

The table below gives the length of various days throughout 1991 for four different latitudes (the length of a day means the time difference between sunrise and sunset)

The number after some of the dates indicates the day of the year, d.

LATITUDE:	52°N		54°N		56°N		58°N	
	hrs	m	hrs	m	hrs	m	hrs	m
Jan 2 (2)	7	52	7	30	7	6	6	37
Jan 14 (14)	8	30	8	14	7	54	7	8
Jan 30 (30)	9	13	8	59	8	45	8	9
Feb 11 (40)	9	42	9	31	9	19	9	5
Feb 27 (58)	10	44	10	37	10	31	10	23
Mar 11	11	32	11	29	11	27	11	24
Mar 23 (80)	12	20	12	19	12	17	12	15
Apr 4 (92)	13	9	13	13	13	19	13	25
Apr 20 (108)	14	11	14	21	14	33	14	45
May 6 (124)	15	10	15	25	15	42	16	1
May 22 (140)	16	0	16	20	16	42	17	8
Jun 7 (156)	16	34	16	58	17	24	17	56
Jun 23 (172)	16	44	17	8	17	38	18	10
Jul 9	16	28	16	51	17	18	17	49
Jul 25	15	52	16	10	16	32	16	56
Aug 10	15	0	15	14	15	30	15	47
Aug 26	14	1	14	10	14	20	14	31
Sep 11	12	58	13	3	13	7	13	12
Sep 27	11	55	11	54	11	53	11	53
Oct 13	10	52	10	46	10	40	10	33
Oct 29	9	50	9	39	9	28	9	15
Nov 14	8	54	8	38	8	22	8	2
Nov 30	8	10	7	50	7	28	7	3
Dec 16	7	47	7	25	6	59	6	30
Dec 24	7	45	7	23	6	57	6	27
Jan 5 (1992)	7	55	7	34	7	10	7	43

Activity 7

For the latitude that most closely matches your home, plot the daylight length against day of the year. Draw a smooth curve as accurately as possible through the data points

What type of equation could represent this data?

It is possible to represent this data with an equation of the form

$$\ell = a\sin\left[\alpha(d - \beta)\right] + b \qquad\qquad (1)$$

This gives a good approximation to the data. You have met equations similar to (1) before in section 11.3, although ' $\alpha(d-\beta)$ ' has replaced ' $\alpha x + \beta$ ' for convenience. A sketch of (1) is shown opposite. This will help you to identify the parameters a, b, α and β.

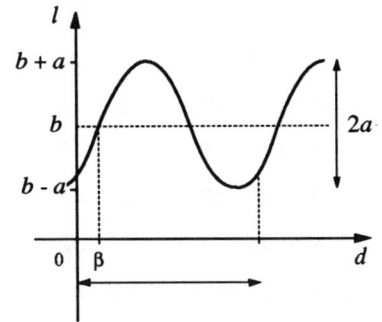

The minimum value of ℓ occurs on December 21st ($d = 355$ or -11) and its maximum value on June 21st ($d = 170$).

What value of d corresponds to β?

Having fixed β, you can now find b, the corresponding value of ℓ.

What is an approximate value of b from your graph?

You can also find an approximate value for a by noting that the difference between the maximum and minimum values of ℓ is $2a$.

What is an approximate value of a for your data?

Finally, you can find α, since you know that the period, $\dfrac{2\pi}{\alpha}$, must be about 365 days.

What is an approximate value for α?

You have now found values for the parameters. Remember that your answer is only an approximation for the model. For example, a possible model for the data at 58°N is given by

$$\ell = 5.9\sin\left(0.0172(d - 78)\right) + 12.1$$

Your values of α and β should be as above $(\alpha = 0.0172,\ \beta = 78)$ but your a and b values will vary.

Activity 8   Testing the model

Using the equation that you have obtained, plot ℓ values for $\frac{1}{2}d = 0,\ 10,\ 20,\dots$ etc on your graph. Draw a smooth curve through the new data points.

How well does it represent the original data?

As you have seen from this last example, fitting an oscillatory equation to a set of data is not straightforward. There are in fact more sophisticated ways of doing this, and computer programs exist that find the curve of best fit, but this is beyond the scope of this text.

Activity 9   Modelling tidal range

Avonmouth is a modern port of Bristol. It is notable for having a very high tidal range. This is illustrated by the data given below for 1991 for highest and lowest tides.

Date	Time	(t)	Water level (m)
March 28th	00:00	(0)	1.6
	05:42	(5.7)	12.5
	12:37	(12.62)	1.3
	18:11	(18.18)	12.7
March 29th	01:01	(25.02)	1.0
	06:31	(30.52)	13.1
	13:29	(37.48)	0.8
	18:56	(42.93)	13.2

Plot these points on a graph of water level against time. Fit an equation of the form

$$h = a\cos(\alpha t) + b$$

for the height of water, where a, b, and α are constants. Determine appropriate values of a, b, and α, and use the model to predict the time of highest and lowest tides for March 30th.

The actual values are

Date	Time	Height
March 30th	01:47	0.7
	07:12	13.4
	14:10	0.7
	19:33	13.4

Compare your theoretical values with these.

Activity 10

A boat grounded in Falmouth needs the water depth to be at least 4 m high before it can float safely. The function below is a model of the water depth at Falmouth on April 16, 1991.

$$h = 2.5\sin(28t - 100) + 3$$

h is the height of the water in metres and t the time in hours after midnight at the start of April 16th. Form an equation and solve it to find the earliest time the boat can be floated off.

11.5 More equations

The previous section should have convinced you that solving trig equations is an important activity, vital to the effective use of equations to model physical phenomenon. Here are some more examples which illustrate the techniques for finding all the solutions.

Example

Find all the solutions between 0 and 2π of the equation

$$\sin\left(x - \tfrac{\pi}{2}\right) + 2 = 1.5$$

giving your answers to 3 d.p.

Solution

Substracting 2 from both sides,

$$\sin\left(x - \tfrac{\pi}{2}\right) = -0.5$$

The figure shows $y = \sin\left(x - \tfrac{\pi}{2}\right)$, together with $y = -0.5$.

This was obtained using a calculator, with the degrees changed to radians.

The graph shows that there are two solutions, A and B, between 0 and 2π. Using a calculator to find the inverse sine of -0.5 gives the value -0.524 radians ($-30°$). As this graph is symmetric about the y axis, the solution at A must be $+0.524$ radian ($+30°$). The symmetry of the graph can be used again to find B. This solution is 0.524 **less** than 2π.

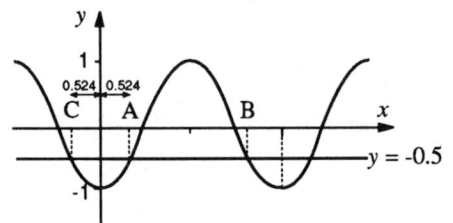

That is,

$$B = 2\pi - 0.524$$
$$= 6.283 - 0.524$$
$$= 5.759 \text{ radians (which is } 330°)$$

The graph of $y = \sin\left(x - \frac{\pi}{2}\right)$ was very useful when dealing with the calculator value of the inverse sine of -0.5, which was **outside** the range of values asked for in the questions. Graphs can also help to find the **number** of solutions an equation may have.

Example

Find the solutions between $0°$ and $360°$ of the equation

$$0.5 = \cos(3x - 45°)$$

Solution

A calculator can be used to make a quick sketch of the graph $y = \cos(3x - 45°)$. The figure below shows this, along with $y = 0.5$.

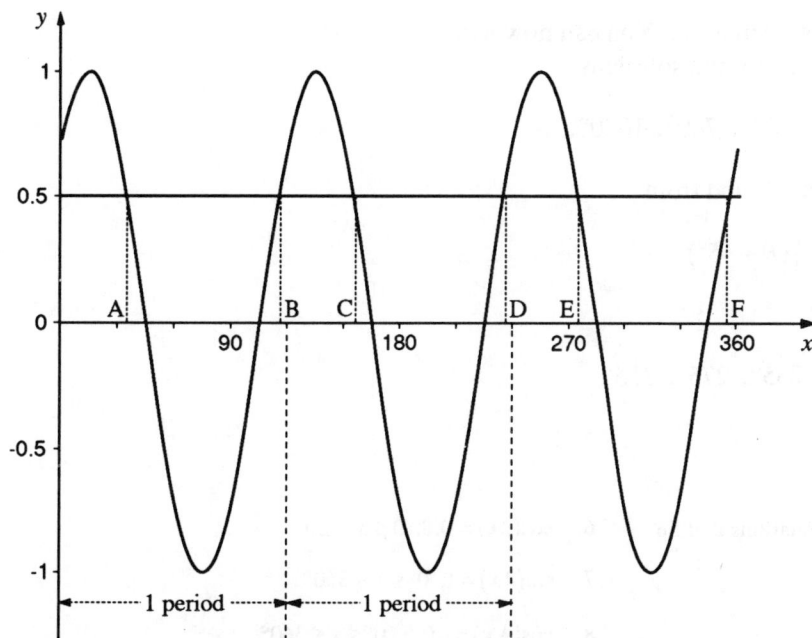

For values of x between $0°$ and $360°$, the graphs show that there are **6** solutions to the equation in the range. A calculator gives inverse cosine of 0.5 as $60°$.

So $3x - 45° = 60°$ will give one possible solution

and $3x = 105°$

$\Rightarrow$ $x = 35°$

It is not so easy to find the other values unless you use the symmetry of the curve. A better way to solve this equation is to define

$$\theta = 3x - 45°$$

and then consider the graphs of $y = 0.5$ and $y = \cos\theta$.

This is illustrated below for an extended range of θ.

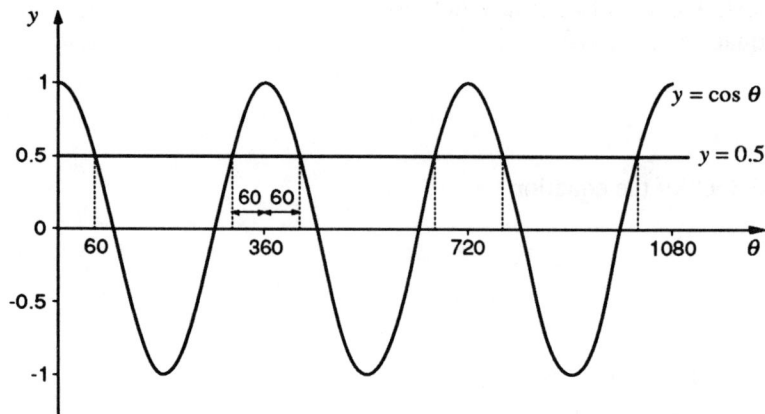

Your calculator will give you the 60° solution. You can now use symmetry about 180°, 360°, 540°,..., to find solutions

$$\theta = 60°, \ 300°, \ 420°, \ 660°, \ 780°, \ 1020°,...$$

The corresponding solutions for x are found from

$$3x - 45° = \theta \ \Rightarrow \ x = \tfrac{1}{3}(\theta + 45°)$$

This gives, in the range 0° to 360°,

$$x = 35°, \ 115°, \ 155°, \ 235°, \ 275°, \ 355°$$

Exercise 11D

Find all the solutions to the following equations that lie in the given range of values of x.

1. $\cos(x - 180°) = 0.7, \ 0° \leq x \leq 360°$

2. $\sin\left(x + \frac{\pi}{2}\right) = -0.2, \ 0 \leq x \leq 2\pi$

3. $\tan(x - 10°) = 1.5, \ 0° \leq x \leq 180°$

4. $\sin(2x) = 0.5, \ 0° \leq x \leq 360°$

5. $2\tan(2x) = -1, \ 0 \leq x \leq 2\pi$

6. $\cos(3x) = 0.8, \ 0 \leq x \leq 2\pi$

7. $\sin(4x) = 1, \ 0° \leq x \leq 360°$

8. $\cos(\tfrac{1}{2}x) = -0.5, \ 0° \leq x \leq 360°$

9. $\tan(2x - 45) = 1, \ 0° \leq x \leq 360°$

10. $\cos(4x + 10°) = -0.5, \ 0° \leq x \leq 180°$

11. $\sin\left(\tfrac{1}{2}x - \frac{\pi}{2}\right) = -1, \ 0 \leq x \leq 2\pi$

11.6 Properties of trig functions

There are a number of important properties, which you have already been using. These are

	Degrees	Radians
1.	$\sin(x+90°)=\cos(x)$	$\sin\left(x+\frac{\pi}{2}\right)=\cos(x)$
2.	$\cos(x-90°)=\sin(x)$	$\cos\left(x-\frac{\pi}{2}\right)=\sin(x)$
3.	$\cos(180°-x)=-\cos(x)$	$\frac{1}{2}\cos(\pi-x)=\cos(x)$
4.	$\sin(180°-x)=+\sin(x)$	$\sin(\pi-x)=\sin(x)$
5.	$\tan x=\dfrac{\sin x}{\cos x}$	
6.	$\sin^2 x+\cos^2 x=1$	

The last result is a very important one and is proved below for $0 < x < 90°$.

Theorem $\qquad \sin^2 x+\cos^2 x=1$

Proof $\qquad$ Using Pythogoras' Theorem,

$$BC^2+AB^2=AC^2$$

Dividing by AC^2,

$$\left(\frac{BC}{AC}\right)^2+\left(\frac{AB}{AC}\right)^2=1$$

But $\qquad \sin x=\dfrac{\text{opp}}{\text{hyp}}=\dfrac{BC}{AC}$

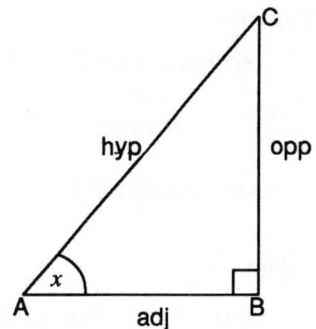

$$\cos x=\dfrac{\text{adj}}{\text{hyp}}=\dfrac{AB}{AC}$$

giving $\qquad \sin^2 x+\cos^2 x=1$

Note $\sin^2 x$ is used for $(\sin x)^2$ so that there is no confusion between $(\sin x)^2$ and $\sin(x^2)$.

Example

Find all solutions between $0°$ and $360°$ of the equation

$$\sin x=2\cos x$$

to 2 d.p.

Solution

Assuming $\cos x \neq 0$, you can divide both sides of the equation by $\cos x$ to give

$$\tan x = 2$$

(since $\dfrac{\sin x}{\cos x} = \tan x$).

The graph of $\frac{1}{2} y = \tan x$ and $y = 2$ is shown opposite. This shows that there are 2 solutions in the range $0°$ to $360°$.

Using a calculator, the inverse tangent of 2 is $63.4°$ (to 1d.p.). This is marked A on the graph. Since the tangent graph has a period of $180°$, B is given by

$$63.4° + 180° = 243.4°$$

Note that at the start of the solution, $\cos x = 0$ was excluded.

When $\cos x = 0$, this gives $\sin x = 0$. These two equations are not both true for any value of x.

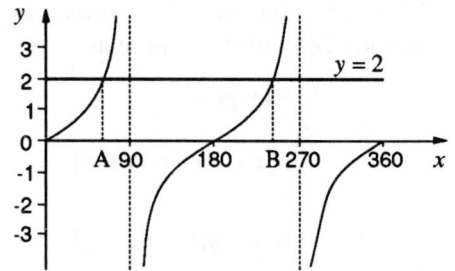

Example

Find all solutions between $0°$ and $360°$ of the equation

$$\cos x \sin x = 3 \cos x$$

giving answers to 1 d.p.

Solution

The equation can be rewritten as

$$\cos x \sin x - 3 \cos x = 0$$

or $\qquad \cos x (\sin x - 3) = 0$

So either $\cos x = 0$ or $\sin x - 3 = 0 \implies \sin x = 3$

The graph of $y = \cos x$ is shown opposite for $0° \leq x \leq 360°$. The solutions are given by $\cos x = 0$, which gives

$$x = 90° \text{ or } 270°$$

The other possibility is $\sin x = 3$. The graph of $y = \sin x$ and $y = 3$ are shown opposite. They do not intersect so there are no solutions.

The only solutions of the equation are therefore $x = 90°$ or $270°$.

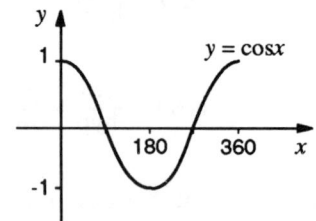

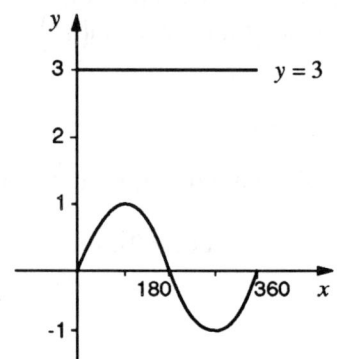

The example above showed that algebraic techniques like factorising can be applied to equations involving trigonometric functions. The next example shows that **quadratics** in sines or cosines can also be solved.

Example

Find all solutions between $0°$ and $360°$ to the equation

$$3\cos x = 2\sin^2 x$$

Solution

This equation contains a mixture of 'cos' and 'sin' terms, but using the identity

$$\cos^2 x + \sin^2 x = 1$$

will make it possible to express the equation in 'cos' terms only.

Now $\qquad \sin^2 x = 1 - \cos^2 x$

so that the equation becomes

$$3\cos x = 2\left(1 - \cos^2 x\right)$$

$$= 2 - 2\cos^2 x$$

Hence $\qquad 2\cos^2 x + 3\cos x - 2 = 0$

(This is a quadratic equation in $\cos x$)

This can be factorised to give

$$(2\cos x - 1)(\cos + 2) = 0$$

and either $\quad 2\cos x - 1 = 0$ or $\cos x + 2 = 0$.

The first equation gives

$$\cos x = \tfrac{1}{2}$$

and $\qquad y = \cos x$ and $y = \tfrac{1}{2}$.

These are shown in the figure opposite.

Using the inverse cosine function on a calculator gives $x = 60^0$, and this corresponds to A on the figure. Using symmetry about 180^0, gives B as

$$270° + 30° = 300°$$

The second equation, $\cos x = -2$, has no solutions. The only possible solutions of the original equation are 60^0 and 300^0.

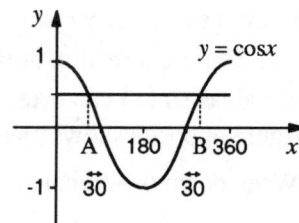

Exercise 11E

1. Find all the solutions to the equation
   $\sin x \cos x + \sin x = 0$ for $0° \le x \le 360°$

2. Solve completely the equation
   $1 - \sin x = 2\cos^2 x$, $0° \le x \le 360°$. (Hint: there are three solutions).

3. Find all solutions between $0°$ and $360°$, to 1 d.p.,
   of the equation $5 - 2\cos x = 8\sin^2 x$.

4. Find all the solutions between 0 and 2π radians of the equation $\sin x = -\cos x$, leaving π in your answers.

5. Solve $3\sin x = \cos x$ for $0° \le x \le 360°$

11.7 Inverse trig functions

Throughout this chapter, you have had to find the 'inverse' sine, cosine or tangent of various numbers in order to solve equations. In Chapter 3, the need for a function to be **one to one** if it is to have an inverse function was made clear. If a function is not one to one, then the inverse mapping has to give more than one 'answer' for some values in its domain. This is not possible for a function, and so the inverse function does not exist.

All the trigonometric functions are many to one, so they cannot have an inverse function. However, if the domains are restricted, each of them can be redefined as a one to one function, within the domain.

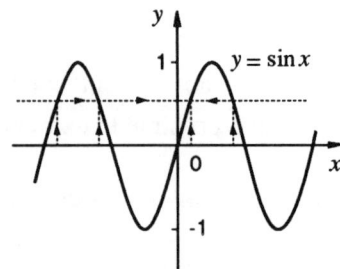

Several values of x are mapped to the same value

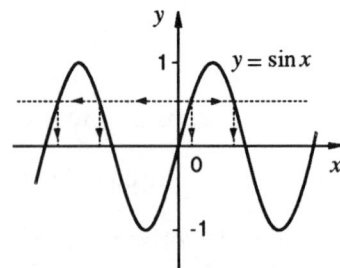

Inverse mapping has to map one value of y to one volume of x.

Activity 11

(a) Draw a sketch of $y = \sin x$, for values of x between -2π and 2π. Choose a section of the graph which spans the full range of values $\sin(x)$ can take, from -1 to $+1$, **once** only. The figure opposite shows one such section.

Write down the domain of $\sin x$ which covers the section you have chosen. For example, $-\frac{3\pi}{2} \le x \le \frac{\pi}{2}$ is the domain of $\sin x$ in the diagram.

(b) Having chosen the domain for which $\sin x$ is one to one, draw a sketch of the graph of the **inverse** function on the same axes. (You will need to use the fact that the graph of a function and its inverse are reflections of each other in the line $y = x$).

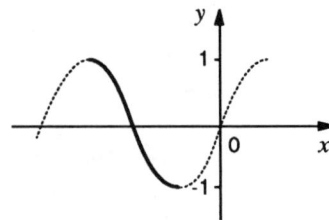

which makes it one to one, and which allows $\cos x$ to 'span' its full range from -1 to $+1$ once only. Then sketch the corresponding inverse functions graph.

Inverse sine function

The figure opposite shows $y = \sin x$, restricted to the domain $-\frac{\pi}{2}$ to $\frac{\pi}{2}$. With this range, $y = \sin x$ is a one to one function and its range is from -1 to 1. It's inverse, denoted by

$$y = \sin^{-1} x$$

is also illustrated.

So, for example, working in radians

$$y = \sin\left(\frac{\pi}{2}\right) = 1 \ \Rightarrow \ \sin^{-1}(1) = \frac{\pi}{2}$$

or

$$y = \sin\left(\frac{\pi}{6}\right) = \tfrac{1}{2} \ \Rightarrow \ \sin^{-1}\left(\tfrac{1}{2}\right) = \frac{\pi}{6}$$

We call $\frac{\pi}{2}$ the **principle** value of $\sin^{-1}(1)$, since for example $\sin\frac{5\pi}{2} = 1$ as well - but remember, the domain is restricted here to $-\frac{\pi}{2}$ to $\frac{\pi}{2}$.

What is the domain of $\sin^{-1} x$?

Also note that the notation $\sin^{-1} x$ for the inverse can be very confusing. It does **not** mean $(\sin x)^{-1} = \frac{1}{\sin x}$ although $\sin^2 x$ does mean $(\sin x)^2$. Sometimes mathematicians have introduced poor notation and this is an example of it. A more recent notation for inverse is 'arcsin'; so for example

$$\arcsin 1 = \frac{\pi}{2}, \ \arcsin \frac{1}{\sqrt{2}} = \frac{\pi}{4}$$

Inverse cosine function

The inverse cosine function, $\cos^{-1} x$ or $\arccos x$ is similarly defined althoug the domain is restricted to 0 to π.

What is the domain of $\cos^{-1} x$?

So, for example since

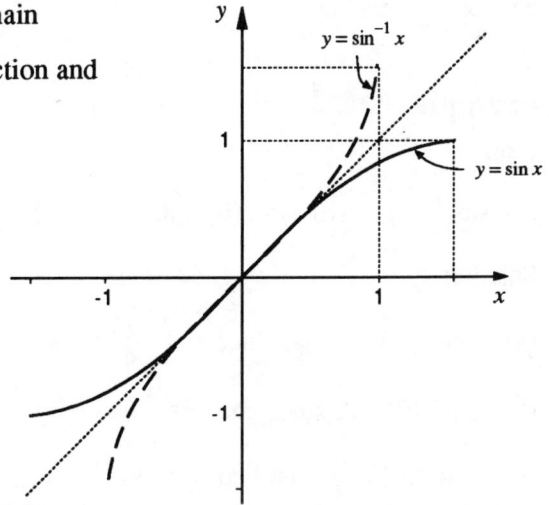

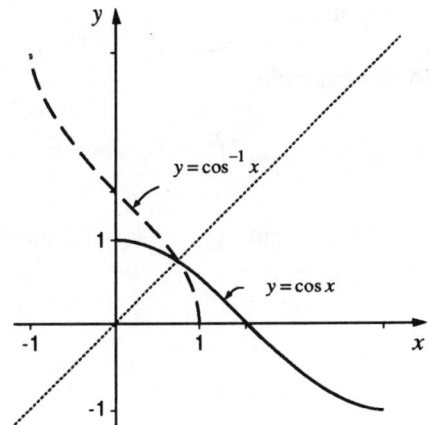

$$\cos 0 = 1 \quad \Rightarrow \quad \cos^{-1}(1) = 0, \text{ and}$$

$$\cos \frac{\pi}{3} = \frac{1}{2} \quad \Rightarrow \quad \cos^{-1}\left(\frac{1}{2}\right) = \frac{\pi}{3}$$

Your calculator will normally always give the principle value of $\sin^{-1} x$ and $\cos^{-1} x$.

Example

Find

(a) $\sin^{-1}\left(-\frac{1}{2}\right)$ (b) $\sin^{-1}(0)$ (c) $\cos^{-1}(0)$ (d) $\cos^{-1}\left(\dfrac{1}{\sqrt{2}}\right)$

Solution

(a) $\sin^{-1}\left(-\frac{1}{2}\right) = -\frac{\pi}{6}$, since $\sin\left(-\frac{\pi}{6}\right) = -\frac{1}{2}$

(b) $\sin^{-1}(0) = 0$, since $\sin 0 = 0$

(c) $\cos^{-1}(0) = \frac{\pi}{2}$, since $\cos \frac{\pi}{2} = 0$

(d) $\cos^{-1}\left(\dfrac{1}{\sqrt{2}}\right) = \frac{\pi}{4}$, since $\cos \frac{\pi}{4} = \dfrac{1}{\sqrt{2}}$

(You can use your calculator to find this or check the answers, but remember to use 'radians' and you will have to use an approximation for $\frac{\pi}{6}$ etc).

Inverse tangent function

What restricted domain is needed for $\tan x$ in order to define its inverse?

To obtain the principle values of $\tan^{-1} x$, the domain of $\tan x$ is restricted to $-\frac{\pi}{2}$ to $\frac{\pi}{2}$ giving a range of all real numbers. The figure opposite illustrates both $y = \tan x$ and $y = \tan^{-1} x$ or $y = \arctan x$.

So, for example

$$\tan^{-1}(0) = 0, \text{ since } \tan 0 = 0, \text{ and}$$

$$\tan^{-1}(1) = \frac{\pi}{4}, \text{ since } \tan \frac{\pi}{4} = 1$$

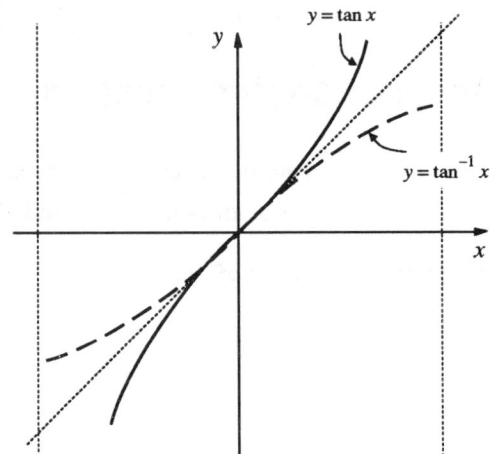

Exercise 11F

1. Find the principle value of

   (a) $\sin^{-1}\left(-\frac{1}{\sqrt{2}}\right)$ (b) $\sin^{-1}\left(\frac{\sqrt{3}}{2}\right)$ (c) $\sin^{-1}(-1)$

2. Find the principle value of

   (a) $\cos^{-1}(-1)$ (b) $\cos^{-1}\left(-\frac{1}{2}\right)$ (c) $\cos^{-1}\left(\frac{\sqrt{3}}{2}\right)$

3. Find the principle value of

   (a) $\tan^{-1}(-1)$ (b) $\tan^{-1}\left(\sqrt{3}\right)$ (c) $\tan^{-1}\left(-\frac{1}{\sqrt{3}}\right)$

11.8 Miscellaneous Exercises

1. Solve $\sin(x) = 0.1$, $0° \le x \le 360°$

2. Solve $\tan(x) = 3$, $-360° \le x \le 360°$

3. Solve $3\sin(x) = -1$, $0 \le x \le 2\pi$

4. Solve $\cos(x) = -\frac{1}{2}$, $0 \le x \le 180°$

5. Solve $\tan(x) = 1.5$, $0 \le x \le 720°$

6. Without the use of a calculator, make sketches of the following graphs for $0 \le x \le 360°$

   (a) $y = \cos\left(x - \frac{\pi}{2}\right)$ (b) $y = \tan\left(x - \frac{\pi}{2}\right)$

   (c) $y = 2\cos(x) + 3$ (d) $y = -\sin(x) + 1$

   (e) $y = \tan(-x) - 2$ (f) $y = \frac{1}{2}\sin\left(x - \frac{\pi}{2}\right)$

   (g) $y = 4\cos(x + \pi)$ (h) $y = 3\sin\left(x + \frac{\pi}{2}\right) + 1$

   (i) $y = \tan\left(x - \frac{\pi}{2}\right) + 1$ (j) $y = 2\cos\left(x - \frac{\pi}{4}\right) - 1$

   (k) $y = \cos(3x)$ (l) $y = \tan\left(\frac{1}{2}x\right)$

   (m) $y = \sin\left(4x - \frac{\pi}{2}\right)$ (n) $y = 3\cos(2x + \pi)$

7. State the period, in degrees, of each of the functions in Question 6. Where appropriate also give the amplitude of the function.

8. A tuning fork produces a note which has a frequency of 300 Hz. This means that the sound wave it generates completes 300 full cycles every second. Given that the amplitude of the sound wave produced is 5Db (decibels), model the sound wave using a sine function.

9. Solve the following equations completely for the values of x shown, to 1 d.p. if necessary.

   (a) $\tan(x + 90°) = 1$, $0° \le x \le 360°$

   (b) $\cos(2x) = \frac{1}{2}$, $0 \le x \le 2\pi$

   (c) $3\cos\left(\frac{1}{2}x + 45°\right) = -1$, $0° \le x \le 360°$

   (d) $\sin(3x) = -\frac{1}{2}$, $0 \le x \le 2\pi$

10. Solve these equations completely giving answers to 1 d.p. where necessary.

    (a) $\sin(x) = \cos(x)$, $0° \le x \le 360°$

    (b) $3\sin(x) + \cos(x) = 0$, $0 \le x \le 2\pi$

    (c) $3\sin^2(x) - 4\sin(x) + 1 = 0$, $0° \le x \le 360°$

    (d) $\frac{1}{2}2\sin^2(x) + 5\cos(x) + 1 = 0$, $0° \le x \le 360°$

*11. Solve these equations completely

    (a) $3\sin(2x) = \cos(2x)$, $0° \le x \le 360°$

    (b) $\cos^2(x) - 1 = 0$, $0° \le x \le 360°$

    (c) $\cos^2(3x) + 4\sin(3x) = 1$, $0° \le x \le 360°$

12 GROWTH AND DECAY

Objectives

After studying this chapter you should

- understand exponential functions;
- be able to construct growth and decay models;
- recognise graphs of exponential functions;
- understand that the inverse of an exponential function is a logarithmic function;
- be able to use logarithms to solve suitable equations;
- be able to differentiate exponential and logarithmic functions.

12.0 Introduction

Amoebae reproduce by dividing after a certain time. Radioactive substances have 'half lives' which are determined by the time it takes the radioactivity to halve. These are examples of systems which are modelled by 'exponential' functions.

The world's human population is growing at about 3% per year. That is after each year the population will be 3% more than it was at the start of the year. In the first activity, you will form a model to describe this population growth, and then use it to find the year when the population will be twice the size it was in 1989.

The population at the end of 1989 was approximately 4.5 billion (4,500,000,000). If the population grew by 3% in 1990, at the end of the year it would be $4.5 \times 1.03 = 4.635$ billion.

The population at the end of 1991 can be found by the calculation

$$4.5 \times 1.03 \times 1.03 = 4.5 \times 1.03^2$$

By the end of 1999, the population would be $4.5 \times (1.03)^3$.

Activity 1   Modelling population

For the model described above, calculate the population at the end of each year, starting at 1989, and continuing until the population has doubled from its value at 1989. Plot the values on a graph.

The model used in this activity can be represented by the equation

$$P = P_o a^x \qquad\qquad (1)$$

when P is the population at the end of the year number x, and P_o is the initial population at year $x = 0$ and a is a constant. So for the world population described above

$$P = 4.5 \times 1.03^x \quad x = 1, 2, 3, \ldots$$

Equations of the form (1) can also be used to describe populations that are declining.

Some species are endangered because they have declining populations, for various reasons like hunting, habitat destruction, new predators or infertility. Many marine species such as whales and some fish give cause for concern. Models are made to help predict future trends in fish stock levels, which take into account many features like fishing techniques and environmental conditions.

In the next activity, you will produce a model of a fish population based on the assumption that it is declining by 15% each year. You then use it to find the number of years before the population becomes so low that it is in danger of being unable to sustain itself.

Activity 2　　Endangered species

Assume that a particular fishery has a population of 100,000 fish and that current fishing methods cause this population to decline by 15% in a year.

Years elapsed	Population
0	100 000
1	100 000 x 0.85 = 85 000
2	...
3	...
4	...
...	...

(a)　Copy and complete the table opposite, and use it to help you form a model of the population p in terms of the years elapsed, $x, x = 1, 2, \ldots, 10$.

(b)　Plot and draw the graph of the fish population for the first 10 years.

(c)　If the population falls below say 25 000, the fish become quite widely separated. In these conditions it becomes difficult to find good catches, and the fish themselves breed at a much reduced rate. Therefore, using 25 000 as an 'action level' use your model to find the number of years before which the population becomes dangerously low.

12.1 Models of growth and decay

The two activities show how mathematics can be used to model growth and decay of populations. Another example bacteria which divides in two every minute. The growth in numbers is illustrated in the following table.

Minutes past introduction of bacteria	Number of bacteria
0	$1 \, (= 2^0)$
1	$2 \left(= 2^1\right)$
2	$2 \times 2 = 2^2 = 4$
3	$2 \times 2 \times 2 = 2^3 = 8$
4	$2 \times 2 \times 2 \times 2 = 2^4 = 16$
...	...
t	2^t

The table shows a 'pattern' for the number of bacteria at any time after the bacteria was introduced. From this pattern, it is easy to see that after t minutes the number of bacteria will be 2^t.

Therefore, $n = 2^t$ is a model for the bacteria growth. The graph for this function is shown here, for values of T between 0 and 4.

Drawing a smooth curve through these discrete points gives a continuous model of growth, although in this example, it only makes sense to use whole numbers for t. However, many situations which have models like this have continuous not discrete domains, so the curve is typical of this type of function.

Any function of the form a^x, where a is a positive constant is called an **exponential** function (as x is the 'exponent' or power of a). Although the example above (and the examples developed in Activities 1 and 2) was only defined for positive values of the exponent $(t \geq 0)$, exponential functions are defined for any real value.

For instance, if $f(x) = 3^x$, then $f(-1) = 3^{-1} = \dfrac{1}{3}$; using the rules for indices, covered in Chapter 9.

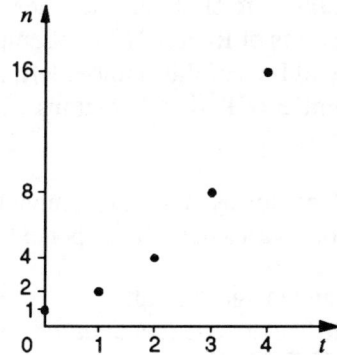

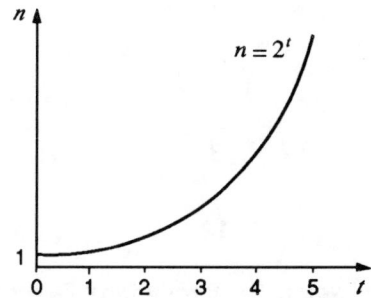

Activity 3

(a) Use a graphic calculator or computer to help you make sketches of these functions, using the same pair of axes. Use a range of the values between -3 and $+3$, and -30 to $+30$ on the y-axis.

(i) $y = 3^x$ (ii) $y = 2^x$ (iii) $y = 1.5^x$ (iv) $y = 1^x$
(v) $y = (0.5)^x$

Do the curves have a common point? What is the relationship between $y = 2^x$ and $y = (0.5)^x$? What happens as x becomes large and positive or large and negative?

(b)   Similarly illustrate the graphs of

(i) $y = 3^{-x}$ (ii) $y = 2^{-x}$ (iii) $y = 1.5^{-x}$

and compare them with $y = 3^x, 2^x, 1.5^x$, and describe their behaviour for large x, positive or negative.

Another example of the use of the exponential function is in the modelling of Radon 219, which is an 'isotope' of the gaseous element Radon. It occurs naturally in some types of rock and its seepage from beneath buildings has been identified as a major concern in some parts of the country. Radon 219 is radioactive, with a **half life** of about 4 seconds. This means that if there are 1000 atoms of Radon 219 in a sample of the gas, 4 seconds later there will be half this number left, 500. 4 more seconds later, and the number of Radon 219 atoms will halve again, to 250, and so on.

This decaying system can be modelled with an exponential function, with a negative exponent.

Time in seconds after sample is collected (seconds)	Number of atoms left
0	1000
4	$1000 \times (\tfrac{1}{2})^1$
8	$1000 \times (\tfrac{1}{2})^2$
12	$1000 \times (\tfrac{1}{2})^3$

Can you write down a formula for N, the number of atoms left after time t seconds?

Since every 4 seconds increases the **power** of the exponent by 1, you can write the model equation as

$$N(t) = 1000 \times (\tfrac{1}{2})^{\frac{1}{4}t}$$

Note that this can also be written as

$$N(t) = 1000 \times (2^{-1})^{\frac{1}{4}t}$$

$$= 1000 \times 2^{-\frac{1}{4}t},$$

using the properties of indices.

Activity 4   U.K. population

The government's statistical service made several predictions
about the United Kingdom's population in 1989. One of these was
that the population would grow by 1% every ten years. If the
population in 1990 was 5.5 millions, form a model for the U.K.'s
future population. Use it to draw a graph of the projected
population, and from this estimate the year when the population
will equal 60 million.

The model used in the activity above is

$$P(t) = P_0 a^{\frac{1}{10}t}$$

when $P_0 = 55$(million) and $a = 1.01$. So far you have used
exponential functions to model various populations and radioactive
decay. This last application can be used to help date
archaeological objects through Carbon dating.

This section is completed with a summary of the general properties
of the exponential functions

$$y = ba^x \text{ and } y = ba^{-x}$$

for $a > 1$. As is shown opposite, both curves pass through $(0, b)$ and
the range of both functions is all real numbers greater than zero.
a is the **base** and x the **exponent** of the function.

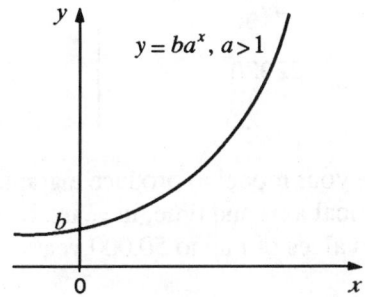

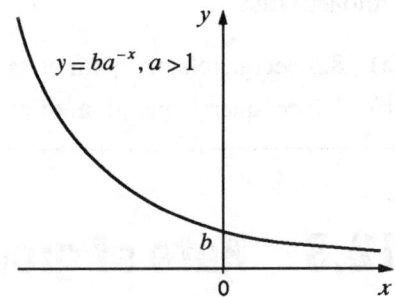

12.2   Carbon dating

Carbon 14 (^{14}C) is an isotope of carbon with a half life of 5730
years. It exists in the carbon dioxide in the atmosphere, and all
living things absorb some Carbon 14 as they breathe. This
remains in an animal or plant, and is constantly added to until the
organism dies. After this time, the Carbon 14 decays, reducing to
half the amount stored in the body after 5730 years. The amount
halves again after another 5730 years, and so on, with no new
Carbon 14 absorbed.

In 1946 an American scientist, *Williard Libby*, developed a way of
'dating' archaeological objects by measuring the Carbon 14
radiation present in them. This radioactivity is compared with that
found in things living now.

For instance, if bones of recently dead animals produce 10
becquerels per gram of bone carbon (a becquerel is the unit of
radioactivity), and an old bone produces only 5 becquerels, the
radioactivity has halved since the animal which had the old bone
died. As the half life of Carbon 14 is 5730 years, this would mean
the animal died in 3740 BC approximately.

Activity 5   Carbon 16

Complete the table below

Age (in years)	Radiation (becs)
0	10
5730	$10 \times \frac{1}{2} = 10 \times 2^{-1}$
11460	$10 \times (\frac{1}{2})^2 = 10 \times 2^{-2}$
17190	...
22920	...
...	...

Use your model to produce a graph, showing radioactivity on the vertical axis and time, in years, on the horizontal. Draw the graph for values of t up to 50,000 years.

From your graph, estimate the ages of bones with these radioactivities

(a)   8.5 becquerels per gram of carbon;

(b)   1.2 becquerels per gram of carbon.

12.3   Rate of growth

Suppose a colony of bacteria doubles in number every minute as every member of the colony divides in two. So if there are 2 bacteria at the start of the colony, there will be 4 a minute later (an increase of 2 in one minute), 8 two minutes later (an increase of 4 in one minute) and so on. As the number of bacteria increases, so the **rate** at which that number increases goes up. So the rate of increase of an exponential function is closely related to the value of the function at any point. This suggests that exponential functions and their derivatives are closely linked.

The next activity you will explore these links.

Activity 6

Plot and draw the curves below for values of x between -2 and $+2$ and on separate axes.

(a) $y = 2^x$ (b) $y = 3^x$ (c) $y = 2.5^x$ (d) $y = 2.9^x$

Using a ruler to draw tangents to each of your curves, calculate the gradient of each one at five different points. The figure opposite illustrates the method.

Note that

$$\text{gradient} = \frac{\text{change in } y}{\text{change in } x}$$

In this case all gradients are positive since a positive change in x results in a positive change in y.

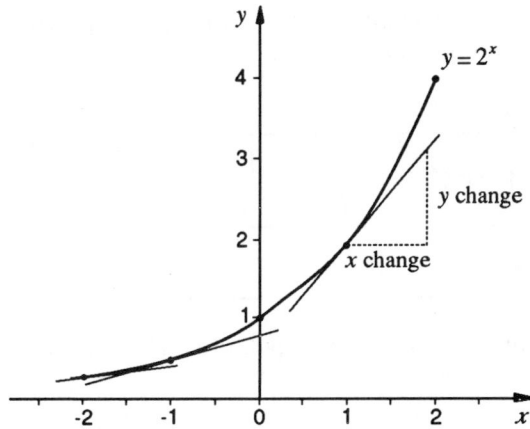

Plot the five values for the gradient of your graph and sketch in the gradient curve. Comment on how the original graph and the gradient curve seem to be related.

If you have access to a computer or calculator that is capable of showing the derivative of a function, then you can find an exponential function whose derivative exactly fits over its own graph by considering $y = a^x$ with a in the range $2.5 < a < 2.9$.

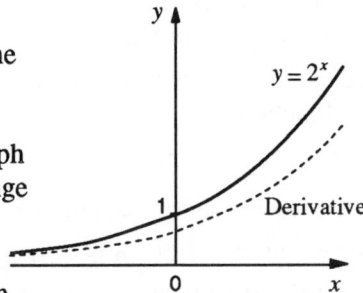

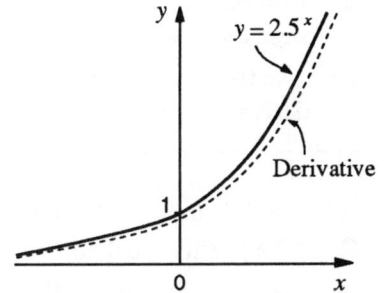

The derivative of 2^x is always less than the value of the function itself. So is the derivative of 2.5^x, although it is a closer fit to the function than that of 2^x. The derivative of 3^x has a greater value than the function. This suggests that there is an exponential function, with a base between 2.5 and 3, which is the same as itself.

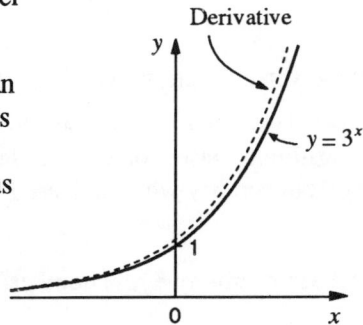

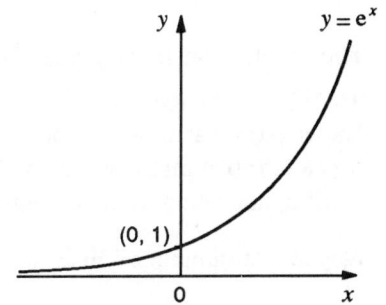

Such a function would therefore be its own derivative. The base required for this to happen is denoted by the letter 'e'.

Unfortunately, its value cannot be given exactly - like π and $\sqrt{2}$ it is irrational, and so it can't be expressed exactly as a fraction or decimal. To five decimal places, it is 2.71828.

239

The function $f(x) = e^x$ is often referred to as the **exponential function**. It is unique in mathematics, in that it is its own derivative. This property makes it extremely important in many branches of the subject.

To summarise

$$y = e^x \quad \Rightarrow \quad \frac{dy}{dx} = e^x$$

Activity 7

Use a graphic calculator or computer to make sketches of these graphs.

(a) $y = e^x$ (b) $y = e^{(x+1)}$ (c) $y = e^{(x-2)}$

(d) $y = e^x + 1$ (e) $y = e^{-x}$

Note that your calculator or computer may use the expression $y = \exp(x)$ for $y = e^x$.

(a) Compare each of your sketches with the graph $y = e^x$, and state the relationship between each graph and that of $y = e^x$.

(b) Use the fact that the derivative of e^x is e^x, to work out the derivatives of each of the other functions.

The function $f(x) = e^x$ is a mapping from the set of real numbers, $\mathbb{R}$, to the positive real numbers. Its graph shows that it is a one to one function. This means that $f(x) = e^x$ has an inverse function. The graph of this inverse function is a reflection in the line $y = x$ of the graph of $y = e^x$.

The graph opposite shows e^x and its inverse function, which is usually written as $\ln(x)$. This function is read as 'the natural (or Naperian) logarithm of x' or 'the logarithm to base e of x'. (**Napier** was a Scottish mathematician of the 16th century who pioneered work connected with this function).

Why is the domain of $\ln(x)$ only the set of positive real number?

The figure shows that $\ln(x)$ is **not** defined for negative values of x (or zero), as there is no graph to the left of the y axis for $\ln(x)$. So $\ln(-2)$, for instance, does not exist. The range of $\ln(x)$, however, is the full set of real numbers.

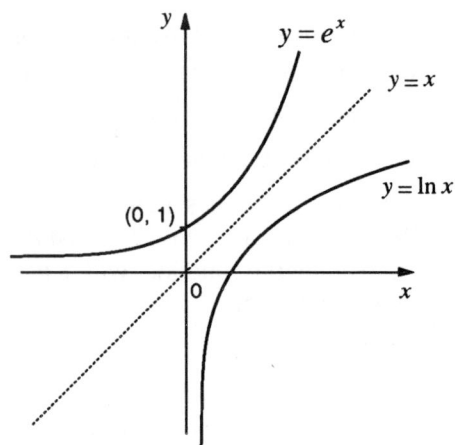

Example

Find x if $e^x = 100$. Give your answer to two d.p.

Solution

Since $e^x = 100$, and $y = \ln x$ is the inverse function of e^x,

$$x = \ln 100$$

Using a calculator to find ln (100), gives $x = 4.61$ to 2 d.p.

To summarise, for $a > 0$,

$$\boxed{e^x = a \Rightarrow x = \ln a}$$

Note that the brackets round 'a' in $\ln a$ have been omitted and will be in future except where it might cause confusion.

Example

Solve, to 3 s.f. the equation $3e^{2x-1} = 5$.

Solution

Since $3e^{2x-1} = 5$, then

$$e^{2x-1} = \tfrac{5}{3}$$

Since e^x and $\ln x$ are inverse functions,

$$2x - 1 = \ln \tfrac{5}{3}$$
$$\Rightarrow \quad 2x = 1 + \ln \tfrac{5}{3}$$
$$\Rightarrow \quad x = \tfrac{1}{2}(1 + \ln \tfrac{5}{3}) = 0.755 \text{ to 3 s.f.}$$

Exercise 12A

1. Solve $e^x = 5$ to 2 d.p.

2. Solve $e^x = \frac{1}{2}$ to 2 d.p.

3. Solve $4e^x = 3$ to 3 s.f.

4. Solve $e^{2x} = 1$ to 2 d.p.

5. Solve $3e^{\frac{1}{2}x} = 4$ to 3 s.f.

6. Solve $e^{-x} = 1.5$ to 2 d.p.

7. Solve $4e^{3x-2} = 16$ to 1 d.p.

8. Solve $7e^{3-x} = 2$ to 3 s.f.

9. Solve $e^x \times e^x = 3$ to 2 d.p.

10. Solve $e^{2x} = 4e^x$ to 3 s.f.

12.4 Solving exponential equations

Earlier in this chapter, you have produced exponential functions as models, and then used graphs to estimate the solution to a problem. The logarithmic function allows you to calculate rather than estimate these solutions as is shown in the example below.

Example

A bacteria colony doubles in number every minute, from a starting population of one. The population model is $P = 2^m$, where P is the population and m the number of minutes since the colony was started. Find the time when the population first equals 1000.

Solution

The problem requires a solution to the equation

$$P = 1000$$

or $\qquad\qquad 2^m = 1000$

Taking log of each side of the equation

$$\ln 2^m = \ln 1000 \qquad\qquad (1)$$

Now, 2 is a positive real number, so there is some number, call it n, such that $e^n = 2$ (see figure opposite). Then $n = \ln(2) \approx 0.693$.

So $2^m = (e^n)^m$ replacing 2 by e^n and since $(e^n)^m = e^{nm}$, using the properties of indices, 2^m in (1) above can be replaced by e^{nm}, where $n = \ln(2)$.

This gives
$$\ln e^{nm} = \ln 1000$$

But $\ln x$ is the inverse of e^x, so $\ln e^{nm} = nm$. Hence

$$mn = \ln 1000$$

Therefore $m = \dfrac{\ln 1000}{n} = \dfrac{\ln 1000}{\ln 2} = 9.97$ minutes.

Hence $m = 9$ minutes, 58 seconds to the nearest second.

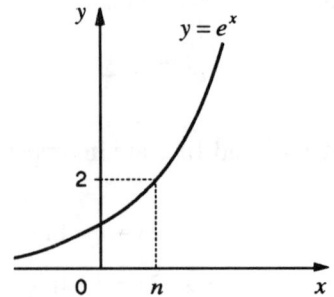

The example illustrates how a general method for solving exponential equations works. This process can be made quicker by using the results developed below.

Consider the function a^x, where $a > 0$.

As a is a number greater than zero, there is a real number, n, such that $e^n = a$ (see figure opposite). This means that $n = \ln a$, since $\ln x$ is the inverse function for the exponential function e^x.

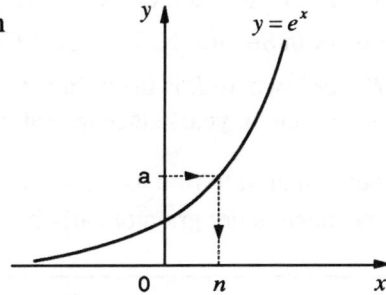

So $a^x = (e^n)^x$ replacing a by e^n

That is, $a^x = e^{xn}$ using laws of indices, and taking logarithms of both sides gives the equation

$$\ln a^x = \ln e^{xn} = xn$$

But $n = \ln a$, so

$$\boxed{\ln a^x = x \ln a}$$

This result is a great help in solving a wide variety of exponential equations.

Example

Solve $3^{2x-1} = 5^x$, giving your answer to 2 d.p.

Solution

Since $\qquad 3^{2x-1} = 5^x$

$\Rightarrow \quad \ln(3^{2x-1}) = \ln 5^x$

$\Rightarrow \quad (2x - 1)\ln 3 = x \ln 5$

$\Rightarrow \quad 2x \ln 3 - \ln 3 = x \ln 5$

$\Rightarrow \quad 2x \ln 3 - x \ln 5 - \ln 3 = 0$

$\Rightarrow \quad 2x \ln 3 - x \ln 5 = \ln 3$

$\Rightarrow \quad x(2 \ln 3 - \ln 5) = \ln 3$

$\Rightarrow \quad x = \dfrac{\ln 3}{2 \ln 3 - \ln 5} = 1.87$ to 2 d.p.

Activity 8

A sample of wood has ^{14}C radioactivity of 6 becquerels per gram. New wood has ^{14}C radioactivity of 6.68 becquerels per gram of Carbon 14. The half life of ^{14}C is 5730 years; form a model based on the work in Section 12.2 for the ^{14}C radiation in wood, of the form $R = ba^t$, where R is the radioactivity, b and a are constants, and t is the time in years since the sample was formed.

Use your equation to find to the nearest year when $R = 6$ becquerels per gram of carbon.

Activity 9

In Activity 4, you used a model for the UK population of the form

$$P = 55 \times 1.01^{\frac{1}{16}t}$$

P is the population in millions, and t the number of years since 1990. You were asked to estimate the year when the population would first equal 60 million. Solve this problem again by substituting $P = 60$ in the equation, and solving for t.

Exercise 12B

1. Solve $2^x = 5$ to 2 d.p.

2. Solve $3^{\frac{1}{2}x} = 1$ to 2 d.p.

3. Solve $4 \times 2^x = 3$ to 3 s.f.

4. Solve $3^x = 5^x$ to 2 d.p.

5. Solve $2^{-x} = 6$ to 3 s.f.

6. Solve $3^{2x} = 4^x$ to 2 d.p.

7. Solve $5^{x-1} = 3$ to 3 s.f.

8. Solve $2^{2x+1} = 4$ to 2 d.p.

9. Solve $5^{x-1} = e^{2x}$ to 1 d.p.

10. Solve $6^{2x+1} = 3^{-x}$ to 2 d.p.

12.5 Properties of logarithms

As well as obeying the rule

$$\ln(a^x) = x\ln a,$$

logarithms also obey, for any real number a, b,

$$\ln(ab) = \ln a + \ln b \qquad\qquad (1)$$

and

$$\ln\left(\frac{a}{b}\right) = \ln a - \ln b \qquad\qquad (2)$$

To prove the first result, (1), note that a and b can be written in the form

$$a = e^m, b = e^n$$

for some real numbers m and n. Then

$$\ln(ab) = \ln(e^m e^n)$$
$$= \ln(e^{m+n})$$
$$= m + n$$

Since $\ln x$ is the inverse function of e^x,

$$\ln a = \ln(e^m) = m$$
$$\ln b = \ln(e^n) = n$$

so that

$$\ln(ab) = \ln a + \ln b$$

How can you deduce equation (2) from (1)?

You will see how useful these results are in the following applications.

Before the theory of gravitation was developed by *Sir Isaac Newton*, the best laws available to describe planetary motion were those formulated by *Johann Kepler*, a German astronomer. His laws were based on his own meticulous observations, and were used later as a 'benchmark test' for Newton's own theory. This activity investigates Keplar's third law.

Activity 10   Kepler's third law

This table shows how the average radius of a planet's orbit around the Sun, R, is related to the period of that orbit in years, T. (The orbits are elliptical, not circular, so an average radius is used here). Only the planets known to Kepler are included.

Planet	Radius, R (millions of km)	Period, T (years)
Mercury	57.9	0.24
Venus	108.2	0.62
Earth	149.6	1
Mars	227.9	1.88
Jupiter	778.3	11.86
Saturn	1427.0	29.46

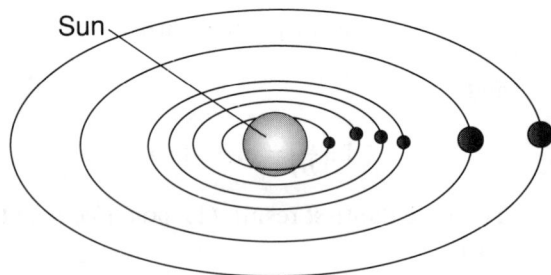

You may assume that T and R are linked by a relationship of the form $T = aR^b$ where a and b are constants to be found.

To fit the model $T = aR^b$ to the data means trying out different values of a and b until you have a good fit with the curve drawn through the data points.

The properties of logarithms will provide us with a better method for finding suitable values of the constants a and b.

Assume a power law of the form

$$T = aR^b$$

Taking logs of each side gives

$$\ln T = \ln(aR^b)$$
$$= \ln a + \ln\left(R^b\right) \quad \text{(using equation (1))}$$
$$= \ln a + b \ln R$$

This equation resembles a straight line equation $y = mx + c$ with y replaced by $\ln T$ and x by R. So a graph of $\ln T$ against $\ln R$ should give a straight line and the constants a and b can be estimated from the graph. The constant b will be the **gradient** of the line, and $\ln a$ will be the **intercept** on the vertical axis.

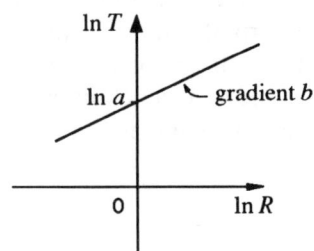

Activity 11

For the data in Activity 10, plot a graph of ln T against ln R, and use it to estimate the values of the constants a and b.

The note produced by a musical instrument is directly related to its frequency (the number of times the air is caused to vibrate every second). The higher the frequency, the higher the note. In order to set the frets on a guitar in the correct place, the maker must know how the length of a string affects the frequency of the note it produces.

Activity 12   Guitar maker's problem

This relationship between length, l (cm), and frequency, f (hz), can be found experimentally. The table shows some data collected by experiment for a particular type of string.

Length, l(cm)	50	60	70	80	90	100
Frequency, f(hz)	410	330	275	255	225	195

The relationship is assumed to be of the form $f = al^b$ where a and b are constant.

Use logarithms to 'linearise' the relationship, as described previously. Plot ln f on a vertical axis and ln l on the horizontal, and draw a line of best fit. Find the gradient and intercept with the vertical axis of this line, and so determine the values of a and b.

The frequencies produced are also effected by the tension in the string and so, even with frets correctly placed, the guitarist must still 'tune' the instrument by changing the tensions in the strings.

'Middle C' has a frequency of 264 Hz.

What length of string gives this frequency?

The last application in this section is based on the method used by forensic scientists to estimate the time of death of a body.

When a person dies, the body's temperature begins to cool. The temperature of the body at any time after death is governed by Newton's Law of Cooling, which applies to any cooling object:

$$D = ae^{-kt}$$

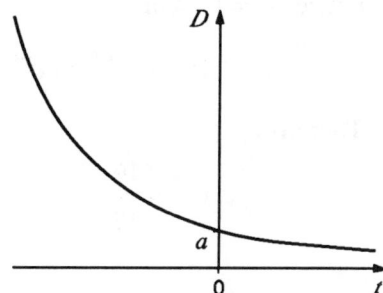

D is the temperature difference between the cooling object and its surrounding, a and k are constants, and t is the time since the object started to cool. The values of a and k depend on the size, shape and composition of the object and the initial temperature difference.

If D is plotted against the time, t, the graph will be similar to the curve shown opposite.

To find out the equation of the curve which applies to a dead body, the values of a and k must be found. This will require two readings of its temperature.

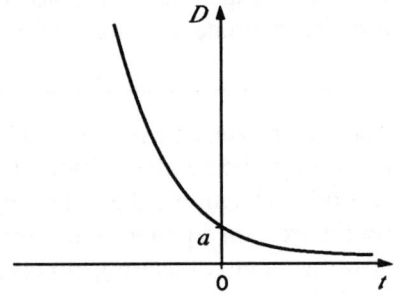

Example

The police arrive at the scene of a murder at 8 a.m.

On arrival, the temperature of the body and its surroundings are measured at $34°C$ and $17°C$ respectively. This was taken to be the moment when the time, t, was equal to zero.

At 9 a.m. when $t = 1$, the body temperature was measured as $33°C$ and the room temperature still as $17°C$.

Estimate the time of death.

Solution

The two sets of data are

$$D = 34 - 17 = 17 \text{ at } t = 0$$

$$D = 33 - 17 = 16 \text{ at } t = 1$$

Substituting in the governing equation

$$D = ae^{-kt}$$

gives

$$17 = ae^{-k.0} = ae^0 = a$$

(since $e^0 = 1$); and

$$16 = ae^{-k.1} = ae^{-k} = 17e^{-k}.$$

Therefore

$$e^{-k} = \frac{16}{17}$$

and taking 'logs',

$$-k = \ln\left(\frac{16}{17}\right) = -0.0606 \Rightarrow k = 0.0606$$

Hence

$$D = 17e^{-0.0606t} \tag{3}$$

Now normal body temperature is given by $36.9°C$, so the corresponding value of D is given by

$$D = 36.9 - 17 = 19.1$$

Substituting this value of D into equation (3) and solving for t will give you the estimated time of death; this gives

$$19.1 = 17e^{-0.0606t}$$

$$\Rightarrow \quad e^{-0.0606t} = \frac{19.1}{17}$$

$$\Rightarrow \quad -0.0606t = \ln\left(\frac{19.1}{17}\right)$$

$$\Rightarrow \quad t = -\frac{1}{0.0606}\ln\left(\frac{19.1}{17}\right)$$

$$= -1.922 \text{ hours}$$

$$\approx -(1 \text{ hour } 55 \text{ mins})$$

So the estimated time of death is estimated at 6.05 am, or about 6.00 am

What important assumptions have been made in this model? Are they reasonable?

Activity 13

A body is found at 11.30 pm. The body temperature at midnight is found to be $33°C$ and at 2.00 am it is $31.5°C$. Assuming the surroundings are at a constant temperature of $30°C$, estimate the time of death.

12.6 Other bases

Many applications of exponential functions do not use the base e. Scientists often use a base of 10 for instance. The graph of $y = a^x$, for $a > 1$, shows that the function of a^x is one to one. This means that it has an inverse function, which is denoted $y = \log_a x$. This is read as "logarithm (or log) to base a of x". The figure opposite also shows the graph $y = \log_a x$. The graph also show that the range of a^x is the positive real numbers, as is the domain of $y = \log_a x$. Provided logarithms use a suitable base, they obey the same laws developed in earlier sections. That is;

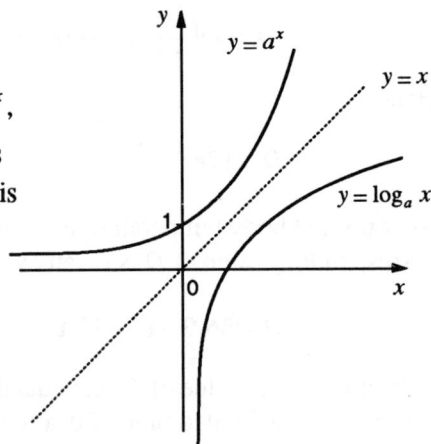

(i) $\log_a p^n = n \log_a p$

(ii) $\log_a pq = \log_a p + \log_a q$ (for any two numbers p and q)

(iii) $\log_a \left(\dfrac{p}{q} \right) = \log_a p - \log_a q$

To summarise; if

$$\boxed{y = a^x, \text{ then } x = \log_a y}$$

Activity 14

Without using a calculator, answer these questions.

(a) For any base a, $a^0 = 1$. Write down $\log_a 1$.

(b) $a^1 = a$. Write down $\log_a(a)$ for any base a.

Also write down $\log_a\left(a^2\right)$, $\log_a\left(\sqrt{a}\right)$

(c) $1000 = 10^3$. Write down $\log_{10}(1000)$

Similarly, find $\log_{10}(100)$, $\log_{10}(\dfrac{1}{10})$, $\log_{10}(0.01)$

(d) $\log_2(8)$ (remember $2^3 = 8$)

Exercise 12C

Without using a calculator, answer these questions

1. $\ln e^2$

2. $\log_{10} 10000$

3. $\log_2 27$

4. $\log_3\left(\frac{1}{16}\right)$

5. $\ln\left(\dfrac{1}{\sqrt{e}}\right)$

6. $\log_{10}\sqrt{10}$

7. $\log_5 125$

8. $\log_{49} 7$

12.7 Derivative of ln x

You have already seen that if

$$y = e^x \Rightarrow \frac{dy}{dx} = e^x$$

Now if $y = \ln x$, and $x > 0$

$$x = e^y$$

so that

$$\frac{dx}{dy} = e^y$$

If δy and δx are corresponding small change in y and x, then

$$\frac{dy}{dx} = \lim_{\delta x \to 0} \left(\frac{\delta y}{\delta x} \right)$$

But

$$\frac{dx}{dy} = \lim_{\delta x \to 0} \left(\frac{\delta x}{\delta y} \right)$$

Hence

$$\frac{dy}{dx} = 1 \bigg/ \frac{dy}{dx}$$

so that, when $y = \ln x$,

$$\frac{dy}{dx} = \frac{1}{e^y} = \frac{1}{x} \, (x > 0)$$

So you have the important result that for $x > 0$

$$\boxed{\frac{d}{dx}(\ln x) = \frac{1}{x}}$$

The function $y = \ln x$ and its derivative are illustrated in the figure opposite. Notice that the graph has only been drawn for values of x greater than zero. This is because $\ln x$ is not defined for negative values of x, so it does not have a derivative when x is less than zero.

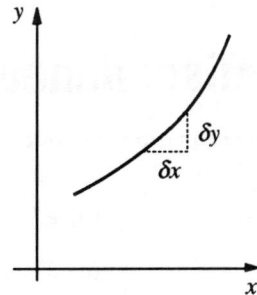

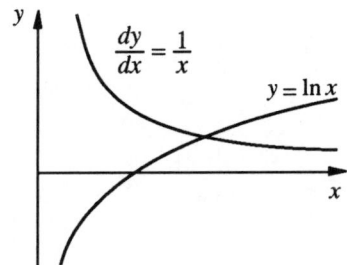

Activity 16

Plot the graphs of the following functions using a calculator or a computer. Now that you know the derivative of $\ln(x)$ is $\dfrac{1}{x}$, try to write down the derivatives of each of these functions, by comparing the curves of each one to $y = \ln(x)$.

(a) $\ln(x+1)$

(b) $\ln(x+2)$

(c) $\ln(x-3)$

(d) $\ln(x)-1$

(e) $-\ln(x)$

If you have a graph plotting package which is capable of displaying the derivatives of each function, you can check your answers.

12.8 Miscellaneous Exercises

1. Solve these equations to three significant figures where appropriate.

   (a) $e^x = 4$

   (b) $e^{3x} = 0.1$

   (c) $e^{2x-1} = 5$

   (d) $3^x = 1$

   (e) $10^x = 5$

   (f) $4^x = 5^{2x+1}$

   (g) $3x^2 = 1$

   (h) $4 \times 7^{2x} = 6$

   (i) $e \times 2^{x-1} = 5^{1-x}$

   (j) $10^x = 1000$

   (k) $5^x = 25$

   (l) $2^x = \dfrac{1}{8}$

2. A physicist conducts an experiment to discover the half life of an element. The radioactivity at one moment from a sample of the element is measured as 30 becquerels. One hour later the radioactivity is just 28 becquerels. Assuming that the radioactivity is governed by a formula of the form

   $$R = a \times 2^{-kt}$$

   where R is the radioactivity in becquerels per gram, t the time in hours, and a and k are constants, find the values of a and k, and hence determine the half life in hours.

13 INTEGRATION

Objectives

After studying this chapter you should

* appreciate why finding the area under a graph is often important;

* be able to calculate the area under a variety of graphs given their equation.

13.0 Introduction

Integration is the process of finding the area under a graph. An example of an area that integration can be used to calculate is the shaded one shown in the diagram. There are several ways of **estimating** the area - this chapter includes a brief look at such methods - but the main objective is to discover a way to find the area **exactly**.

Finding the area under a graph is not just important for its own sake. There are a number of problems in science and elsewhere that need integration for a solution. Section 13.1 starts by looking at an example of such a problem.

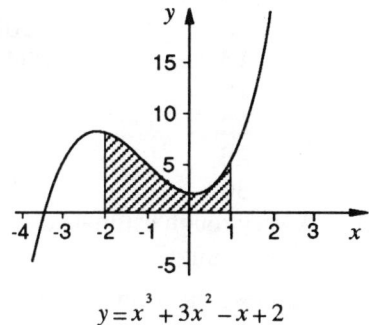

$$y = x^3 + 3x^2 - x + 2$$

13.1 Estimating future populations

Shortly after the Second World War, it was decided to establish several 'new towns'. Two well known examples are Hemel Hempstead, near London, and Newton Aycliffe in the north-east; in these cases the 'new towns' were based on existing small communities. More recently, the city of Milton Keynes was built up where once there was practically nothing.

Clearly such ambitious developments require careful planning and one factor that needs to be considered is population growth. Apart from anything else, the services and infrastructure of a new community need building up in accordance with the projected population: chaos would ensue, for instance, if there were 5000 children but only enough schools for 3000.

Imagine you are planning a new town. You have been advised that planned population growth will conform to the following model:

Initial growth rate 6000 people per year;
thereafter the rate of growth will decrease
by 30% every five years.

Your task is to estimate what the total population will be 30 years later. One approach to this problem is detailed in the following activity.

Activity 1   Rough estimates

(a)   Start by setting up a mathematical model. Let t stand for the time in years; when $t = 0$ the growth rate is 6000 people per year. When $t = 5$ (after 5 years) the growth rate is 6000 less 30%, i.e., $0.7 \times 6000 = 4200$.

Copy and complete this table of growth rates :

Time (years)	Growth Rate
0	6000
5	4200
10	2940
15	...
20	...
25	...
30	706

(b)   A very rough estimate can be obtained by following this line of reasoning:

Suppose the growth rate remains fixed at 6000 throughout the first 5 years. Then after 5 years the population will be

$$5 \times 6000 = 30\ 000.$$

Now suppose the growth rate over the next five years is a constant 4200 per year. Then after 10 years the population will be

$$30\ 000 + 5 \times 4200 = 51\ 000.$$

Continue this process to obtain an estimate for the population after 30 years.

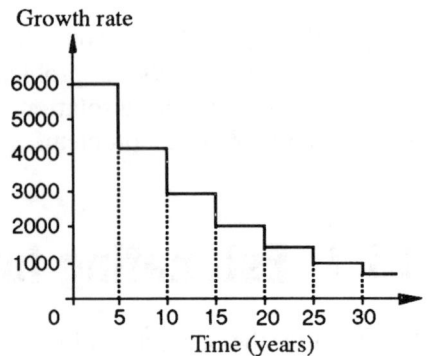

(c)   Your figure for (b) will clearly be an over-estimate. A similar process can be applied to give an under-estimate. The process starts as follows:

Suppose the growth rate over the first five years is fixed at 4200. Then after five years the population will be

$$5 \times 4200 = 21\ 000$$

After 10 years the population will be

$$21\ 000 + 5 \times 2940 = 35\ 700.$$

Continue this process to obtain a second estimate.

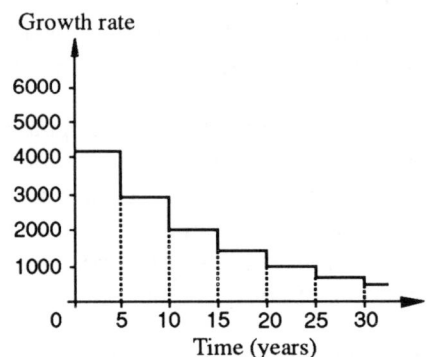

(d)   Use your two estimates to make a third, more realistic estimate of the population after 30 years' growth.

The two diagrams alongside the above activity should give a good idea of what the processes you have just completed actually mean. In parts (b) and (c) what you essentially did was to find the area under a bar chart; one estimate was too large, the other too small, with the 'true' answer lying somewhere in the middle.

The graph of growth rate against time will not, in truth, be a 'bar chart' at all. More realistically it will resemble the curve shown opposite, which shows the growth rate decreasing continuously. The figure below shows the same curve with the two bar charts superimposed.

The 'true' population estimate will be the **exact** area under the curve in the figure above.

Activity 2   More accurate estimate

A more refined estimate can be obtained by using growth rates calculated at yearly intervals rather than 5-yearly ones. A decrease of 30% every **five** years is roughly equivalent to a decrease of 6.9% every year.

Hence the table of growth rates starts like this:

Time (years)	Growth Rate
0	6000
1	5587
2	5202
etc.	etc.

Use these figures to get a closer estimate of the population. Follow the same sort of process as in Activity 1.

Why was the figure 6.9% used in this activity?

With a little ingenuity you might be able to save tedious calculation by efficient use of a computer. If you do this, try and refine the estimate still further by using smaller intervals.

The graphs alongside Activity 2 illustrate why using smaller intervals leads to increased accuracy. The smaller the intervals, the more the 'bar chart' resembles the continuous curve. Hence the area of the 'bar charts' approaches the exact value of the area under the curve as the interval size is decreased.

The next two activities provide further examples of finding estimates for the area under a curve.

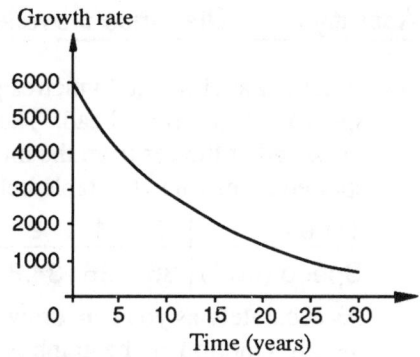

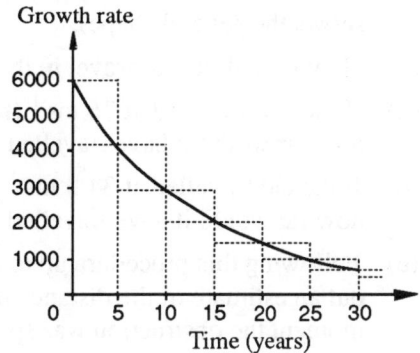

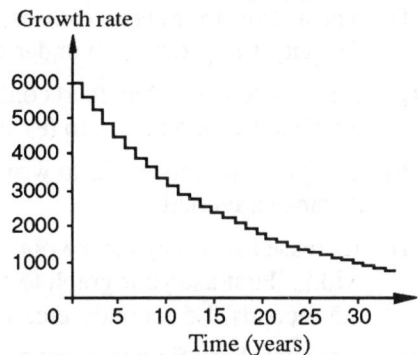

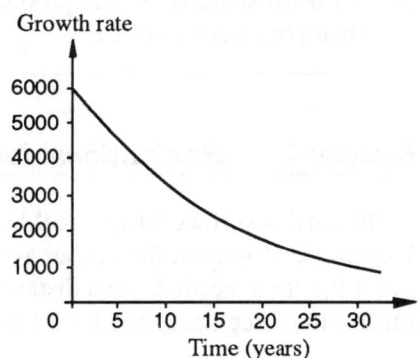

Activity 3   Distance travelled

(a) A car is travelling at 36 metres per second when the driver
     spots an obstruction ahead. The car does an 'emergency stop';
     the speed of the car from the moment the obstruction is
     spotted is shown in the table below.

Time (s)	0	1	2	3	4	5	6	7
Speed (ms^{-1})	36	36	34.8	29.9	23.2	15.2	4.8	0

As accurately as you can, draw a speed-time curve describing
the car's motion. (The graph opposite is not accurate but
shows the general shape).

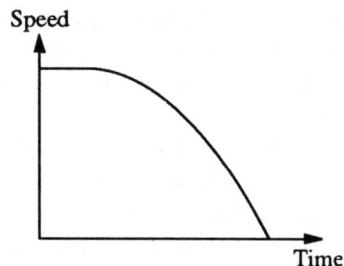

(b) How far did the car travel in the first second?

(c) If the car travelled at 36 ms^{-1} **throughout** the 2nd second,
     how far would it have travelled in this second?

(d) If the car travelled at 34.8 ms^{-1} throughout the 3rd second,
     how far would it have travelled in this second?

(e) Following this procedure up to the 7th second inclusive; work
     out an estimate of the distance travelled by the car from the
     moment the obstruction was spotted.

(f) The answer to (e) is an over-estimate. Use the ideas of
     Activity 1 to produce an under-estimate.

(g) Superimpose two bar charts onto your graph, the areas under
     which are your answers to (e) and (f).

(h) Use your answers so far to write down a better estimate of the
     distance travelled.

(i) Increased accuracy can be obtained by halving the interval
     width. First use your graph to estimate the car's speed after
     0.5 seconds, 1.5 seconds, etc. Record your answers clearly.

(j) Now use these figures to produce a more accurate estimate
     than your answer to (h).

Activity 4   Gravitational force

At the earth's surface a force of 9.81 Newtons is required to lift a
1 kg mass. However, the further away from the surface you go, the
lower the force needed. At a distance of x kilometres from the
surface the force needed to lift a 1 kg mass is given by the formula

$$F = \frac{4.02 \times 10^8}{(6400 + x)^2}$$

(a)   Use this formula to complete this table :

Distance from surface (km)	Lifting force required (N)
0	9.81
500	8.44
1000	...
1500	...
2000	...
2500	...
3000	4.55

(b)   Draw a graph of force against distance. Make it as accurate as you can.

The question to be answered is this: how much **energy** is required to lift the 1 kg mass to a point 3000 km above the surface?

For a constant force,

energy = force ×distance travelled.

In this situation, though, the force is changing the further away the mass is taken.

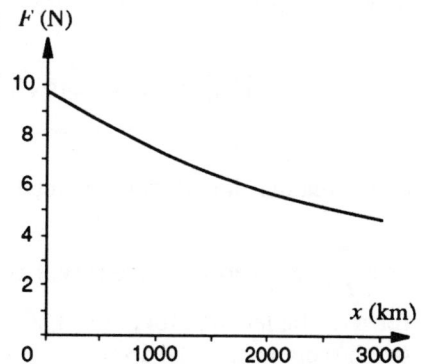

(c)   Suppose the force remained at 9.81 N for the first 500 km. What energy (in kilojoules (kJ)) would be required for this part of the journey?

(d)   Continue the argument along these lines to estimate the total energy needed to lift the 1 kg mass through 3000 km. You might need to refer back to the procedure used in Activity 3.

(e)   Work out the lifting forces required at 250 km, 750 km etc. Use these smaller intervals to produce a better estimate of the energy required.

Exercise 13A

1.   The sketch opposite shows the graph of the curve with equation

$$y = 4x^3 - 15x^2 + 12x + 5.$$

(a) Copy and complete this table of values :

x	1.0	1.1	1.2	1.3	1.4	1.5	1.6	1.7	1.8	1.9	2.0
y	6	5.37..	..	..	..	2.18..	..	..	..	..	1

(b) Use your table to estimate the shaded area in the diagram. Employ methods like those in the earlier questions and activities.

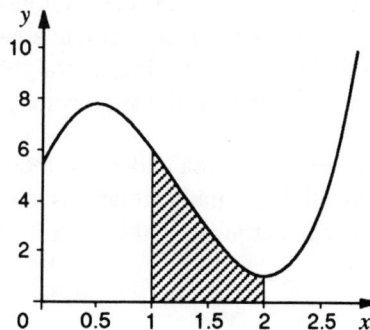

257

13.2 Notation for area

Three examples have now been encountered in which the area under a graph has been of significance. Mathematicians have for many centuries appreciated the importance of areas under curves. The process of working out such areas is called 'integration' and early approaches to the problem were similar to the method investigated so far, namely the splitting of areas into several thin rectangles.

The notation developed by the German mathematician *Leibniz* quickly became widely adopted and is still used today. The area under the curve

$$y = 4x^3 - 15x^2 + 12x + 5$$

between $x = 1$ and $x = 2$ is written as

$$\int_1^2 \left(4x^3 - 15x^2 + 12x + 5\right) dx,$$

and is read as the integral between 1 and 2 of $4x^3 - 15x^2 + 12x + 5$.

The $\int$ sign, known as the **integral sign**, derives from the ancient form of the letter S, for sum. The 'dx' represents the width of the small rectangles. The above notation denotes the sum of lots of very thin rectangles between the limits $x = 1$ and $x = 2$.

Hence the area in Activity 4 would be written

$$\int_0^{3000} \frac{4.02 \times 10^8}{(64000 + x)^2} \, dx.$$

The $\int$ and the 'dx' enclose the function to be integrated.

Area under straight line

The activities in the previous section highlighted the fact that, at present, the methods used to calculate areas have only been approximate. When the graph is a straight line, however, it is a simple matter to work out the area exactly.

As you have seen, the area under a velocity-time graph gives the distance travelled. In many situations, particularly when a body is moving freely under gravity, the velocity-time graph is a straight line.

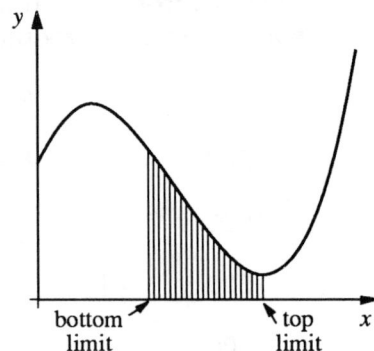

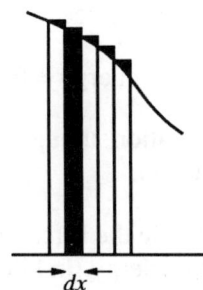

Consider the case of a stone thrown vertically downwards from a cliff-top. If its initial velocity is 5 metres per second then its velocity v meters per second after t seconds will be given approximately by the formula

$$v = 5 + 10t$$

How far will it have travelled after 5 seconds, assuming it hasn't hit anything by then? The answer to this question can be worked out by finding the area under, that is integrating, the velocity-time graph.

The problem thus boils down to finding the shaded area in the diagram opposite. Using the integral sign, this area can be written

$$\int_0^5 (5 + 10t)\, dt$$

(Note that dt is used here rather than dx as the variable along the horizontal axis is t).

The shaded area is a trapezium, the area of which can be worked out using the general rule

$$\text{area} = \frac{(\text{sum of parallel sides}) \times (\text{distance between them})}{2}$$

In this case the calculation yields $\dfrac{(55 + 5) \times 5}{2} = 150.$

So in 5 seconds the stone will have fallen 150 metres.

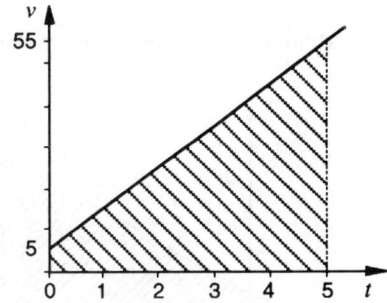

Example

Calculate the integral $\int_4^6 (20 - 3x)\, dx.$

Solution

The area required is shown in this sketch. Again, it is a trapezium.

To work out the lengths of the parallel sides you need to know the y-coordinates of A and B. These can be worked out from the equation of the line, $y = 20 - 3x$.

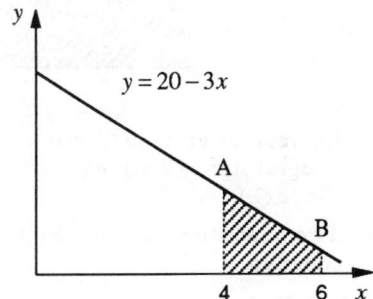

At A: $y = 20 - (3 \times 4) = 8.$

At B: $y = 20 - (3 \times 6) = 2.$

So the area is $\dfrac{(8+2) \times 2}{2} = 10$ square units.

Exercise 13B

1. Calculate the shaded areas in these diagrams.

   (a)

   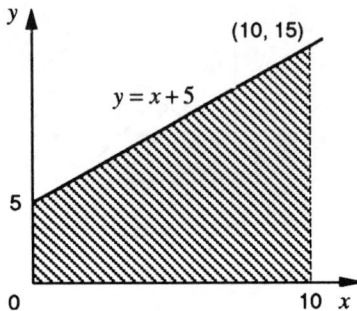

   (b)

   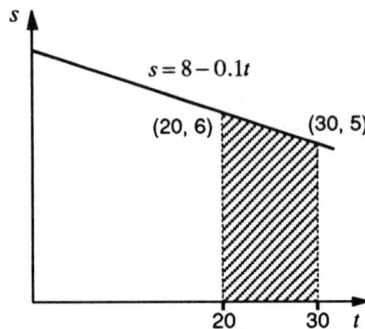

   (c)

   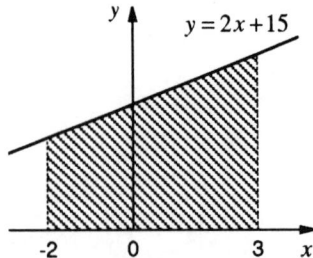

2. Express the areas in Question 1 using the integral sign. (Warning : don't automatically use '*dx*'.)

3. Two quantities p and q are related by the equation

   $$q = 12p - 45.$$

   (a) Sketch the graph of this relationship for values of p between 0 and 10.

   (b) Calculate

   $$\int_5^8 (12p - 45)\ dp.$$

   and shade on your graph the area this represents.

4. Calculate these integrals. Draw a sketch diagram if it helps.

   (a) $\int_1^6 (6x + 3)\, dx$

   (b) $\int_{-1}^1 (25 - 7t)\, dt$

   (c) $\int_{20}^{75} (-p + 100)\, dp$

   (d) $\int_{2.7}^{6.5} (3.5s + 17.1)\, ds$

5. The speed of a car as it rolls up a hill is given by $v = 20 - 3t$ where t is the time in seconds and v is measured in metres per second.

   (a) Draw a sketch graph showing speed against time between $t = 0$ and $t = 5$.

   (b) Integrate to find how far the car travels in the first 5 seconds.

6. The rate at which a city's population grows is given by the formula

   $$R = 1000 + 700t$$

   where R is the rate of increase in people per year and t is the number of years since 1st January 1990.

   (a) How fast will the population be growing on 1st January 1993, according to this model?

   (b) The population starts from zero on 1st January 1990. Calculate, by integration, what the population should be on 1st January 1993. (Draw a sketch graph.)

13.3 General formula

It is straight forward to find the area under any straight line graph, say

$$y = mx + c$$

between $x = a$ and $x = b$, as shown in the figure opposite.

What are the y coordinates of A and B?

The area of the trapezium is given by

$$\tfrac{1}{2}((mb+c)+(ma+c))(b-a)$$

$$= \tfrac{1}{2}(mb+ma+2c)(b-a)$$

$$= \tfrac{1}{2}(mb^2+mab+2cb-mab-ma^2-2ca)$$

$$\text{area} = \left(\tfrac{1}{2}mb^2+cb\right)-\left(\tfrac{1}{2}ma^2+ca\right)$$

You can use this formula to check your answer to Question 1 in Exercise 13B.

The formula is usually written $\left[\tfrac{1}{2}mx^2+cx\right]_a^b$

This is short for the 'function $\tfrac{1}{2}mx^2+cx$ evaluated at the top limit, $x = b$, minus the value of the same function at the lower limit, $x = b$'.

The function $\tfrac{1}{2}mx^2+cx$ can thus be used to find areas efficiently. It is called an **indefinite integral** of the function $mx+c$; one way of thinking of it is as the 'area function' for the graph of $y = mx+c$. For the moment the 'area function', or indefinite integral, will be denoted by $A(x)$, whereas the area between $x = a$ and $x = b$ is given by

$$\text{Area} = \int_a^b (mx+c)dx = \left[\tfrac{1}{2}mx^2+cx\right]_a^b,$$

for the straight line function $y = mx+c$, and this is called a **definite integral** .

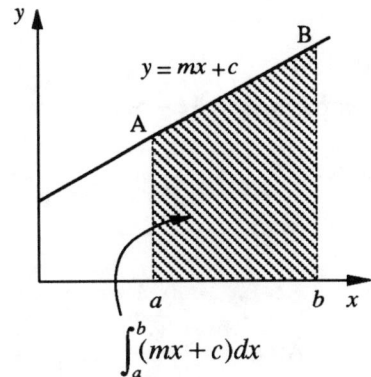

$$\int_a^b (mx+c)dx$$

261

Example

Find $A(x)$ for the straight line $y = 3x - 5$. Use it to work out the area under $y = 3x - 5$ between $x = 2$ and $x = 10$.

Solution

Comparing $y = 3x - 5$ with $mx + c$, you see that $m = 3$ and $c = -5$.

Hence

$$A(x) = \frac{3x^2}{2} - 5x$$

$$\text{Area} = \left[\frac{3x^2}{2} - 5x \right]_2^{10}$$

$$= \left(\frac{3 \times 10^2}{2} - 5 \times 10 \right) - \left(\frac{3 \times 2^2}{2} - 5 \times 2 \right)$$

$$= 100 - (-4) = 104 \text{ units}$$

Activity 5

a) Write down an indefinite integral of the function $12 - 8x$

b) Evaluate the area under the graph of $y = 12 - 8x$ between $x = 1$ and $x = 3$.

c) Draw a sketch graph and interpret your answer.

Activity 6

If $y = 2x - 7$ then the indefinite integral is $A(x) = x^2 - 7x$.

Find $\dfrac{dA}{dx}$. Investigate further for different straight line formulas.

Exercise 13C

1. Find the area function A for these straight lines :

    (a) $y = 2x + 7$

    (b) $s = 10 - t$

    (c) $z = 2.8 + 11.4w$

    (d) $y = -14 - 11x$

2. Use the area function method to calculate the following areas. Try to set out your working as in the worked example above.

    (a) The area under $y = 6x + 1$ between $x = 0$ and $x = 3$.

    (b) The area under $s = 13 - 5t$ between $t = -2$ and $t = 1$.

    (c) $\displaystyle\int_1^{10} (x + 0.5)\, dx$

    (d) $\displaystyle\int_{20}^{30} \left(\frac{y}{2} - 1 \right) dy$

13.4 The reverse of differentiation

Activity 6 gives a vital clue in the search for a method of integrating more difficult functions. For straight lines, the formula of the line and its indefinite integral are connected by the following rule.

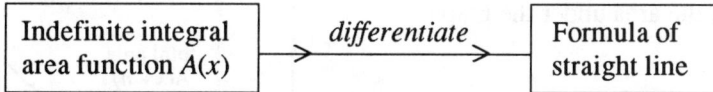

Indefinite integral area function $A(x)$	*differentiate* $\longrightarrow$	Formula of straight line

For example, in the worked example just before Activity 5, the equation of the line was $y = 3x - 5$. The corresponding indefinite integral was

$$A(x) = \frac{3x^2}{2} - 5x.$$

It is clear that in this case $\dfrac{dA}{dx} = y$. If this connection were true for more complex functions than straight lines, the process of finding indefinite integrals and consequently area for functions like $x^2 - 5x + 7$ would automatically be simplified.

In fact, there is one important piece of evidence supporting just such a conclusion. In Section 13.1 the population was estimated by considering the area under the graph of **rate of change** of population. In Activity 3 the distance travelled by a car was calculated by finding the area under the graph of **rate of change** of distance (i.e. velocity).

In general you should appreciate that

> the area under a graph showing the **rate of change** of some quantity will give the quantity itself.

But the process of finding rates of change is differentiation, hence integration must be the reverse process.

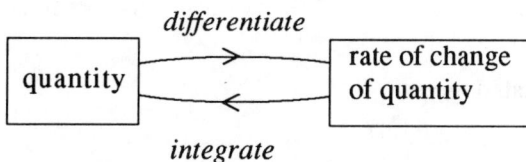

For any function, therefore, it is said that

integration is the reverse of differentiation.

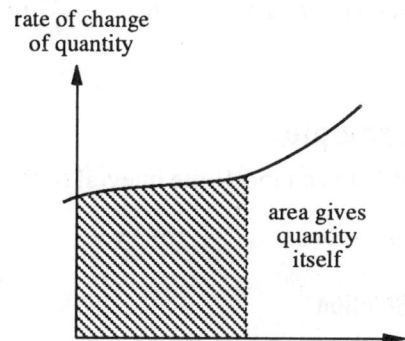

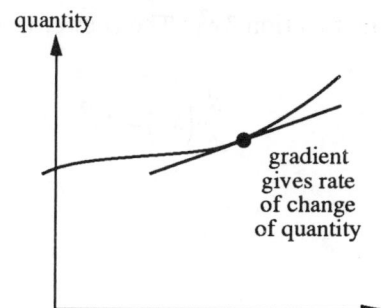

rate of change of quantity

area gives quantity itself

quantity

gradient gives rate of change of quantity

You may feel that the justification for this conclusion is rather vague: too many words and not enough mathematics! This is the way mathematicians often operate, get an instinctive 'feel' for a result and then prove it rigorously. So here is a more mathematical justification for this crucial result.

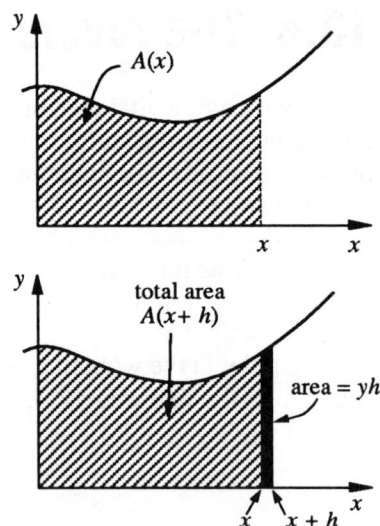

This graph represents y as a function of x. It does not matter at all what sort of function it is. Let $A(x)$ be the area under the graph between $x = 0$ and $x = x$.

Now suppose an extra strip is added to the area. It has width h and is shaded black in the diagram. The area of both shaded regions together is $A(x + h)$. The area of the black strip is approximately yh, since it is roughly a rectangle with height y.

Hence

$$A(x + h) \approx A(x) + yh$$

$$A(x + h) - A(x) \approx yh$$

$$y \approx \frac{A(x + h) - A(x)}{h}.$$

This is only approximately true (hence the '$\approx$') but the equation becomes more and more exact the smaller h becomes. But as h approaches 0,

$$\frac{A(x + h) - A(x)}{h} \to \frac{dA}{dx}.$$

Hence it can be seen that $y = \dfrac{dA}{dx}$ for any function y.

Example

What is an indefinite integral of the following?

(a) $y = 3x^2$ (b) $y = x^2$

Solution

To answer (a) consider which function, when differentiated, gives the function $3x^2$. The answer is x^3, since

$$\frac{d}{dx}\left(x^3\right) = 3x^2$$

Part (b) follows from this; x^2 is one third of $3x^2$. Hence the indefinite integral for x^2 must be

$$\frac{x^3}{3}, \text{ since } \frac{d}{dx}\left(\frac{x^3}{3}\right) = x^2.$$

The usual way of writing the two answers above, using $\int$ notation, is

$$\int 3x^2 dx = x^3 \text{ and } \int x^2 dx = \tfrac{1}{3}x^3$$

Activity 7   Standard functions

Following the procedure of the example above, work out these indefinite integrals :

(a) $\int x^3 dx, \ \int x^4 dx, \ \int x^5 dx, \ \text{etc}$

(b) $\int x^{-2} dx, \int x^{-3} dx, \int x^{-4} dx \ \text{etc}$

(c) $\int x^{-1} dx$

(d) $\int e^x dx$

Can you formulate general rules for finding indefinite integrals before turning the page and going on to the next section? Carefully consider the pattern emerging each time.

Is integration unique?

You already know that $\int 3x^2 \ dx = x^3$, and that in reverse this is equivalent to saying $\frac{d}{dx}\left(x^3\right) = 3x^2$.

But what about $\frac{d}{dx}\left(x^3 + 1\right), \ \frac{d}{dx}\left(x^3 + 4\right)$ and $\frac{d}{dx}\left(x^3 - 7\right)$?

They all give the answer $3x^2$. So, in general, you can write

$$\int 3x^2 \ dx = x^3 + K$$

where K is any constant.

Activity 8   Integral of a constant

Sketch the graph of $y = 2$. What is $\int 2\,dx$?

Generalise to find $\int k\,dx$, where k is any number.

Standard Integrals

Below is a summary of what you should have found out from Activities 7 and 8. These results are most important and although ready available should be memorised, so that they can be recalled instantly when needed.

$$\int x^n\,dx = \frac{x^{n+1}}{n+1} + K \text{ for any integer } n, \text{ except } -1.$$

$$\int x^{-1}\,dx = \int \frac{1}{x}\,dx = \log\,x + K$$

$$\int e^x\,dx = e^x + K$$

Note that each of the results includes the term '$+K$'. K is known as the **arbitrary** constant, or constant of integration. You must **always** include the arbitrary constant when working out **indefinite** integrals.

Examples

(1) $\displaystyle\int 5x^3\,dx = 5\int x^3\,dx = 5\frac{x^4}{4} + K = \frac{5x^4}{4} + K$

(2) $\displaystyle\int \frac{7}{x^3}\,dx = 7\int x^{-3}\,dx$ (because $\dfrac{7}{x^3} = 7x^{-3}$)

$$= 7\frac{x^{-2}}{(-2)} + K = -\frac{7}{2}x^{-2} + K = -\frac{7}{2}\cdot\frac{1}{x^2} + K$$

$$= -\frac{7}{2x^2} + K$$

(3) $\displaystyle\int\left(3x^5-\frac{2}{x}\right)dx=3\int x^5\,dx-2\int\frac{1}{x}\,dx$

$$=3\frac{x^6}{6}-2\ln x+K$$

$$=\frac{x^6}{2}-2\ln x+K$$

(4) $\displaystyle\int\frac{x+x^2}{3}\,dx=\frac{1}{3}\int\left(x+x^2\right)dx$

$$=\frac{1}{3}\left\{\int x\,dx+\int x^2\,dx\right\}$$

$$=\frac{1}{3}\left(\frac{x^2}{2}+\frac{x^3}{3}\right)+K$$

$$=\frac{x^2}{6}+\frac{x^3}{9}+K$$

In the worked examples above, it has been assumed that integration is a **linear** process. For example, in Example (1), the first step was

$$\int 5x^3\,dx=5\int x^3\,dx.$$

In Example (4) the second step assumed that

$$\int\left(x+x^2\right)dx=\int x\,dx+\int x^2\,dx.$$

Why are these assumptions valid?

If $u(x)$ and $v(x)$ are any functions of x and a and b are any constant numbers, then

$$\boxed{\int\left(au(x)+bv(x)\right)dx=a\int u(x)dx+b\int v(x)dx}$$

Exercise 13D

Evaluate these indefinite integrals.

Numbers 1 to 10 are relatively straightforward.

Numbers 11 to 20 might need more care.

Tidy each answer as much as you can; for example

$-\dfrac{3}{4x^4}$ is better than $3\dfrac{x^{-4}}{(-4)}$.

1. $\displaystyle\int x^9\,dx$

2. $\displaystyle\int x^{-8}\,dx$

3. $\displaystyle\int \dfrac{1}{x^5}\,dx$

4. $\displaystyle\int \left(x^4+x^7\right)dx$

5. $\displaystyle\int \left(\dfrac{1}{x}+\dfrac{1}{x^2}\right)dx$

6. $\displaystyle\int 6x^7\,dx$

7. $\displaystyle\int \dfrac{3}{t^3}\,dt$

8. $\displaystyle\int \dfrac{2}{w}\,dw$

9. $\displaystyle\int \left(e^p-3p\right)dp$

10. $\displaystyle\int \left(2-\dfrac{1}{q^3}\right)dq$

11. $\displaystyle\int \dfrac{x^3}{2}\,dx$

12. $\displaystyle\int \dfrac{4}{3}x^7\,dx$

13. $\displaystyle\int \dfrac{3}{2y^3}\,dy$

14. $\displaystyle\int 2\left(x-\dfrac{1}{3x}\right)dx$

15. $\displaystyle\int \left(\dfrac{e^k}{4}+k^{-3}\right)dk$

16. $\displaystyle\int \dfrac{3-2x^5}{4}\,dx$

17. $\displaystyle\int \left(e^m-\dfrac{2}{5m^4}\right)dm$

18. $\displaystyle\int \dfrac{5x^2-x+1}{2}\,dx$

19. $\displaystyle\int \dfrac{1}{5}\left(z^2-\dfrac{1}{z}\right)dz$

20. $\displaystyle\int \dfrac{2x-1}{x}\,dx$

13.5 Finding areas

The original aim of this chapter was to calculate exact values for areas under curves. The groundwork for this has now been done. The procedure is best explained through a worked example, but before you read through it you might like to remind yourself how indefinite integrals were used to work out areas under straight lines in Section 13.2.

Example

Calculate the area under the curve $y = x^2 + 2$ between $x = 1$ and $x = 4$.

Solution

The diagram shows the area required.

Using integral notation this area would be denoted

$$\int_1^4 \left(x^2+2\right)dx.$$

The indefinite integral of x^2+2 is the function

$$\dfrac{x^3}{3}+2x+K$$

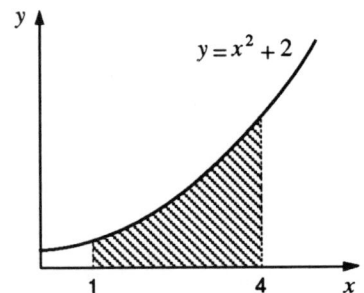

Hence

$$\text{Area} = \left[\frac{x^3}{3} + 2x + K \right]_1^4$$

$$= \left(\frac{4^3}{3} + 2 \times 4 + K \right) - \left(\frac{1^3}{3} + 2 \times 1 + K \right)$$

$$= \left(29\tfrac{1}{3} + K \right) - \left(2\tfrac{1}{3} + K \right)$$

$$= 27 \text{ units.}$$

An important point to note in this calculation is that the arbitrary constant **cancels** out. Integrals with limits, such as

$$\int_1^4 \left(x^2 + 2 \right) dx$$

are called **definite** integrals, and it is customary when evaluating these to omit the arbitrary constant altogether.

Example

Work out the area under the graph of $y = 10e^x + 3x$ between $x = -1$ and $x = 3$, to one decimal place.

Solution

$$\text{Area} = \int_{-1}^3 \left(10e^x + 3x \right) dx$$

$$= \left[10e^x + \frac{3x^2}{2} \right]_{-1}^3$$

$$= \left(10e^3 + \frac{3 \times 3^2}{2} \right) - \left(10e^{-1} + \frac{3 \times (-1)^2}{2} \right)$$

$$= 214.35537 - 5.1787944$$

$$= 209.2 \text{ to 1 d.p.}$$

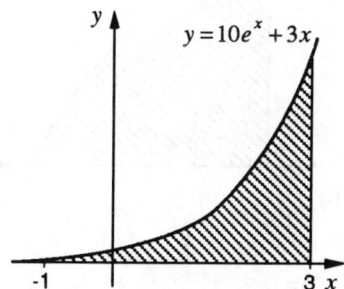

$y = 10e^x + 3x$

Activity 9   Negative areas

(a)   Evaluate $\int_0^6 \left(x^2 - 2x - 8\right) dx$.

(b)   Sketch the curve $y = x^2 - 2x - 8$ and interpret the value of the integral.

(c)   Calculate the total shaded area in this diagram.

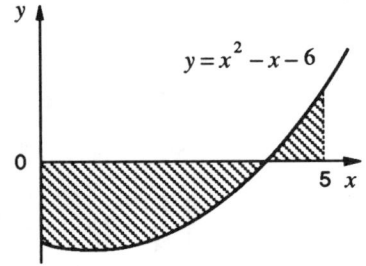

Activity 10   Simple products

Is it true that

$$\int (x+3)(x+5)\, dx = \left(\int (x+3)\, dx\right) \times \left(\int (x+5)\, dx\right)?$$

What is the best way to evaluate $\int (x+3)(x+5)\, dx$?

Exercise 13E

1. Work out the shaded areas below to 3 significant figures. Use integral notation when setting out your solutions.

(a)

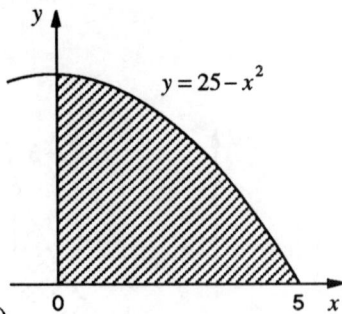

(b)

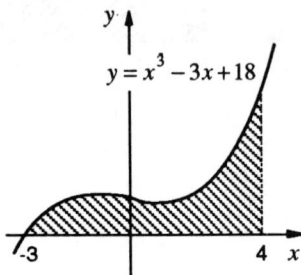

(c)

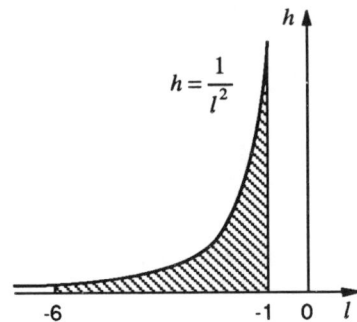

(d)

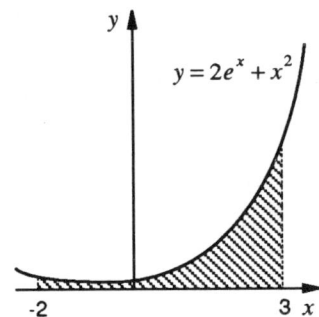

(e)

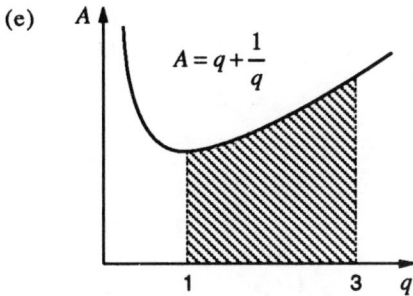

$A = q + \dfrac{1}{q}$

(f)

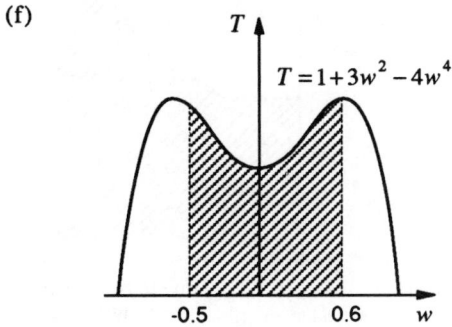

$T = 1 + 3w^2 - 4w^4$

2. Evaluate these to 3 significant figures.

(a) $\displaystyle\int_1^5 \left(x^2 - \dfrac{1}{x^2} \right) dx$ (b) $\displaystyle\int_{-2}^{-1} \left(6z^2 - 1 \right) dz$

(c) $\displaystyle\int_0^1 \dfrac{1 + 5m^3}{6}\, dm$ (d) $\displaystyle\int_{2.5}^3 \left(4 + \dfrac{2}{3x} \right) dx$

3. The diagram below shows a sketch of $y = x - x^2$. Find the shaded area to 3 significant figures.

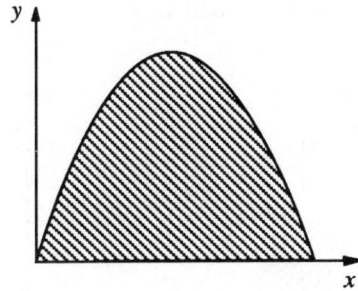

4. A pig-trough has a cross-section in the shape of the curve

$$y = x^{10},$$

for x between -1 and $+1$.

(a) Calculate $\displaystyle\int_{-1}^1 x^{10}\, dx$.

(b) Work out the cross-sectional area of the trough, given that one unit on the graph represents one metre.

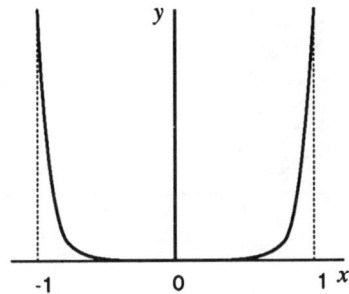

13.6 Using integration

This chapter began with illustrations of particular situations where the area under a graph had significance. Activity 11 should remind you of some of them and introduce you to some more.

Activity 11   What does the area mean?

For each of these graphs, describe in words what quantity is represented by the shaded area.

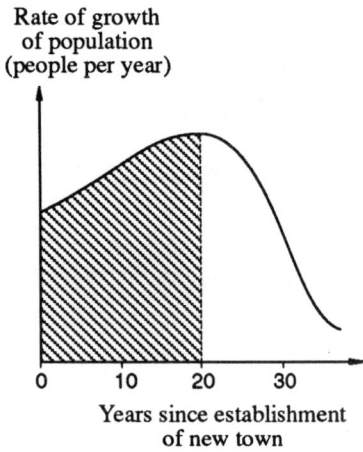

A

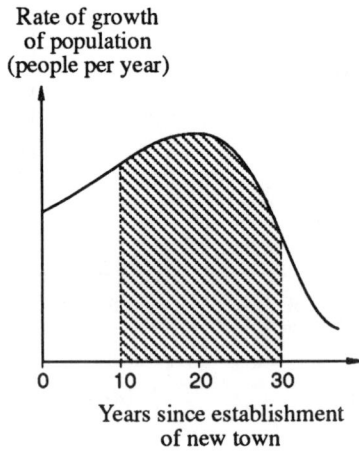

B

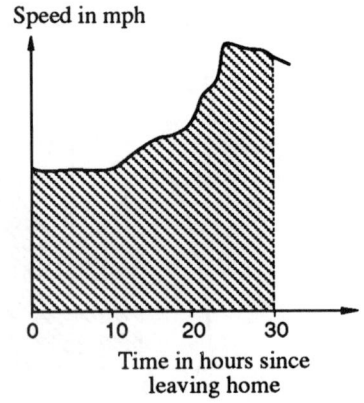

C

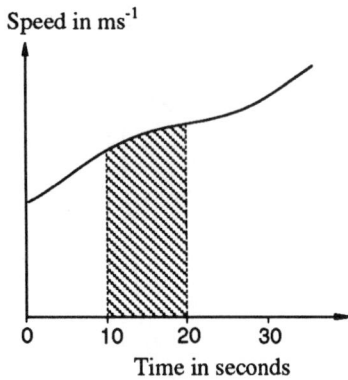

D

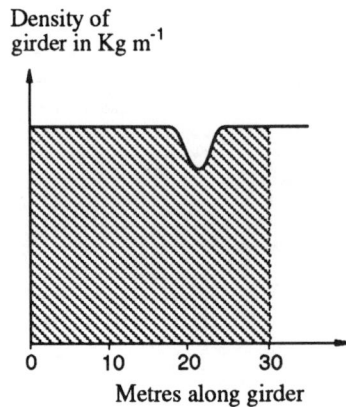

E

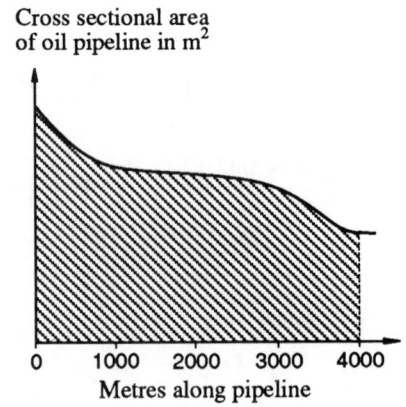

F

Activity 12   The rogue car

A motorist parks his car next to a telephone box. He makes a call from the box but, during the conversation, suddenly sees the car rolling down the hill; he clearly forgot to apply the handbrake. When he first noticed the car it was 20 metres away moving with speed $0.7 + 0.2t$, where t is the time in seconds after the first time he noticed the car rolling away.

(a)   How far does the car travel between $t = 0$ and $t = 5$?

(b)   Find a formula for the car's distance from the phone box, in terms of t.

13.7 Initial conditions

Activity 12 gave an example of how integration can be used to find a formula, not just a numerical value. To find the answer to part (b) properly, you needed to apply what are called initial conditions. Integrating the velocity formula produced a constant, which could be evaluated by knowing the distance when t was zero. Here are two further examples.

Example

A particle P is moving along a straight line with velocity $4 + 6t + t^2$. When $t = 0$, P is a distance of 8 metres from a fixed point F. Find an expression for the distance FP.

Solution

The distance can be found by integrating the velocity formula.

$$FP = \int (4 + 6t + t^2)dt$$

$$= 4t + 3t^2 + \frac{t^3}{3} + \text{constant}$$

$FP = 8$ when $t = 0$, so the constant $= 8$. Hence

$$FP = 4t + 3t^2 + \frac{t^3}{3} + 8.$$

Example

A town planning committee notes that the rate of growth of the town's population since 1985 has followed the formula

$$(1500 + 200t) \text{ people per year}$$

where t is the number of years since 1st January 1985. On 1st January the population was 25 000. Find a formula for the city's population valid from 1985 onwards.

Solution

$$\text{Population} = \int (1500 + 200t)\, dt$$

$$= 1500t + 100t^2 + K$$

When $t = 7$, $P = 25\ 000$. Putting this information into the formula gives

$$25\ 000 = 10\ 500 + 4900 + K$$

$$\Rightarrow \quad K = 9600$$

Population $= 100t^2 + 1500t + 9600$.

Activity 13   Integration to find volumes

The upper part of a cocktail glass is a cone. The cone can be thought of as the line $y = 1.3x$ rotated around the x–axis through $360°$. The second of the two diagrams opposite shows this idea.

(a)   The cross-sectional area of the cone varies with x. Show that it does so according to the approximate formula

$$A = 5.3093x^2.$$

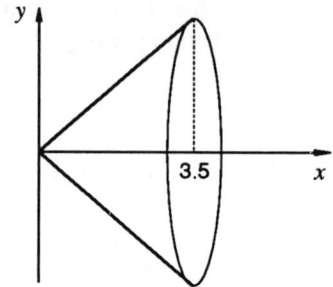

(b)   The cone is 3.5 cm high. This means that the above formula applies only for $0 \le x \le 3.5$. Sketch the graph of A against x for this range.

(c)   Find the volume of the glass in cm³. Give your answer to 3 s.f.

Example

The line $y = x + 6$ between $x = -1$ and $x = 5$ is rotated $360°$ about the y-axis. What is the volume of the solid this creates?

Solution

The cross-sectional area of the solid is given by the formula

$$A = \pi(x+6)^2.$$

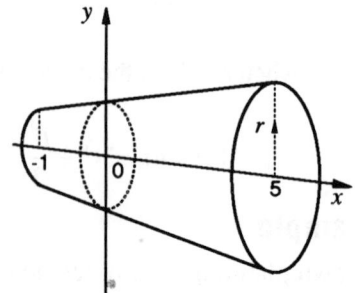

The graph opposite shows a sketch of this function between $x = -1$ and $x = 5$. The volume of the solid is given by the area under this graph.

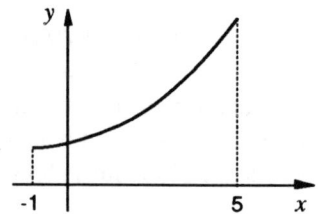

$$\text{Volume} = \int_{-1}^{5} A\ dx$$

$$= \int_{-1}^{5} \pi(x+6)^2\ dx$$

$$= \pi \int_{-1}^{5} (x^2 + 12x + 36)\ dx$$

$$= \pi \left[\frac{x^3}{3} + 6x^2 + 36x \right]_{-1}^{5}$$

$$= \pi(371\tfrac{2}{3} - (-30\tfrac{1}{3}))$$

$$= 402\pi$$

13.8 Solids of revolution

Activity 34 and the worked example that followed both involved volumes or **solids of revolution**. As their name suggests, these are formed by rotating a straight line or curve around a fixed axis. Many objects can be thought of as solids of revolution; a few of them are pictured opposite, and you may be able to think of more. Integration can be used to find the **volumes** of such solids, the general principle being outlined in the following activity.

Cooling tower   Lampshade

Activity 14

The curve $y = f(x)$ is rotated 360° about the x-axis. Consider how to find the cross-sectional area of the solid as a function of x, and hence explain why the formula for the volume is

$$\text{Volume} = \pi \int_a^b y^2 \, dx \text{ or } \pi \int_a^b [f(x)]^2 \, dx$$

where a and b are as in the diagram.

Exercise 13F

1. The speed of a falling object in ms^{-1} is given by the formula $2 - 10t$, where t is the time in seconds. When $t = 0$ its height h above the ground is 1000 m.

   (a) Find a formula for h in terms of t.

   (b) When does the object hit the ground?

2. The speed of a particle P moving along a straight line is given by the formula

   $$5 + t - \frac{t^2}{4}.$$

   When $t = 6$, the particle is at the point A.

   (a) Find a formula for the distance AP.

   (b) Verify that the particle is again at A when t is between 7.7 and 7.8 seconds.

3. Find the volumes swept out when the following curves are rotated 360° about the x-axis.

   (a) $y = 7x^3$ between $x = 0$ and $x = 2$;

   (b) $y = \dfrac{2}{\sqrt{x}}$ between $x = 1$ and $x = 5$.

   (c) $y = x^2 + 1$ between $x = -3$ and $x = 3$.

4. You have encountered the total cost function $C(Q)$ in previous questions in this book. A related economic concept is that of the marginal cost function $M(Q)$. This gives the change in total cost if the level of output (Q) is increased by 1 unit. $M(Q)$ and $C(Q)$ are thus related as follows

   $$M(Q) = \frac{dC}{dQ}$$

   (a) The marginal cost function for a particular firm is $Q^2 - 3Q + 5$. Find the total cost function $C(Q)$, given that $C(0) = 5$. [$C(0)$ is the total cost when the level of output is zero, and hence gives the firm's fixed costs.]

   (b) For a different firm, $M(Q) = 2Q^2 - 10Q + 17$. Find the increase from 5 to 8 units.

5. A parabolic bowl can be thought of as the curve
$y = 2\sqrt{x}$ rotated around the x-axis.

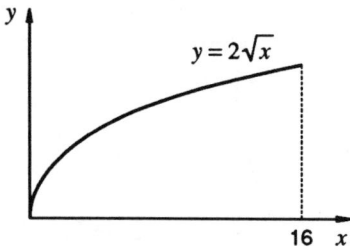

The height of such a vessel is 16 cm and the
radius of the top is 8 cm. Show that the volume
of the bowl is 1.61 litres, to 3 s.f.

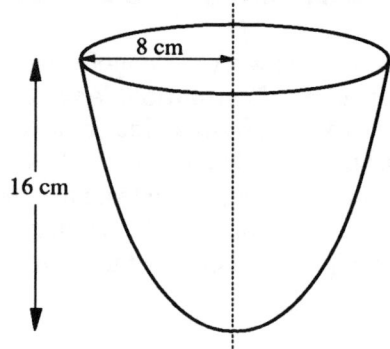

Activity 15   Some well-known formulas

(a) You already know that the volume of a cone is given by the

formula $\dfrac{\pi r^2 h}{3}$, where r is the base radius and h is the height.

Rotating the line segment OP about the x-axis will produce a
cone. Use this fact to prove the above formula. Start by
obtaining the equation of the line OP in terms of x, y, r, and h.

(b) The diagram opposite shows a semicircle, radius r, centre
$(0, 0)$. The point (x, y) lies on the semicircle.

Find a relationship between x, y and r, and use this to find a
formula for the volume of a sphere with radius r.

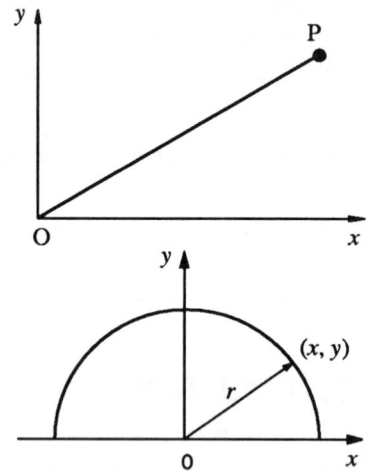

13.9   Miscellaneous Exercises

1. Work out these indefinite integrals:

(a) $\int (5x + 2)\, dx$

(b) $\int \left(\dfrac{t^3}{2} + \dfrac{2t^2}{3} \right) dt$

(c) $\int \left(\dfrac{7}{p} - p \right) dp$

(d) $\int \dfrac{5}{4s^6}\, ds$

(e) $\int \left(x - \dfrac{2}{x} \right)^2 dx$

2. Evaluate these definite integrals:

(a) $\int_0^{15} (e^x + 2x^2)\, dx$

(b) $\int_{-3}^{3} (e^x + 2x^2)\, dx$

(c) $\int_1^2 \left(\dfrac{2}{x} + \dfrac{3}{5x^2} \right) dx$

(d) $\int_2^3 (x + 5)(x - 2)\, dx$

(e) $\int_{-2}^{-1} (x^2 - 2)(x + 1)\, dx$

3. The graph shows the function
$f(x) = x(x + 1)(x - 2)$.

Find the areas labelled A and B.

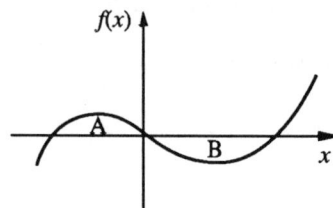

4. This graph shows a function $p(x)$.

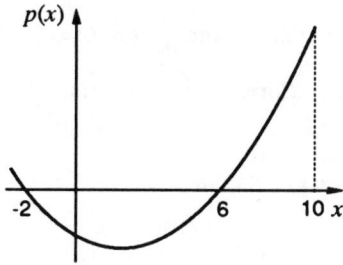

If $\int_{-2}^{6} p(x)\,dx = -10$ and $\int_{-2}^{10} p(x)\,dx = 1$, write

down the value of $\int_{6}^{10} p(x)\,dx$.

5.

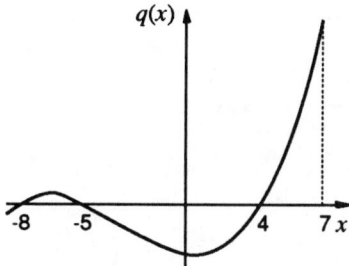

Given that

$\int_{-8}^{7} q(x)\,dx = -4, \int_{4}^{7} q(x)\,dx = 4$ and $\int_{-5}^{4} q(x)\,dx = -11$,

find the value of $\int_{-8}^{-5} q(x)\,dx$.

6. Find the shaded area in these diagrams.

(a)

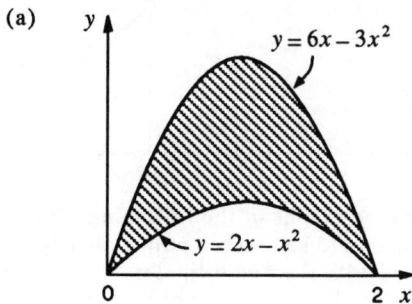

(b)

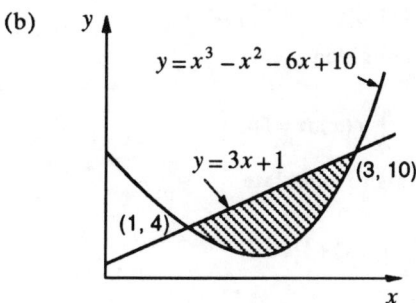

(c)

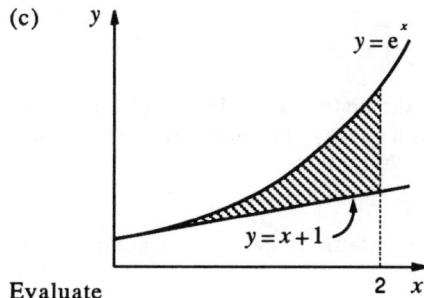

7. Evaluate

(a) $\int_{1}^{3} \frac{x-1}{x}\,dx$ (b) $\int_{-4}^{-1} \frac{xe^{x}-3}{2x}\,dx$

8. A petrol tank, when full, contains 36 litres of petrol. It develops a small hole which widens as time goes by. The rate at which fuel leaks out (in litres per day) is given by the formula

$$0.009t^2 + 0.08t + 0.01$$

where t is the time in days. When $t = 0$ the tank is full.

(a) Find the formulas for

   (i) the amount of fuel lost
   (ii) the amount of fuel left in the tank after t days.

(b) How many cm³ does the tank lose on

   (i) the first day;
   (ii) the tenth day?

(c) How much fuel is left in the tank after

   (i) 5 days;
   (ii) 15 days?

9. A gas is being kept in a large cylindrical container, the height of which can be altered by means of a piston.

The pressure of the gas (p), volume in which it is kept (V) and temperature (T) are related by the equation

$$pV = 5430T.$$

(a) If $T = 293$ (degrees Kelvin) and the radius of the base of the cylinder is 1 metre, show that

$$p \approx \frac{5.06 \times 10^5}{h}$$

where h is the height of the piston above the base of the cylinder.

(b) The energy (in joules) required to compress the gas from a height h_1, to a height h_2 is given by

277

$$\int_{h_2}^{h_1} p\, dh$$

Initially the piston is at a height of 5 metres. How much energy is required to push the piston down to a height of

(i) 4 m   (ii) 1 m?

*10.   A pewter tankard can be thought of as a solid of revolution.

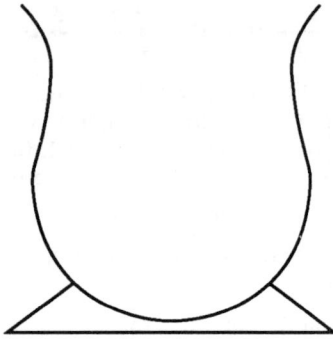

The shape that has been rotated is that shown in the graph below.

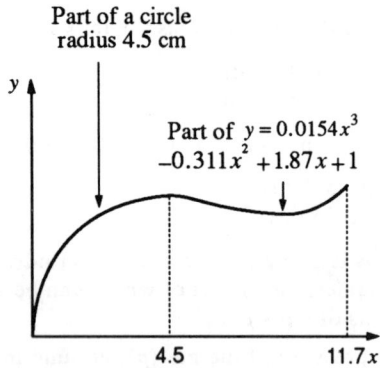

Part of a circle radius 4.5 cm

Part of $y = 0.0154x^3 -0.311x^2 + 1.87x + 1$

(a) The bottom 4.5 cm of the shape is a hemisphere. Find its volume. (Use the formula $\frac{4}{3}\pi r^3$ for the volume of a sphere.)

(b) From $x = 4.5$ to $x = 11.7$ the curve has equation

$$y = 0.0154x^3 - 0.311x^2 + 1.87x + 1.$$

The cross-sectional area of the tankard has the following form. Explain how it was calculated and work out what numbers go in the gaps;

$$A = 0.00074506x^6 - 0.030093x^5 + 0.4848x^4$$
$$-3.5574x^3 + 9.0318x^2 + ?x + ?.$$

(c) Calculate the volume of the tankard to the nearest whole number.

11. $u(x)$ and $v(x)$ are two functions of x.

$$\int_0^3 u(x)\, dx = 5 \quad \text{and} \quad \int_0^3 v(x)\, dx = 8.$$

Use this information to calculate these, where possible.

(a) $\int_0^3 (u(x) + v(x))dx$

(b) $\int_0^3 u(x)v(x)dx$

(c) $\int_0^3 xu(x)dx$

(d) $\int_0^3 (2u(x) - 3v(x))\, dx$

(e) $\int_0^3 \left[(u(x))^2 + (v(x))^2 \right] dx$

(f) $\int_0^6 u(x)dx$

(g) $\int_0^3 \dfrac{u(x)}{v(x)}\, dx.$

12. The graph of a function $f(x)$ looks like this:

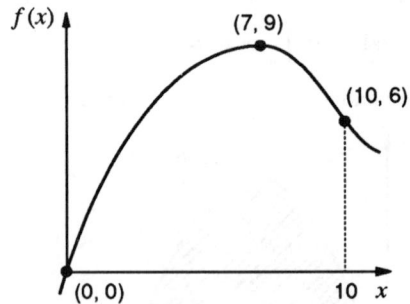

(a) Sketch the graphs of the following. Each sketch should have three points clearly labelled with their coordinates.

(i) $f(x) + 1$ (ii) $f(x - 4)$

(iii) $2f(x)$ (iv) $f(2x)$

(b) You are given

$$\int_0^{10} f(x)dx = 56.$$

Use this to calculate

(i) $\int_0^{10} [f(x) + 1]dx$

(ii) $\int_0^{10} 2f(x)dx$

*13(a) The graph shows $y = \dfrac{1}{x}$. Explain why

$$\int_1^p \frac{1}{x}\, dx = \ln p.$$

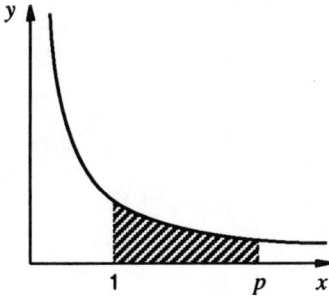

(b) Now imagine you don't know how to calculate

$$\int_1^p \frac{1}{x}\, dx$$

exactly. Use a numerical method to evaluate these approximately.

$$\int_1^2 \frac{1}{x}\, dx; \quad \int_1^3 \frac{1}{x}\, dx; \quad \int_1^6 \frac{1}{x}\, dx.$$

(c) Why do your answers suggest that $\int_1^p \dfrac{1}{x}\, dx$ might be a logarithmic function?

(d) Find the values of a and b in these equations:

(i) $\displaystyle \int_1^4 \frac{1}{x}\, dx + \int_1^5 \frac{1}{x}\, dx = \int_1^a \frac{1}{x}\, dx.$

(ii) $\displaystyle \int_1^2 \frac{1}{x}\, dx + \int_1^b \frac{1}{x}\, dx = \int_1^{50} \frac{1}{x}\, dx.$

14 SEQUENCES AND SERIES

Objectives

After studying this chapter you should

- be able to recognise geometric and arithmetic sequences;
- understand Σ notation for sums of series;
- be familiar with the standard formulas for Σr, Σr^2 and Σr^3;
- know how to find the limits of several kinds of sequence.

14.0 Introduction

Suppose you go on a sponsored walk. In the first hour you walk 3 miles, in the second hour 2 miles and in each succeeding hour $\frac{2}{3}$ of the distance the hour before. How far would you walk in 10 hours? How far would you go if you kept on like this for ever?

This gives a sequence of numbers: 3, 2, $1\frac{1}{3}$, .. etc. This chapter is about how to tackle problems that involve sequences like this and gives further examples of where they might arise. It also examines sequences and series in general, quick methods of writing them down, and techniques for investigating their behaviour.

Legend has it that the inventor of the game called **chess** was told to name his own reward. His reply was along these lines.

'Imagine a chessboard.

Suppose 1 grain of corn is placed on the first square,

2 grains on the second,
4 grains on the third,
8 grains on the fourth,

and so on, doubling each time up to and including the 64th square. I would like as many grains of corn as the chessboard now carries.'

It took his patron a little time to appreciate the enormity of this request, but not as long as the inventor would have taken to use all the corn up.

Number of grains of corn shown

Activity 1

(a)   How many grains would there be on the 64th square?

(b)   How many would there be on the *n*th square?

(c)   Work out the numerical values of the first 10 terms of the sequence.

$$2^0, \ 2^0 + 2^1, \ 2^0 + 2^1 + 2^2 \ \text{etc.}$$

(d)   How many grains are there on the chessboard?

14.1 Geometric sequences

The series of numbers 1, 2, 4, 8, 16 ... is an example of a **geometric sequence** (sometimes called a geometric progression). Each term in the progression is found by **multiplying** the previous number by 2.

Such sequences occur in many situations; the multiplying factor does not have to be 2. For example, if you invested £2000 in an account with a fixed interest rate of 8% p.a. then the amounts of money in the account after 1 year, 2 years, 3 years etc. would be as shown in the table. The first number in the sequence is 2000 and each successive number is found by multiplying by 1.08 each time.

Number of years	Money in account (£)
0	2000
1	2160
2	2332.80
3	2159.42
4	2720.98

Accountants often work out the residual value of a piece of plant by assuming a fixed depreciation rate. Suppose a piece of equipment was originally worth £35 000 and depreciates in value by 10% each year. Then the values at the beginning of each succeeding year are as shown in the table opposite. Notice that they too form a geometric progression.

Year	Value (£)
0	35 000
1	31 500
2	28 350
3	25 515
4	22 963.50

The chessboard problem in Activity 1 involved adding up

$$2^0 + 2^1 + 2^2 + \ldots \ldots + 2^{64}$$

The sum of several terms of a sequence is called a **series**. Hence the sum $2^0 + 2^1 + 2^2 + \ldots \ldots + 2^{64}$ is called a **geometric series** (sometimes geometric progression, GP for short)

Activity 2   Summing a GP

In Activity 1 you might have found a formula for

$$1 + 2 + 2^2 + \ldots + 2^{n-1}$$

(a)   Work out the values of

$$3^0, \ 3^0 + 3^1, \ 3^0 + 3^1 + 3^2$$

(b) Find a formula for

$$1+3+3^2+ \ldots +3^{n-1}$$

(c) Find a formula for $1+4+4^2 \ldots + 4^{n-1}$

(d) Now find a formula for

$$1+r+r^2+ \ldots +r^{n-1}$$

where r is any number. Test your theory.

(e) In practice, geometric series do not always start with 1. Suppose the first term is a. How is the series in part (d) altered? How can you adapt your formula for the total of all terms?

The general form of a geometric sequence with n terms is

$$a, ar, ar^2, \ldots, ar^{n-1}$$

The ratio r of consecutive terms, is known as the **common ratio**. Notice that the nth term of is ar^{n-1}.

In the chessboard problem the solution involved adding up the first 64 terms. The sum of the first n terms of a series is often denoted by S_n, and there is a formula for S_n which you may have found in Activity 2. Here is a way of proving the formula, when $r \neq 1$.

$$S_n = a+ar+ar^2+ \ldots +ar^{n-1} \qquad (1)$$

Multiply both sides by r :

$$rS_n = ar+ar^2+ \ldots +ar^{n-1}+ar^n \qquad (2)$$

Notice that the expressions for S_n and rS_n are identical, with the exception of the terms a and ar^n. Subtracting equation (1) from equation (2) gives

$$rS_n - S_n = ar^n - a$$
$$\Rightarrow \quad S_n(r-1) = a(r^n-1)$$
$$\Rightarrow \quad S_n = \frac{a(r^n-1)}{r-1}$$

Activity 3   Understanding and using the formula

(a)   Sometimes it is useful to write

$$S_n = \frac{a(1-r^n)}{1-r} \text{ instead of } S_n = \frac{a(r^n-1)}{r-1}$$

Why are these formulae identical? When might it be more convenient to use the alternative form?

(b)   For what value of r do these formulas not hold? What is S_n in this case?

Example:

Find

(a) $4+6+9+ \ldots + 4\times(1.5)^{10}$

(b) $8+6+4.5+ \ldots + 8\times(0.75)^{25}$

Solution

(a)   First term $a=4$, common ratio $r=1.5$, number of terms $n=11$;

$$S_{11} = \frac{4(1.5^{11}-1)}{1.5-1} = 684.0 \text{ to 4 s.f.}$$

(b)   First term $a=8$, common ratio $r=0.75$, number of terms $n=26$;

$$S_{26} = \frac{8(1-0.75^{26})}{1-0.75} = 31.98 \text{ to 4 s.f.}$$

Example:

A plant grows 1.67 cm in its first week. Each week it grows by 4% more than it did the week before. By how much does it grow in nine weeks, including the first week?

Solution

The growths in the first 9 weeks are as follows :

$$1.67, \ 1.67\times1.04, \ 1.67\times1.04^2, \ldots$$

Total growth in first nine weeks is

$$S_9 = \frac{1.67(1.04^9 - 1)}{1.04 - 1} = 17.67 \text{ cm to 4 s.f.}$$

Example:

After how many complete years will a starting capital of £5000 first exceed £10 000 if it grows at 6% per annum?

Solution

After n years, the capital sum has grown to

$$5000 \times (1.06)^{n-1}$$

When is this first greater than 10 000, n being a natural number? In other words, the smallest value of n is required so that

$$5000 \times (1.06)^{n-1} > 10\,000, \ n \in \mathbb{N}$$

$$\Rightarrow \quad (1.06)^{n-1} > 2$$

Now take logs of both sides:

$$(n-1)\log 1.06 > \log 2$$

$$\Rightarrow \quad n - 1 > \frac{\log 2}{\log 1.06}$$

$$\Rightarrow \quad n - 1 > 11.9$$

$$\Rightarrow \quad n > 12.9, n \in \mathbb{N}$$

After 13 years, the investment has doubled in value.

£5000

⇓ ?

£10, 000

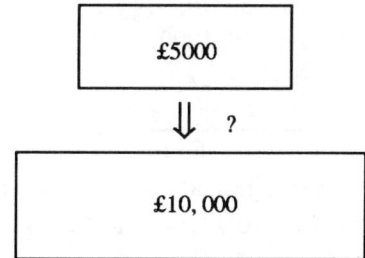

How many years later?

Activity 4   GP in disguise

(a)   Why is this a geometric sequence?

$$1, -2, 4, -8, 16, \dots ?$$

What is its common ratio? What is its nth term? What is S_n?

(b)   Investigate in the same way, the sequences

$$1, -1, 1, -1, \dots$$

Exercise 14A

1. Write down formulas for the nth term of these sequences:

   (a) 3, 6, 12, 24, ...

   (b) 36, 18, 9, 4.5, ...

   (c) 2, −6, 18, −54, ...

   (d) 90, −30, 10, $-3\frac{1}{3}$, ...

   (e) 10, 100, 1000, ...

   (f) 6, −6, 6, −6, ...

   (g) $\dfrac{1}{4}, \dfrac{1}{12}, \dfrac{1}{36}, \dfrac{1}{108}$, ...

2. Use the formula for S_n to calculate to 4 s.f.

   (a) 5 + 10 + 20 + ... to 6 terms

   (b) 4 + 12 + 36 + ... to 10 terms

   (c) $\dfrac{1}{3} + \dfrac{1}{6} + \dfrac{1}{12}$ +...to 8 terms

   (d) 100 − 20 + 4 − ... to 20 terms

   (e) 16 +17.6 + 19.36 + ... to 50 terms

   (f) 26 − 16.25 + 10.15625 ... to 15 terms

3. Give the number (e.g. 12th term) of the earliest term for which

   (a) the sequence 1, 1.5, 2.25, ... exceeds 50;

   (b) the sequence 6, 8, $10\frac{2}{3}$, ... exceeds 250;

   (c) the sequence $\dfrac{2}{5}, \dfrac{1}{5}, \dfrac{1}{10}$,... goes below $\dfrac{1}{1000}$

4. (a) For what value of n does the sum $50 + 60 + 72 + ... + 50 \times (1.2)^{n-1}$ first exceed 1000?

   (b) To how many terms can the following series be summed before it exceeds 2 000 000?

   $$2 + 2.01 + 2.02005 + ...$$

5. Dave invests £500 in a building society account at the start of each year. The interest rate in the account is 7.2% p.a. Immediately after he invests his 12th instalment he calculates how much money the account should contain. Show this calculation as the sum of a GP and use the formula for S_n to evaluate it.

14.2 Never ending sums

Many of the ideas used so far to illustrate geometric series have been to do with money. Here is one example that is not. If you drop a tennis ball, or any elastic object, onto a horizontal floor it will bounce back up part of the way. If left to its own devices it will continue to bounce, the height of the bounces decreasing each time.

The ratio between the heights of consecutive bounces is constant, hence these heights follow a GP. The same thing is true of the times between bounces.

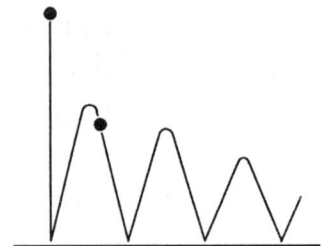

Activity 5   Bouncing ball

(a) A tennis ball is dropped from a height of 1 metre onto a concrete floor. After its first bounce, it rises to a height of 49cm. Call the height after the nth bounce h_n. Find a formula for h_n and say what happens to h_n as n gets larger.

(b) Under these circumstances the time between the first and second bounces is 0.6321 seconds. Call this t_1. The next time t_2, is $0.7t$, and each successive time is 0.7 times the previous one. Find a formula for t_n.

(c)   If $S_2 = t_1 + t_2$, what does S_2 represent What does S_n mean?
Calculate S_{10}, S_{20} and S_{50}. How long after the first bounce
does the ball take to stop bouncing altogether, to the nearest
tenth of a second?

Activity 5 gave an example of a **convergent** sequence.
Convergence, in this context, means that the further along the
sequence you go, the closer you get to a specific value. For
example, in part (a) the sequence to the nearest 0.1 cm is

$$100, 49.0, 24.0, 11.8, 5.8, 2.8, 1.4, ...$$

and the numbers get closer and closer to zero. Zero is said to be
the **limit** of the sequence.

Part (b) also gave a sequence that converged to zero. In part (c),
the sequence of numbers $S_1, S_2, S_3, ...$ start as follows :

$$0.6321, 1.0746, 1.3844, 1.6012, 1.7530, ...$$

You should have found that this sequence did approach a limit, but
that this was not zero. Hence the series has a convergent **sum**, that
is, the sum S_n of the series also converges.

The series 1, 2, 4, 8 ...... is a **divergent** sequence. It grows without
limit as the number of terms increases. The same is true, in a
slightly different sense, of the sequence 1, –2, 4, –8 ...... Any
sequence that does not converge is said to be **diverge**.

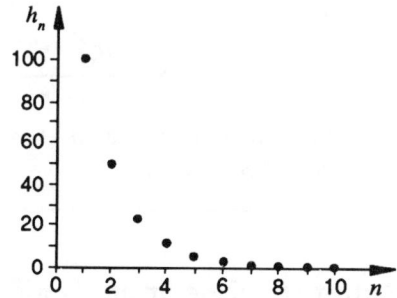

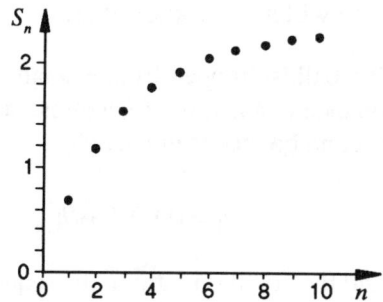

Activity 6   Convergent or divergent?

For each of these sequences

> (i)   write a formula for the nth term;
> (ii)   find whether the sequence converges;
> (iii) find whether the sum S_n converges.

(a) $6, 2, \frac{2}{3}, \frac{2}{9}, ...$

(b) $1, 1.5, 2.25, 3.375, ...$

(c) $4, -3, \frac{9}{4}, -\frac{27}{16} ...$

(d) $1, 1.01, 1.012, 1.013, ...$

(e) $8, -9.6, 11.52, -13.824 ...$

Activity 7 Behaviour of r^n

(a) What happens to r^n as n gets larger (i.e. as $n \to \infty$)? (You might need to see what happens for a variety of different values of r, positive and negative, large and small.)

(b) The sum of the first n terms of a geometric sequence is given by

$$S_n = \frac{a\left(1 - r^n\right)}{1 - r}$$

If r^n converges to 0 as $n \to \infty$, what can you say about the limit of S_n as $n \to \infty$?

Activity 8 Experimental verification

Conduct an experiment with a bouncing ball. Calculate the theoretical time from the first bounce until it stops bouncing. Then use a stop-watch to see how close the answer is to your calculation. You will need to know that:

If a ball is dropped from a height of 1 metre and rises after the first bounce to a height of h metres, then the time between the first and second bounce is given by

$$t_1 = 0.90305\sqrt{h},$$

and the common ratio in the sequence $t_1, t_2, t_3 \ldots$ is $\sqrt{h}$.

A geometric series, $a + ar + ar^2 + \ldots + ar^{n-1}$ converges when $|x| < 1$;

i.e. for $-1 < x < 1$. Since if $|x| < 1, r^n \to 0$ as $n \to \infty$ and

$$\boxed{S_n \to \frac{a}{1 - r} \text{ as } n \to \infty}$$

The limit $\dfrac{a}{1 - r}$ is known as the 'sum to infinity' and is denoted S_∞.

Example

Find

(a) $8 + 4 + 2 + 1 + \ldots$

(b) $20 - 16 + 12.8 - 10.24 + \ldots$

Solution

(a) This is a geometric series with first term 8 and common

ratio $\frac{1}{2}$, so

$$S_\infty = \frac{8}{1-\frac{1}{2}} = 16$$

(b) This is a geometric series with first term 20 and common
ratio -0.8, so

$$S_\infty = \frac{20}{1-^-0.8} = \frac{20}{1.8} = \frac{100}{9} \ (= 11.1 \text{ to 3 s.f.})$$

Exercise 14B

1. Find these sums to infinity, where they exist.

(a) $80 + 20 + 5 + 1.25 + \ldots$

(b) $180 - 60 + 20 - 20/3 + \ldots$

(c) $2 + 1.98 + 1.9602 + \ldots$

(d) $-100 + 110 - 121 + \ldots$

(e) $1/10 + 1/100 + 1/1000 + \ldots$

2. (a) What is $1/10 + 1/100 + 1/1000 + \ldots$ as a
recurring decimal?

(b) Express $0.37373737\ldots$ as an infinite
geometric series and find the fraction it
represents.

3. What fractions do these decimals represent?

(a) $0.52525252\ldots$

(b) $0.358358358\ldots$

(c) $0.194949494\ldots$

4. (a) A GP has a common ratio of 0.65. Its sum to
infinity is 120. What is the first term?

(b) Another GP has 2.8 as its first term and its
sum to infinity is 3.2 Find its common ratio.

5. Rosita is using a device to
extract air from a bottle of wine.
This helps to preserve the wine
left in the bottle.

The pump she uses can extract a
maximum of 46 cm³. In practice
what happens is that the first
attempt extracts 46 cm³ and
subsequent extractions follow a
geometric sequence.

Rosita's second attempt extracts
36 cm³. What is the maximum
amount of air she can remove in
total?

6. A rubber ball is dropped from a
height of 6 metres and after the first bounce rises
to a height of 4.7 m. It is left to continue
bouncing until it stops.

(a) A computerized timer is started when it first
hits the ground. The second contact with the
ground occurs after 1.958 second and the
third after 3.690 seconds. Given that the
times between consecutive contacts with the
ground follow a geometric sequence, how
long does the ball take to stop bouncing?

(b) The heights to which the ball rises after each
impact also follow a geometric sequence.
Between the release of the ball and the
second bounce the ball travels $6 + 2 \times 4.7 =$
15.4 m. How far does the ball travel
altogether?

14.3 Arithmetic sequences

Geometric sequences involve a constant ratio between consecutive terms. Another important type of sequence involves a constant **difference** between consecutive terms; such series are called **arithmetic** sequence.

In an experiment to measure the descent of a trolley rolling down a slope a 'tickertape timer' is used to measure the distance travelled in each second. The results are shown in the table.

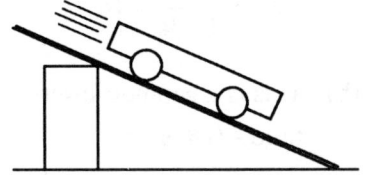

The sequence 3, 5, 7, 9, 11, 13 is an example of an arithmetic sequence. The sequence starts with 3 and thereafter each term is 2 more than the previous one. The difference of 2 is known as the **common difference**.

It would be useful to find the total distance travelled in the first 6 seconds by adding the numbers together. A quick numerical trick for doing this is to imagine writing the numbers out twice, once forwards once backwards, as shown below

Second	cm travelled in second
1	3
2	5
3	7
4	9
5	11
6	13

$$3 \quad 5 \quad 7 \quad 9 \quad 11 \quad 13$$

$$13 \quad 11 \quad 9 \quad 7 \quad 5 \quad 3$$

Each pair of vertical numbers adds up to 16.

6×16 between them. Hence the sum of the original series is

$$\frac{1}{2} \times (6 \times 16) = 48$$

The sum of terms of an arithmetic sequence is called an **arithmetic series** or **progression**, often called AP for short.

Activity 9   Distance travelled

Use the example above of a trolley rolling down a slope to answer these questions.

(a)   Work out the distance travelled in the 20th second.

(b)   Calculate S_{20}, the distance travelled in the first 20 seconds, using the above method.

(c)   What is the distance travelled in the nth second?

(d)   Show that the trolley travels a distance of $n(n+2)$ cm in the first n seconds.

Example

Consider the arithmetic sequence 8, 12, 16, 20 ...

Find expressions

(a) for u_n, (the nth term)   (b)   for S_n.

Solution

In this AP the first term is 8 and the common difference 4.

(a) $u_1 = 8$

$u_2 = 8 + 4$

$u_3 = 8 + 2 \times 4$

$u_4 = 8 + 3 \times 4$ etc.

u_n is obtained by adding on the common difference $(n-1)$ times.

$$\Rightarrow \qquad u_n = 8 + 4(n-1)$$

$$= 4n + 4$$

(b)   To find S_n, follow the procedure explained previously:

8	12	....	$4n$	$4n+4$
$4n+4$	$4n$	....	12	8

Each pair adds up to $4n + 12$. There are n pairs.

So $2S_n = n(4n + 12)$

$$= 4n(n+3)$$

giving $S_n = 2n(n+3)$.

Exercise 14C

1. Use the 'numerical trick' to calculate
    (a) $3 + 7 + 11 + ... + 27$

    (b) $52 + 46 + 40 + ... + 4$

    (c) the sum of all the numbers on a traditional clock face;

    (d) the sum of all the odd numbers between 1 and 99.

2. Find formulas for u_n and S_n in these sequences :
    (a) 1, 4, 7, 10, ...

    (b) 12, 21, 30, 39, ...

    (c) 60, 55, 50, 45, ...

    (d) 1, $2\frac{1}{2}$, 4, $5\frac{1}{2}$, ...

3. A model railway manufacturer makes pieces of track of lengths 8 cm, 10 cm, 12 cm, etc. up to and including 38 cm. An enthusiast buys 5 pieces of each length. What total length of track can be made?

The general arithmetic sequence is often denoted by

$$a, a+d, a+2d, a+3d, \text{etc. ...}$$

To sum the series of the first n terms of the sequence,

$$S_n = a + (a+d) + (a+2d) + \ldots + (a+(n-1)d)$$

Note that the order can be reversed to give

$$S_n = (a+(n-1)d) + (a+(n-2)d) + \ldots + a$$

Adding the two expressions for S_n gives

$$2S_n = (2a+(n-1)d) + (2a+(n-1)d) + \ldots + (2a+(n-1)d)$$
$$= n[2a+(n-1)d]$$

So

$$\boxed{S_n = \tfrac{n}{2}(2a+(n-1)d)}$$

An alternative form for S_n is given in terms of its first and last term, a and l, where

$$l = a+(n-1)d$$

The nth term of the sequence is given by

$$u_n = a+(n-1)d$$

Thus

$$S_n = \tfrac{n}{2}(a+l)$$

Example

Sum the series $5 + 9 + 13 + \ldots$ to 20 terms.

Solution

This is an arithmetic sequence with first term 5 and common difference 4; so

$$S_{20} = \frac{20}{2}(2 \times 5 + 19 \times 4) = 860$$

Example

The sum of the series $1 + 8 + 15 + ...$ is 396. How many terms does the series contain?

Solution

This is an arithmetic sequence with first term 1 and common difference 7. Let the number of terms in the sequence be n.

$$S_n = 396 \Rightarrow \frac{n}{2}(2 + 7(n-1)) = 396$$

$$\Rightarrow n(7n - 5) = 792,$$

$$\Rightarrow 7n^2 - 5n - 792 = 0,$$

$$\Rightarrow (7n + 72)(n - 11) = 0,$$

$$\Rightarrow n = 11 \text{ since } -\frac{72}{7} \text{ is not a integer}$$

The number of terms is 11.

Activity 10 Ancient Babylonian problem

Ten brothers receive 100 shekels between them. Each brother receives a constant amount more than the next oldest. The seventh oldest brother receives 7 shekels. How much does each brother receive?

Exercise 14D

1. Find the sum of
   (a) $11 + 14 + 17 + ...$ to 16 terms
   (b) $27 + 22 + 17 + ...$ to 10 terms
   (c) $5 + 17 + 29 + ... + 161$
   (d) $7.2 + 7.8 + 8.4 + ...$ to 21 terms
   (e) $90 + 79 + 68 + ... -20$
   (f) $0.12 + 0.155 + 0.19 + ...$ to 150 terms

2. The last three terms of an arithmetical sequence with 18 terms is as follows : $...67, 72, 77$. Find the first term and the sum of the series.

3. How many terms are there if
   (a) $52 + 49 + 46 + ... = 385?$
   (b) $0.35 + 0.52 + 0.69 + ... = 35.72?$

4. The first term of an arithmetic series is 16 and the last is 60. The sum of the arithmetic series is 342. Find the common difference.

5. New employees joining a firm in the clerical grade receive an annual salary of £8500. Every year they stay with the firm they have a salary increase of £800, up to a maximum of £13300 p.a. How much does a new employee earn in total, up to and including the year on maximum salary?

14.4 Sigma notation

Repeatedly having to write out terms in a series is time consuming. Mathematicians have developed a form of notation which both shortens the process and is easy to use. It involves the use of the greek capital letter Σ (sigma), the equivalent of the letter S, for sum.

The series $2 + 4 + 8 + \ldots + 2^{12}$ can be shortened to $\sum\limits_{r=1}^{12} 2^r$.

This is because every term in the series is of the form 2^r, and all the values of 2^r, from $r=1$ to $r=12$ are added up. In this example the '2^r' is called the **general term**; 12 and 1 are the top and bottom **limits** of the sum.

Similarly, the series

$$60 + 60 \times (0.95) + \ldots + 60 \times (0.95)^{30}$$

can be abbreviated to

$$\sum\limits_{r=0}^{30} 60 \times (0.95)^r.$$

Often there is more than one way to use the notation. The series

$$\frac{1}{2} + \frac{2}{3} + \frac{3}{4} + \ldots + \frac{99}{100}$$

has a general term that could be thought of as either

$\frac{r}{r+1}$ or as $\frac{r-1}{r}$. Hence the series can be written as either

$$\sum\limits_{r=1}^{99} \frac{r}{r+1} \quad \text{or} \quad \sum\limits_{r=2}^{100} \frac{r-1}{r}.$$

Example

Write out what $\sum\limits_{r=11}^{9}(10-r)^2$ means and write down another way of expressing the same series, using Σ notation.

Solution

$$\sum_{r=1}^{9}(10-r)^2 = (10-1)^2 + (10-2)^2 + \ ... \ + (10-9)^2$$

$$= 9^2 + 8^2 + \ ... \ + 1^2$$

An alternative way of writing the same series is to think of it in reverse:

$$1^2 + 2^2 + \ ... \ + 8^2 + 9^2 = \sum_{r=1}^{9}r^2$$

Example

Express in Σ notation 'the sum of all multiples of 5 between 1 and 100 inclusive'.

Solution

All multiples of 5 are of the form $5r, r \in \mathbb{N}$.

$100 = 5 \times 20$, so the top limit is 20. The lowest multiple of 5 to be included is 5×1. The sum is therefore

$$5 + 10 + 15 + \ ... \ + 100 = \sum_{r=1}^{20} 5r$$

Example

Express in Σ notation 'the sum of the first n positive integers ending in 3'.

Solution

Numbers ending in 3 have the form $10r+3$, $r \in \mathbb{N}$. The first number required is 3 itself, so the bottom limit must be $r = 0$. This means that the top limit must be $n-1$. Hence the answer is

$$\sum_{r=0}^{n-1}(10r+3) \quad (= 3 + 13 + \ ... \ + (10n-7))$$

[An alternative answer is $\sum_{r=1}^{n}(10r-7)$]

Exercise 14E

1. Write out the first three and last terms of:

    (a) $\displaystyle\sum_{r=5}^{15} r^2$ (b) $\displaystyle\sum_{r=1}^{10}(2r-1)$

    (c) $\displaystyle\sum_{r=1}^{n} r$ (d) $\displaystyle\sum_{r=3}^{10}\frac{r-2}{r}$

    (e) $\displaystyle\sum_{r=6}^{100}(r-2)^2$

2. Shorten these expressions using Σ notation.

    (a) $1+\dfrac{1}{2}+\dfrac{1}{3}+...+\dfrac{1}{25}$

    (b) $10+11+12+...+50$

    (c) $1+8+27+...+n^3$

    (d) $1+3+9+27+...+3^{12}$

    (e) $6+11+16+...+(5n+1)$

    (f) $14+17+20+...+62$

    (g) $5+50+500+...+5\times10^n$.

    (h) $\dfrac{1}{6}+\dfrac{2}{12}+\dfrac{3}{20}+...+\dfrac{20}{21\times22}$

3. Use Σ notation to write:

    (a) the sum of all natural numbers with two digits;

    (b) the sum of the first 60 odd numbers;

    (c) the sum of all the square numbers from 100 to 400 inclusive;

    (d) the sum of all numbers between 1 and 100 inclusive that leave remainder 1 when divided by 7.

4. Find alternative ways, using Σ notation, of writing these:

    (a) $\displaystyle\sum_{r=1}^{19}(20-r)$ (b) $\displaystyle\sum_{r=2}^{41}\frac{1}{r-1}$ (c) $\displaystyle\sum_{r=-3}^{3} r^2$

14.5 More series

Activity 11

(a) Write down the values of $(-1)^0$, $(-1)^1$, $(-1)^2$, $(-1)^3$ etc. Generalise your answers.

(b) Write down the first three terms and the last term of

   (i) $\displaystyle\sum_{r=0}^{10}(-1)^r\frac{1}{2^r}$ (ii) $\displaystyle\sum_{r=0}^{10}(-1)^{r+1}\left(\frac{r^2}{r^2+1}\right)$

(c) How can you write the series

   $100-100\times(0.8)+100\times(0.8)^2$ ... to n terms

   using Σ notation?

Activity 12   Properties of Σ

(a) Calculate the numerical values of

   $\displaystyle\sum_{r=1}^{5} r$ $\displaystyle\sum_{r=1}^{5} r^2$ $\displaystyle\sum_{r=1}^{5} r^3$ $\displaystyle\sum_{r=1}^{5}\left(r+r^2\right)$ $\displaystyle\sum_{r=1}^{5} 3r$

(b) If $u_1, u_2, ... u_n$ and $v_1, v_2, ... v_n$ are two sequences of numbers, is it true that

$$\sum_{r=1}^{n}(u_r + v_r) = \sum_{r=1}^{n}u_r + \sum_{r=1}^{n}v_r ?$$

Justify your answer.

(c) Investigate the truth or falsehood of these statements:

(i) $\sum_{r=1}^{n}u_r v_r = \left(\sum_{r=1}^{n}u_r\right)\left(\sum_{r=1}^{n}v_r\right)$

(ii) $\sum_{r=1}^{n}u_r^2 = \left(\sum_{r=1}^{n}u_r\right)^2$

(iii) $\sum_{r=1}^{n}\alpha u_r = \alpha\left(\sum_{r=1}^{n}u_r\right)$ [α is any number.]

Again, justify your answers fully.

(d) What is the value of

$$\sum_{r=1}^{5}(r+1)?$$

What is $\sum_{r=1}^{n}1$? and $\sum_{r=1}^{n}\alpha$?

Exercise 14F

1. Work out the numerical value of

(a) $\sum_{r=1}^{10}1$

(b) $\sum_{r=1}^{25}4$

(c) $\sum_{r=0}^{16}(3+5r)$

(d) $\sum_{r=0}^{30}3\times(3.5)^r$

(e) $\sum_{r=1}^{\infty}(0.7)^r$

(f) $\sum_{r=0}^{\infty}5\times(-\tfrac{2}{3})^r$

2. Use $\sum$ notation to write these:

(a) $1-\dfrac{1}{2}+\dfrac{1}{3}-\dfrac{1}{4}+ ... -\dfrac{1}{6}$

(b) $-1+4-9+16- ... +144$

(c) $12-12\times0.2+12\times0.04- ... +12\times(0.2)^{50}$

3. If you know that

$$\sum_{r=1}^{n}u_r = 20 \text{ and } \sum_{r=1}^{n}v_r = 64$$

calculate where possible:

(a) $\sum_{r=1}^{n}(u_r + v_r)$

(b) $\sum_{r=1}^{n}u_r v_r$

(c) $\sum_{r=1}^{n}u_r^2$

(d) $\sum_{r=1}^{n}\tfrac{1}{2}v_r$

(e) $\sum_{r=1}^{n}(v_r - u_r)$

(f) $\sum_{r=1}^{n}(5u_r - v_r)$

(g) $\sum_{r=1}^{2n}u_r$

(h) $\sum_{r=1}^{n}(-1)^r v_r$

14.6 Useful formulae

You will find it useful to know three important results

- a formula for $\sum_{r=1}^{n} r$ $(1+2+3+ \dots +n)$

- a formula for $\sum_{r=1}^{n} r^2$ $\left(1^2 +2^2 +3^2 + \dots +n^2\right)$

- a formula for $\sum_{r=1}^{n} r^3$ $\left(1^3 +2^3 +3^3 + \dots +n^3\right)$

The following few activities are designed to illustrate cases where these series arise, and to work out what these formulas are.

Activity 13 Sum of first n natural numbers

Find a formula by $\sum_{r=1}^{n} r$ by

(a) treating the sum as arithmetic series:

(b) using a geometrical argument based on the diagram on the right;

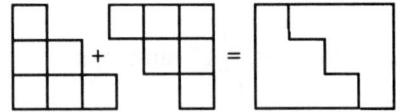

(c) starting from the statement

$$\sum_{r=1}^{n} r = \sum_{r=1}^{n} \left((n+1)-r\right).$$

Activity 14 Another chess board problem

(a) How many squares are there in each of the 'mini chessboards' on the right?

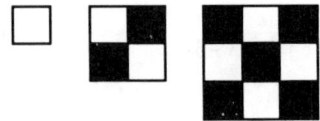

(b) How many squares are there on a chessboard? Express your answer using $\sum$ notation.

(c) What is a corresponding three dimensional problem?

Activity 15 Sum of squares

(a) Work out $\sum_{r=1}^{n} r(r+1)$ for $n = 1, 2, 3, \ldots 8$.

Copy and complete the table of results opposite.

(b) Prepare formula for $\sum_{r=1}^{n} r(r+1)$.

(c) Use your formulae for $\sum_{r=1}^{n} r(r+1)$ and $\sum_{r=1}^{n} r$ to obtain a formula

for $\sum_{r=1}^{n} r^2$.

n	$\sum_{r=1}^{n} r(r+1)$
1	2
2	8
3	20
4	...
5	...
6	...
7	168
8	...

Activity 16 Sum of cubes

(a) For $n = 1, 2, 3$ and 4, work out $\sum_{r=1}^{n} r$ and $\sum_{r=1}^{n} r^3$.

Conjecture a formula for $\sum_{r=1}^{n} r^3$.

(b) Prove this formula by starting from the statement

$$\sum_{r=1}^{n} r^3 = \sum_{r=1}^{n} \left((n+1)-r\right)^3.$$

Why could a similar approach not be used to prove the formula for $\sum r^2$?

The results of the last few activities can be summarised as follows.

$$\sum_{r=1}^{n} r = \frac{1}{2}n(n+1)$$

$$\sum_{r=1}^{n} r^2 = \frac{1}{6}n(n+1)(2n+1)$$

$$\sum_{r=1}^{n} r^3 = \frac{1}{4}n^2(n+1)^2$$

The useful fact that $\sum r^3 = (\sum r)^2$ is a coincidence (if there is such a thing in maths). It is not possible to extend this to find $\sum r^4, \sum r^5$ etc. Formulas do exist for sums of higher powers, but they are somewhat cumbersome and seldom useful.

Exercise 14G

1. Write down the general term, and hence evaluate:

   (a) $1+2+3+ \ldots +20$

   (b) $1^2 +2^2 +3^2 + \ldots +10^2$

   (c) $2+8+18+ \ldots +(2 \times 15^2)$

   (d) $2+4+6+ \ldots +100$

   (e) $1+3+5+ \ldots +25$

   (f) $1+8+27+ \ldots +1000$

2. Work out $\sum\limits_{r=10}^{20} r$. Use the fact that

   $$\sum_{r=10}^{20} r = \sum_{r=0}^{20} r - \sum_{r=0}^{9} r.$$

3. Use techniques similar to that in Question 2 to calculate

   (a) $11^2 +12^2 + \ldots +24^2$

   (b) $7^3 +8^3 + \ldots +15^3$

   (c) $21+23+ \ldots +61$

4. Calculate $21+23+25 \ldots +161$

5. Prove that $\sum\limits_{r=0}^{2n} r = n(2n+1)$

   and hence that $\sum\limits_{r=n+1}^{2n} r = \frac{1}{2}n(3n+1)$

14.7 Combining series

The results established above enable a formula to be worked out for any sum of the form

$$\sum_{r=1}^{n} \left(ar^3 + br^2 + cr + d \right)$$

The algebra required to do this neatly may be quite involved. Two worked examples are given below. Pay particular attention to the way in which the complicated expressions are simplified by the use of factors.

Example

$\sum\limits_{r=1}^{n} (r + 5)(r - 3)$

$= \sum\limits_{r=1}^{n} \left(r^2 + 2r - 15 \right)$

$= \sum\limits_{r=1}^{n} r^2 + 2\sum\limits_{r=1}^{n} r - \sum\limits_{r=1}^{n} 15$

$= \dfrac{1}{6}n(n+1)(2n+1) + n(n+1) - 15n$

$$= \frac{1}{6}n\{(n+1)(2n+1)+6(n+1)-90\}$$

$$= \frac{1}{6}n\{2n^2+3n+1+6n+6-90\}$$

$$= \frac{1}{6}n(2n^2+9n-83)$$

The quadratic $2n^2 + 9n{-}83$ cannot be factorised, so this is the simplest form of the answer.

Example

$$1^2 \times 0 + 2^2 \times 1 + 3^2 \times 2 + \ldots + n^2 \times (n-1)$$

$$= \frac{1}{12}n(n+1)(n-1)(3n+2)$$

Question 1 below will give you practice in tidying up complicated formulas. This should prepare you for Questions 2 and 4, which are similar to the worked examples.

Exercise 14H

1. Copy and complete:

   (a) $\frac{1}{6}n(n+1)(2n+1)-\frac{3}{2}n(n+1)$

   $$= \quad \frac{1}{6}n(n+1)\{ \ldots \}$$

   $$= \quad \frac{1}{3}n(n+1)\{ \ldots \}$$

   (b) $\frac{1}{4}n^2(n+1)^2 - 3n(n+1)$

   $$= \quad \frac{1}{4}n(n+1)\{ \ldots \} = \ldots$$

   (c) $\frac{3}{4}n^2(n+1)^2 - \frac{5}{6}n(n+1)(2n+1)$

   $$= \quad \frac{1}{12}n(n+1)(9n^2 - \ldots)$$

   (d) $\frac{1}{6}n(n+1)(2n+1)-\frac{5}{2}n(n+1)+4n$

   $$= \quad \frac{1}{6}n\{ \ldots \}$$

   $$= \quad \frac{1}{3}n(\ldots)(\ldots)$$

   (e) $\frac{n^2(n+1)^2}{4} - \frac{4n(n+1)(2n+1)}{3} + \frac{7n(n+1)}{2}$

   $$= \quad \frac{n(n+1)}{12}\{ \ldots \}$$

2. Prove that :

(a) $\sum_{r=1}^{n}(r+4)=\dfrac{n(n+9)}{2}$

(b) $1\times3+2\times4+3\times5+\ ...\ +n(n+2)$

$$=\frac{1}{6}n(n+1)(2n+7)$$

(c) $\sum_{r=1}^{n}(r+3)(r-7)=\dfrac{n\left(2n^2-9n-137\right)}{6}$

(d) $1^2\times2+2^2\times3+\ ...\ $ to n terms

$$=\frac{n(n+1)(n+2)(3n+1)}{12}$$

(e) $\sum_{r=1}^{n}r(r-1)(r+3)=\dfrac{n(n-1)(n+1)(3n+14)}{12}$

3. Use your answers to Question 2 to calculate

(a) $5+6+7+\ ...\ +25$

(b) $4\times(-6)+5\times(-5)+6\times(-4)+\ ...\ +20\times10$

(c) $(1\times3)+(2\times4)+(3\times5)+\ ...\ +(30\times32)$

(d) $\left(1^2\times2\right)+\left(2^2\times3\right)+\left(3^2\times4\right)+\ ...\ +\left(50^2\times51\right)$

(e) $(0\times1\times4)+(1\times2\times5)+(2\times3\times+6)\,...\,(9\times10\times13)$

4. Work out formulas for the following. Simplify, and check your answers.

(a) $\sum_{r=1}^{n}r(3r+1)$

(b) $\sum_{r=1}^{n}(r-5)(r+2)$

(c) $0\times3+1\times4+2\times5+\ ...\ +(n-1)(n+2)$

(d) $\sum_{r=1}^{n}\left(r^3-10r\right)$

(e) $(1\times2\times3)+(2\times3\times4)+(3\times4\times5)+\ ...\ $ to n terms

14.8 Sequences and limits

An amateur dramatic society uses a printing firm for its publicity material. For printing standard posters the firm charges a fixed charge of £30 plus £12 per hundred copies of the poster. The Committee wishes to know the average cost per hundred posters.

If the society has n hundred posters printed the total cost of the printing will be £$(30 + 12n)$. The average cost per hundred will be

$$a_n = \frac{30+12n}{n}$$

Clearly the average cost per poster depends on the number of posters printed. The numbers a_n represent a sequence.

What is the limit of a_n as $n\to\infty$?

Mathematically, this would be expressed by writing

$$\left(\frac{30+12n}{n}\right)=\lim_{n\to\infty}\left(\frac{30}{n}+12\right)=12$$

Example

Find $\lim\limits_{n \to \infty}\left(\dfrac{4n+7}{n-3}\right)$

Solution

The numerator can be re-written

$$4n+7 = 4(n-3)+12+7$$
$$= 4(n-3)+19$$

This enables the formulae to be rewritten

$$\frac{4n+7}{n-3} = \frac{4(n-3)+19}{n-3}$$
$$= 4+\frac{19}{n-3}$$

This form makes it clear that the required limit, as $n \to \infty$, is 4.

An equivalent of finding the answer would have been to divide both numerator and denominator by n :

$$\frac{4n+7}{n-3} = \frac{4+\dfrac{7}{n}}{1-\dfrac{3}{n}}$$

As $n \to \infty$, the numerator tends to 4 and the denominator tends to 1. Hence the original expression tends to 4.

Example

Find $\lim\limits_{n \to 0}\left(\dfrac{n^3-7n^2+12n}{n}\right)$

Solution

This function is not defined when $n = 0$, but the question is really seeking a description of its behaviour **near** zero.

$$\frac{n^3-7n^2+12n}{n} = n^2-7n+12 \text{ when } n \neq 0.$$

As $n \to 0$, $n^2-7n+12 \to 12$.

Hence the limit required is 12.

Activity 17 Another look at integration

You already know how to work out the area in the top diagram. Here is a way of verifying the answer.

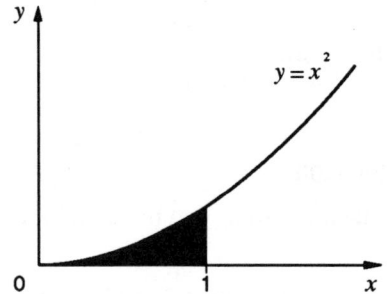

(a) Suppose the area is approximated by n vertical strips as shown.

  Show that the combined area of the strips is $\dfrac{1}{n^3} \sum_{r=1}^{n} r^2$

(b) Use this formula to express the area exclusively in terms of n. What is the limit as $n \to \infty$?

(c) Adapt this method to work out

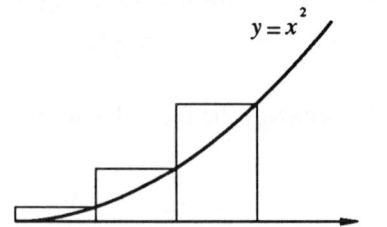

  (i) $\displaystyle\int_0^1 x^3 dx$ (ii) $\displaystyle\int_0^2 x^2 dx$ (iii) $\displaystyle\int_0^a x^2 dx$

Exercise 14I

1. Find these limits :

  (a) $\displaystyle\lim_{n \to \infty} \frac{7n+3}{2n-1}$

  (b) $\displaystyle\lim_{n \to \infty} \frac{(n-1)^2}{n^2}$

  (c) $\displaystyle\lim_{n \to \infty} \frac{1-10n}{n^2}$

  (d) $\displaystyle\lim_{n \to 0} \frac{n^2-5n}{2n}$

  (e) $\displaystyle\lim_{n \to 0} \frac{(n+1)^3-1}{n}$

  (f) $\displaystyle\lim_{n \to 2} \frac{n^2-3n+2}{n-2}$

  (g) $\displaystyle\lim_{n \to 0} \frac{(n+3)^2-(n-3)^2}{5n}$

  (h) $\displaystyle\lim_{n \to -3} \frac{n^3+3n^2-9n-27}{(n+3)^2}$

14.9 Miscellaneous Exercises

1. A piece of paper is 0.1 mm thick. Imagine it can be folded as many times as desired. After one fold, for example, the paper is 0.2 mm thick, and so on.

  (a) How thick is the folded paper after 10 folds?

  (b) How many times should the paper be folded for its thickness to be 3 metres or more? (About the height of a room.)

  (c) How many more times would it need to be folded for the thickness to reach the moon? (400 000 km away)

2. Find formulas for the nth terms of each of these sequences:

  (a) 4, 6, 9, 13.5, ...

  (b) 250, 244, 238, 232, ...

  (c) 10, 2, 0.4, 0.08, ...

  (d) 0.17, 0.49, 0.81, 1.13, ...

3. Evalate these sums:

(a) $12+15+18+21+ \ldots$ to 20 terms;

(b) $4+12+36+108+ \ldots$ to 12 terms;

(c) $5-2-9-16- \ldots -65$;

(d) $240+180+135+ \ldots$ to infinity.

4. The first term of a geometric sequence is 7 and the third term is 63. Find the second term.

5. Consider the arithmetic series $5+9+13+17+\ldots$

(a) How many terms of this series are less than 1000?

(b) What is the least value of n for which $S_n < 1000$?

6. Abbreviate the following using Σ notation:

(a) $1+3+5+7+ \ldots 23$

(b) $\dfrac{1}{3}+\dfrac{2}{6}+\dfrac{3}{11}+\dfrac{4}{18}+ \ldots +\dfrac{25}{225+2}$.

7. Evaluate

(a) $\displaystyle\sum_{r=1}^{\infty} \frac{1}{r}$

(b) $\displaystyle\sum_{r=1}^{\infty} 6\times(0.3)^r$

(c) $\displaystyle\sum_{r=1}^{100} (4-r)$

8. Prove that

(a) $\displaystyle\sum_{r=1}^{n} (r^2 +3r+3) = \dfrac{n(n^2 +6n+14)}{3}$

(b) $\displaystyle\sum_{r=1}^{n} r^2(r+1) = \dfrac{n(n+1)(n+2)(3n+1)}{12}$

9. Find formulas for

(a) $\displaystyle\sum_{r=1}^{n} r(r+6)$

(b) $\displaystyle\sum_{r=1}^{n} r(r^2 -6)$

10. Find these limits

(a) $\displaystyle\lim_{n\to\infty} \left(2+\left(\frac{1}{2}\right)^n\right)$

(b) $\displaystyle\lim_{n\to 0} \frac{1}{4(4+3n)}$;

(c) $\displaystyle\lim_{n\to\infty} \frac{6n+5}{n}$;

(d) $\displaystyle\lim_{n\to 1} \frac{n^2 -1}{n-1}$

11.

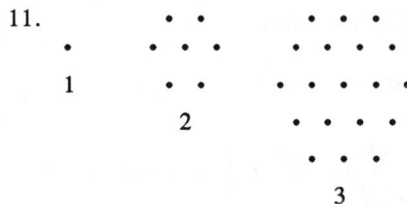

1
2
3

The diagram shows the first three hexagonal numbers

(a) The rth hexagonal number is a quadratic function of r. Find this function.

(b) Show algebraically that the sum of the first n hexagonal numbers is n^3.

(c) Can you give a non-algebraic explanation for the result in (b)? Use the diagrams below as a clue.

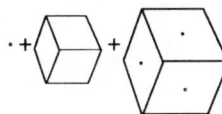

$1+9+17=27$

12. (a) You know that $\displaystyle\sum_{r=1}^{n} r = \frac{1}{2}n(n+1)$,

show that $\displaystyle\sum_{r=1}^{n} r(r+1) = \frac{n(n+1)(n+2)}{3}$.

Prepare a formula for $\displaystyle\sum_{r=1}^{n} r(r+1)(r+2)$

*(b) Prepare a formula for $\displaystyle\sum_{r=1}^{n} r(r+1)(r+2)(r+3)$

Use this to show that

$$\sum_{r=1}^{n} r^4 = \frac{n(n+1)(2n+1)(3n^2 +3n-1)}{30}$$

13. (a) Show that $\dfrac{1}{r(r+1)} = \dfrac{1}{r} - \dfrac{1}{r+1}$

(b) Write out in full $\displaystyle\sum_{r=1}^{4} \left(\frac{1}{r} - \frac{1}{r+1}\right)$

(c) Show that $\displaystyle\sum_{r=1}^{n} \frac{1}{r(r+1)} = \frac{n}{n+1}$

14. Here is another proof that

$$\sum_{r=1}^{n} r^2 = \frac{1}{6}n(n+1)(2n+1).$$

(a) Consider $\sum_{r=1}^{n}\left[(r+1)^3 - r^3\right]$ and show that it

equals $(n+1)^3 - 1$.

(b) Show that $(r+1)^3 - r^3 = 3r^2 + 3r + 1$ and use it

to prove the formula for $\sum_{r=1}^{n} r^2$.

15. (a) Show that $\sum_{r=1}^{2n} r^2 = \frac{n(2n+1)(4n+1)}{3}$ and hence

that $\sum_{r=n+1}^{2n} r^2 = \frac{n(2n+1)(7n+1)}{6}$.

(b) Use a similar method to show that

(i) $\sum_{n+1}^{2n} r^3 = \frac{n^2(3n+1)(5n+3)}{4}$.

(ii) $\sum_{n}^{3n} r = 2n(2n+1)$.

(iii) $\sum_{2n+1}^{4n} r(r+1) = \frac{4n(2n+1)(7n+1)}{3}$.

(iv) $\sum_{n}^{\infty} 2^{-r} = 2^{-(n-1)}$.

15 FURTHER CALCULUS

Objectives

After studying this chapter you should

* be able to find the second derivatives of functions;
* be able to use calculus to find maximum or minimum of functions;
* be able to differentiate composite functions or 'function of a function';
* be able to differentiate the product or quotient of two functions;
* be able to use calculus in curve sketching;
* be able to produce a power series for certain functions.

15.0 Introduction

You should have already covered the material in Chapters 8, 12 and 13 before starting this chapter. By now, you will already have met the ideas of calculus, both differentiation and integration, and you will have used the techniques developed in the earlier chapters to solve problems. This section extends the range of problems you can solve, including finding the greatest or least value of a function and differentiating complicated functions.

15.1 Rate of change of the gradient

The sketch opposite shows the graph of

$$y = x^3 + 3x^2 - 9x - 4$$

You have already seen that the derivative of this function is given by

$$\frac{dy}{dx} = 3x^2 + 6x - 9$$

This is also illustrated opposite.

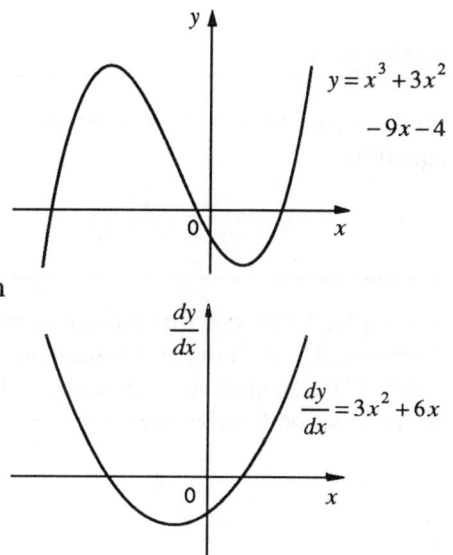

$$y = x^3 + 3x^2 - 9x - 4$$

$$\frac{dy}{dx} = 3x^2 + 6x$$

The gradient function, $\dfrac{dy}{dx}$, is also a function of x, and can be
differentiated again to give the **second** differential

$$\frac{d^2 y}{dx^2} = 6x + 6$$

Again this is illustrated opposite.

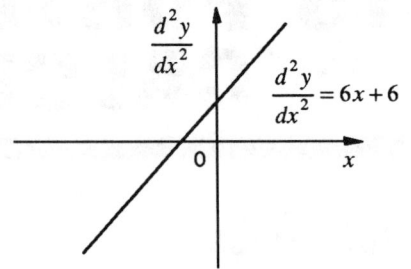

Exercise 15A

1. Find the second derivative of these functions :

    (a) $y = x^3$ (b) $y = x^4$ (c) $y = x^2$

    (d) $y = x$ (e) $y = \dfrac{1}{x}$ (f) $y = 4x^3 - 12x^2 + 5$

    (g) $y = 3x + 1$ (h) $y = e^x$ (i) $y = \ln x$

2. Find $\dfrac{d^3 y}{dx^3}$ if y is

    (a) x^4 (b) $5x^2 + \dfrac{3}{x^2}$ (c) e^x

    (d) $\ln x$ (e) $2x$

15.2 Stationary points

In this section you will consider the curve with equation

$$y = 2x^3 + 3x^2 - 12x$$

This is cubic, and a rough sketch of its graph is shown opposite. It has two stationary points (sometimes called turning points) at which the gradient is zero.

How can you find the coordinates of the stationary points?

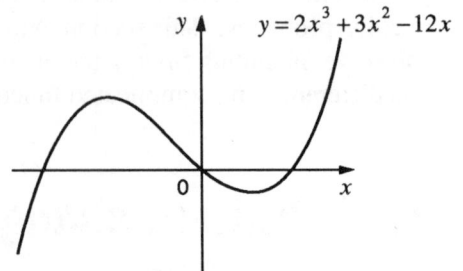

Activity 1

Find the coordinates of the stationary points for the curve with equation

$$y = 2x^3 + 3x^2 - 12x$$

In Chapter 6 you saw that the nature of the stationary points can be determined by looking at the gradient on each side of the stationary point. Here an alternative more formal method is developed, based on using second derivatives.

Activity 2

Draw an accurate sketch of the curve with equation

$$y = 2x^3 + 3x^2 - 12x$$

between $x = -3$ and $+2$. Choose the y axis to show values between -10 and $+20$.

For every x value $-3, -2.8, -2.6, -2.4, ..., 2$, note the gradient on the diagram.

Plot a graph of the gradient function and note how it behaves near the stationary point of the function.

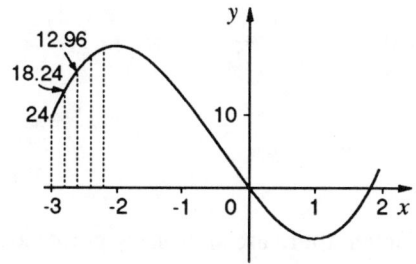

For a **maximum** value of a function, note that the gradient is decreasing in value as it passes through the value zero at the stationary point, whereas for a **minimum** value this gradient is increasing.

The result can be summarised as

For **stationary points** of a function $y(x)$

$$\frac{dy}{dx} = 0$$

If $\dfrac{d^2y}{dx^2} < 0$ at a stationary point, it corresponds to a **maximum** value of y.

If $\dfrac{d^2y}{dx^2} > 0$ at a stationary point, it corresponds to a **minimum** value of y.

Example

Find maxima and minima of the curve with equation

$$y = \frac{1}{4}x^4 + \frac{1}{3}x^3 - 6x^2 + 3$$

Hence sketch the curve.

Solution

For stationary points, $\dfrac{dy}{dx} = 0$ which gives

$$\frac{dy}{dx} = x^3 + x^2 - 12x$$

So $\qquad \dfrac{dy}{dx} = 0 \Rightarrow x^3 + x^2 - 12x = 0$

$$x(x^2 + x - 12) = 0$$

$$x(x+4)(x-3) = 0$$

Hence there are stationary points at $x = 0, -4$ and 3. To find out their nature, the second derivative is used. Now

$$\frac{d^2y}{dx^2} = 3x^2 + 2x - 12$$

At $x = 0$, $\dfrac{d^2y}{dx^2} = -12 < 0 \Rightarrow$ a maximum at $x = 0$ of value $y = 3$

At $x = -4$, $\dfrac{d^2y}{dx^2} = 3(-4)^2 + 2(-4) - 12$

$$= 28 > 0 \Rightarrow \text{minimum at } x = -4 \text{ of value}$$

$$y = \frac{1}{4}(-4)^4 + \frac{1}{3}(-4)^3 - 6(-4)^2 + 3 = -\frac{151}{3}$$

At $x = 3$, $\dfrac{d^2y}{dx^2} = 3(3)^2 + 2(3) - 12$

$$= 21 > 0 \Rightarrow \text{minimum at } x = 3 \text{ of value}$$

$$y = \frac{1}{4}(3)^4 + \frac{1}{3}(3)^3 - 6(3)^2 + 3 = -\frac{57}{3}$$

The information so far found can be sketched on a graph (not to scale). You also know that as $x \to \pm\infty$, $y \to +\infty$. It is now clear how to sketch the shape - shown dashed on the diagram.

There is one further type of stationary point to be considered and that is **point of inflection**. An example is given in the next activity.

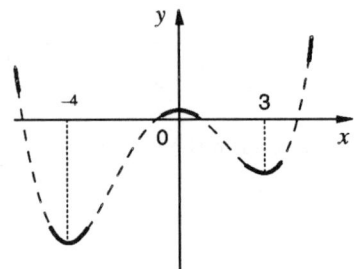

Activity 3

Find the stationary point of $y = x^3$.

What is the value of $\dfrac{d^2y}{dx^2}$ at the stationary point?

Sketch the graph of $y = x^3$.

For a **horizontal** point of inflection, not only does $\dfrac{dy}{dx} = 0$, but also

$\dfrac{d^2y}{dx^2} = 0$, and $\dfrac{d^3y}{dx^3} \neq 0$, at the point. These are **sufficient** but not

necessary conditions, as can be seen by considering $y = x^5$.

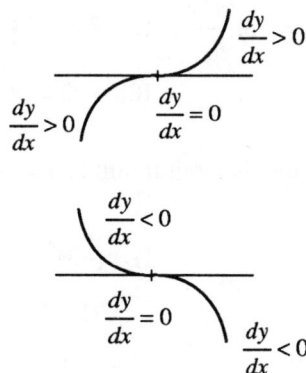

Activity 4

Find the stationary point of $y = x^5$.

What is the value of $\dfrac{d^2y}{dx^2}$ and $\dfrac{d^3y}{dx^3}$ at its stationary point?

Sketch the graph of $y = x^5$.

Example

Find the nature of the stationary points of the curve with equation

$$y = x^4 + 4x^3 - 6$$

Sketch a graph of the curve.

Solution

Now

$$\frac{dy}{dx} = 4x^3 + 12x^2$$

$$= 0 \text{ when } 4x^3 + 12x^2 = 0$$

$$\Rightarrow \quad 4x^2(x+3) = 0$$

$$\Rightarrow \quad x = 0, -3 \text{ for stationary points}$$

But

$$\frac{d^2y}{dx^2} = 12x^2 + 24x$$

At $x = 0$, $\dfrac{d^2y}{dx^2} = 0$, but $\dfrac{d^3y}{dx^3} = 24x + 24 = 24 > 0$ at $x = 0$

So there is a point of inflection at $(0, -6)$

At $x = -3$, $\dfrac{d^2y}{dx^2} = 12(-3)^2 + 24(-3)$

$$= 108 - 72 = 36 > 0$$

So there is a minimum at $x = -3$ of value

$$y(-3) = (-3)^4 + 4(-3)^3 - 6$$

$$= 81 - 4 \times 27 - 6$$

$$= -33$$

To sketch the curve, also note that

$$y \to \infty \text{ as } x \to \pm \infty$$

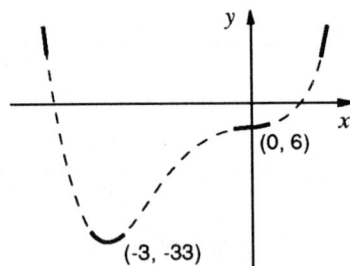

You can then deduce the form of curve, shown dashed opposite.

Finally, in this section, it should be noted that the conditions given earlier are **sufficient** conditions to guarantee max/min values, but not **necessary**.

Activity 5

Draw the graph of $y = x^4$.

Does it have any maximum or minimum values?

Do the conditions hold?

By now you are probably getting confused since there is a rule to determine the nature of stationary points, yet not all functions satisfy it. So let it be stressed that

(a)   Stationary points are always given by $\dfrac{dy}{dx} = 0$.

(b)   If, at the stationary point, $\dfrac{d^2y}{dx^2} > 0$, there is a minimum,

whereas if $\dfrac{d^2y}{dx^2} < 0$, there is a maximum.

(c) If, at the stationary point, $\dfrac{d^2y}{dx^2} = 0$, then there is a point of

inflection provided $\dfrac{d^3y}{dx^3} \neq 0$.

Conditions (b) and (c) are sufficient to guarantee the nature of the stationary point, but, as you have already seen **not** necessary.

When this analysis does not hold, that is when $\dfrac{d^2y}{dx^2} = \dfrac{d^3y}{dx^3} = 0$ at a

stationary point, it is easier to consider the **sign** of the gradient each side of the stationary point.

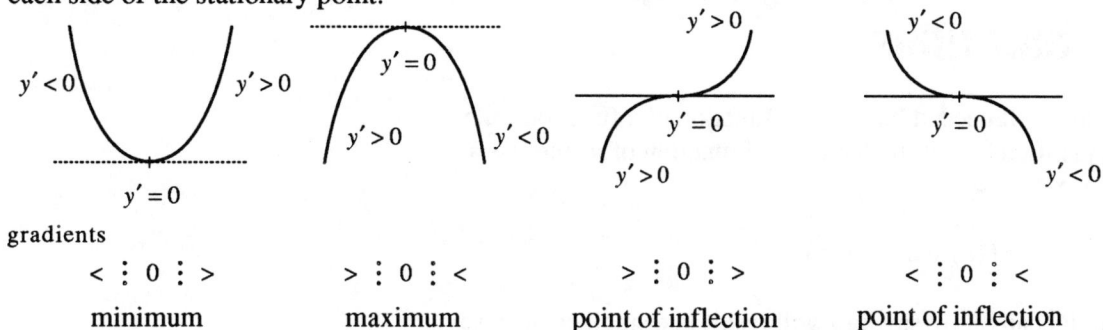

gradients

$<\ \vdots\ 0\ \vdots\ >$	$>\ \vdots\ 0\ \vdots\ <$	$>\ \vdots\ 0\ \vdots\ >$	$<\ \vdots\ 0\ \vdots\ <$
minimum	maximum	point of inflection	point of inflection

Example

Show that $y = x^4$ has a minimum at $x = 0$.

Solution

For stationary points $\qquad \dfrac{dy}{dx} = 4x^3$

$$= 0 \Rightarrow x = 0$$

But, if $x = 0$, $\qquad \dfrac{d^2y}{dx^2} = 12x^2 = 0$

and $\qquad \dfrac{d^3y}{dx^3} = 24x = 0$

So you cannot use the usual results, but at, for example,

$$x = -0.1 \Rightarrow \dfrac{dy}{dx} = 4(-0.1)^3 = -0.004 < 0$$

$$x = 0.1 \Rightarrow \dfrac{dy}{dx} = 4(0.1)^3 = 0.004 > 0$$

So there is a **minimum** at $x = 0$.

Exercise 15B

1. Find the turning points of these curves using differentiation. In each case, find out whether the points found are maxima, minima or points of inflection.

   (a) $y = 2x^2 + 3x - 1$ (b) $y = x^3 - 12x + 6$

   (c) $y = x^3 + 2x^2 - 5x - 6$ (d) $y = x^2 - 4x + 2\ln(x)$

   (e) $y = e^x - 4x$ (f) $y = e^x + 5$

2. Investigate the nature of the stationary points of the curve

$$y = (x-1)(x-a)^2$$

   when (a) $a > 1$ (b) $0 < a < 1$. In each case give a sketch of the curve.

15.3 Differentiating composite functions

The crucial result needed here is the 'function of a function' rule. It can be justified by noting that if y is a function of u, and u is a function of x

i.e. $y = f(u)$ and $u = g(x)$

then a small change, δx say, in x will result in a small change, say δu in u, which in turn results in a small change, δy in y. Now

$$\frac{\delta y}{\delta x} = \frac{\delta y}{\delta u} \times \frac{\delta u}{\delta x}$$

and if you let $\delta x \to 0$, then $\delta u \to 0$ and $\delta y \to 0$, and the equation becomes

$$\frac{dy}{dx} = \frac{dy}{du} \times \frac{du}{dx}$$

For example, suppose

$$y = (3x + 2)^2$$

then one way to differentiate this function is to multiply out the brackets and differentiate term by term;

$$y = (3x + 2)(3x + 2)$$

$$= 9x^2 + 12x + 4$$

and $\dfrac{dy}{dx} = 18x + 12$

Another method, which will be even more useful as the functions get more complicated, is to introduce a new variable u defined by

$$u = 3x + 2$$

so that

$$y = u^2$$

Now $\dfrac{dy}{du} = 2u$ and $\dfrac{du}{dx} = 3$, so using the result above

$$\frac{dy}{du} = (2u) \times 3 = 6(3x + 2), \text{ as before.}$$

Activity 6

Use both methods described above to differentiate the functions

(a) $y = (5x - 1)^2$ (b) $y = (3 - 2x)^3$

Example

If $y = e^{x^2}$, find $\dfrac{dy}{dx}$

Solution
As before introduce a new variable u defined by

$$u = x^2$$

so that

$$y = e^u$$

Now

$$\frac{dy}{du} = e^u \text{ and } \frac{du}{dx} = 2x$$

giving

$$\frac{dy}{dx} = \frac{dy}{du} \times \frac{du}{dx} = (e^u)2x = 2xe^{x^2}$$

What is the value of $\displaystyle\int xe^{x^2} dx$?

Another important result needed is given by

$$\boxed{\frac{dy}{dx} = \frac{1}{\dfrac{dx}{dy}}}$$

but note that this is only true for **first** derivatives.

For example, if $y = x^2$

then $\dfrac{dy}{dx} = 2x$

But, expressing x as a function of y,

$$x = y^{\frac{1}{2}}$$

and $\dfrac{dx}{dy} = \dfrac{1}{2}y^{-\frac{1}{2}} = \dfrac{1}{2y^{\frac{1}{2}}} = \dfrac{1}{2x}$

So, as expected $\dfrac{dy}{dx} = \dfrac{1}{\frac{dx}{dy}}$

Activity 7

Verify the result $\dfrac{dy}{dx} = \dfrac{1}{\frac{dx}{dy}}$ when

(a) $y = x^3$ (b) $y = \frac{1}{2}(x+3)$

The proof of the result is based on using small increments (increase) δx and δy, noting that

$$\dfrac{\delta y}{\delta x} = \dfrac{1}{\frac{\delta x}{\delta y}}$$

and taking the limit as $\delta x \to 0$ (and $\delta y \to 0$).

This section is concluded by using this result in finding the derivative of the function

$$y = a^x$$

To achieve this, you must first take 'logs' of both sides to give

$$\ln y = \ln(a^x) = x \ln a \text{ (properties of logs)}$$

Defining $u = (\ln a)x \Rightarrow \dfrac{du}{dx} = \ln a$

and $u = \ln y$

Hence $\dfrac{du}{dy} = \dfrac{1}{y}$

giving $\dfrac{dy}{dx} = \dfrac{dy}{du}\dfrac{du}{dx}$

$$= \left(\dfrac{1}{\dfrac{du}{dy}}\right)\ln a$$

$$= y \ln a$$

$$\dfrac{d}{dx}(a^x) = (\ln a)a^x$$

so, for example

$$\dfrac{d}{dx}(2^x) = (\ln 2)2^x$$

What happens if a = e in the formula above?

Exercise 15C

1. Differentiate these functions :

    (a) $y = (2x-5)^8$ $\quad (u = 2x-5)$

    (b) $y = (x^2 + x^3)^{10}$ $\quad (u = x^2 + x^3)$

    (c) $y = \dfrac{1}{x-2}(u = x-2, y = u^{-1})$

    (d) $y = \dfrac{1}{(3x+1)^2}$ $\quad (u = 3x+1, y = u^{-2})$

    (e) $y = \sqrt{x+1}$ $\quad (u = x+1, y = u^{\frac{1}{2}})$

    (f) $y = (e^x - x)^6$ $\quad (u = e^x - x)$

    (g) $y = \ln(3x+4)$ $\quad (u = 3x+4)$

    (h) $y = e^{\sqrt{x}}$ $\quad (u = -\sqrt{x})$

2. Find any turning points on these curves :

    (a) $y = e^{x^2}$

    (b) $y = e$ $\quad (u = -(x+4)^2)$

    (c) $y = (\ln(x))^2$ $\quad (u = \ln(x))$

    (d) $y = \left(x + \dfrac{1}{x}\right)^4$ $\quad \left(u = x + \dfrac{1}{x}\right)$

3. A kettle contains water which is cooling according to the equation $W = 80e^{-0.04t} + 20$

    where W is the temperature of the water in °C at time t (in minutes) after the kettle was switched off. Find the rate at which the water is cooling in °C/min when $t = 30$ and when $t = 60$.

4. The population of a country is modelled by the function

    $$P = 12 \times (1.03)^t$$

    where P is the population in millions and t is the time in years after the start of 1990. Find the rate at which the population is increasing in millions per year at the start of the year 2000.

5. (a) Differentiate e^{2x}. Hence write down $\int e^{2x} dx$.

    (b) Differentiate e^{-x}, and so write down $\int e^{-x} dx$.

6. Differentiate e^{x^2}, and use your result to find

    $$\int x e^{x^2} dx.$$

7. The derivative of $\ln x$ is $\dfrac{1}{x}$.

    By differentiating a suitable logarithmic function, find $\int \dfrac{1}{x+2} dx$.

15.4 Integration again

The results that have been developed in the last section are, as you will see, very useful in integration. For example, if

$$y = (x+5)^4$$

then $\dfrac{dy}{dx} = 4(x+5)^3$ (using the methods in the last section).

Hence $\displaystyle\int 4(x+5)^3 dx = (x+5)^4 + C$ (C is arbitrary constant)

or $\displaystyle\int (x+5)^{dx} = \frac{1}{4}(x+5)^4 + C'$ $(C' = \frac{1}{4}C)$

Activity 8

Differentiate $y = (ax+b)^{n+1}$ and hence deduce the value of

$$\int (ax+b)^n\, dx$$

Use your results to find

(a) $\displaystyle\int (3x-1)^9\, dx$ (b) $\displaystyle\int \frac{1}{(2x+4)^2}\, dx$ (c) $\displaystyle\int \sqrt{x-1}\, dx$

Similarly, it can be seen that if

$$y = \ln(ax+b)$$

then $\dfrac{dy}{dx} = \dfrac{a}{(ax+b)}$

Hence $\displaystyle\int \frac{a\,dx}{(ax+b)} = \ln(ax+b) + C$

or $\boxed{\displaystyle\int \frac{dx}{(ax+b)} = \frac{1}{a}\ln(ax+b) + C'}$ $C' = \dfrac{1}{a}C$

Activity 9

Use the result above to evaluate

(a) $\displaystyle\int \frac{dx}{(x+1)}$ (b) $\displaystyle\int \frac{dx}{4-x}$ (c) $\displaystyle\int \frac{dx}{1-2x}$

Another important method of integration is based on the result that
if

$$y = \ln f(x)$$

then defining $u = f(x)$, $y = \ln u$ and

$$\frac{dy}{dx} = \frac{dy}{du} \times \frac{du}{dx}$$

$$= \frac{1}{u} \times f'(x)$$

$$= \frac{f'(x)}{f(x)}$$

Hence

$$\int \frac{f'(x)}{f(x)} \, dx = \ln f(x) + C$$

Example

Find $\qquad \int \frac{(3x^2 + x)}{(2x^3 + x^2)} \, dx$

Solution

If $f(x) = 2x^3 + x^3$, then $f'(x) = 6x^2 + 2x$, so the integral can be
written as

$$I = \int \frac{(3x^2 + x)}{(2x^3 + x^2)} \, dx$$

$$= \frac{1}{2} \int \frac{2(3x^2 + x)}{(2x^3 + x^2)} \, dx$$

$$= \frac{1}{2} \int \frac{f'(x)}{f(x)} \, dx \qquad (\text{when } f(x) = 2x^3 + x^2)$$

$$= \frac{1}{2} \int \ln f(x) + C$$

$$= \frac{1}{2} \ln(6x^2 + 2x) + C$$

Exercise 15D

1. Find

(a) $\int (2x+1)^4 \, dx$ (b) $\int \dfrac{1}{(x-5)^2} \, dx$

(c) $\int \dfrac{1}{\sqrt{x+1}} \, dx$ (d) $\int \sqrt{4x-1} \, dx$

2. Evaluate

(a) $\int_0^1 \dfrac{dx}{(x+1)}$ (b) $\int_1^2 \dfrac{dx}{(2x-1)}$

3. Find

(a) $\int \dfrac{dx}{(3x+2)}$ (b) $\int \dfrac{x}{x^2+1} \, dx$

(c) $\int \dfrac{e^x}{1+e^x} \, dx$

4. Evaluate $\int_0^1 xe^{-x^2} \, dx$

15.5 Differentiating products and quotients

These are very useful formula for differentiating both products and quotients. For example, if y is defined as the product of the functions, u and v of x, then

$$y(x) = u(x)\, v(x)$$

Using the basic definition of a derivative, let δx be a small change in x, then consider

$$y(x+\delta x) - y(x) = u(x+\delta x)\, v(x+\delta x) - u(x)\, v(x)$$

$$= (u(x+\delta x) - u(x))\, v(x+\delta x)$$

$$+ u(x)(v(x+\delta x) - v(x))$$

(the middle two terms cancel)

Diving both sides by δx gives

$$\frac{y(x+\delta x) - y(x)}{\delta x} = \frac{(u(x+\delta x) - u(x))}{\delta x}\, v(x+\delta x) + u(x)\frac{(v(x+\delta x) - v(x))}{\delta x}$$

and taking the limit as $\delta x \to 0$, gives

$$\boxed{\frac{dy}{dx} = \frac{du}{dx}\, v + u\,\frac{dv}{dx}}$$

(since $v(x+\delta x) \to v(x)$ as $\delta x \to 0$)

Example

If $y = x^2(x-1)$ find $\dfrac{dy}{dx}$ by

(a) using the formula above;

(b) multiplying out and differentiating term by term.

Solution

(a) Here $u = x^2, v = (x-1)^2$, so

$$\frac{dy}{dx} = 2x(x-1)^2 + x^2\, 2(x-1)$$

$$= 2x(x-1)(x-1+x)$$

$$= 2x(x-1)(2x-1)$$

(b) $y = x^2(x^2 - 2x + 1)$

$$= x^4 - 2x^3 + x^2$$

$$\frac{dy}{dx} = 4x^3 - 6x^2 + 2x$$

$$= 2x(2x^2 - 3x + 1)$$

$$= 2x(x-1)(2x-1) \qquad \text{(as before)}$$

Example

If $y = (x+1)e^{-x}$, find $\dfrac{dy}{dx}$.

Solution

This time, you must use the formula, with $u = x+1, v = e^{-x}$

So $\quad \dfrac{dy}{dx} = 1 \times (e^{-x}) + (x+1)(-e^{-x})$

$$= -xe^{-x}$$

Turning to the equivalent formula for a quotient, with

$$y = \frac{u}{v} = u\left(\frac{1}{v}\right)$$

which is the product of u and $\dfrac{1}{v}$.

Now

$$\frac{d}{dx}\left(\frac{1}{v}\right) = \frac{d}{dv}\left(\frac{1}{v}\right)\frac{dv}{dx} \quad \text{('function of a function')}$$

$$= -\frac{1}{v^2}\frac{dv}{dx}$$

and

$$\frac{dy}{dx} = \frac{du}{dx} \times \frac{1}{v} + u\frac{d}{dx}\left(\frac{1}{v}\right)$$

$$= \frac{du}{dx} \times \frac{1}{v} + u\left(-\frac{1}{v^2}\frac{dv}{dx}\right)$$

i.e.

$$\frac{dy}{dx} = \frac{\left(v\dfrac{du}{dx} - u\dfrac{dv}{dx}\right)}{x^2}$$

Example

Use the quotient formula to find $\dfrac{d}{dx}((1+x)e^{-x})$

Solution

Here $u = (1 + x)$, $v = e^x$, so that

$$y = \frac{u}{v} = \frac{(1+x)}{(e^x)} = (1+x)e^{-x}$$

Thus

$$\frac{dy}{dx} = \frac{e^x \times 1 - (1+x)e^x}{(e^x)^2}, \text{ since } \frac{d}{dx}(e^x) = e^x$$

$$= -\frac{xe^x}{(e^x)^2}$$

$$= -\frac{x}{e^x}$$

$$= -xe^{-x} \text{ (as before)}$$

So you can see that the quotient formula is just another form of the product formula. It is though sometimes very convenient to use.

Example

Differentiate $y = \dfrac{x-1}{2x-3}$ with respect to x.

Solution

Here $u = x - 1$, $v = 2x - 3$ and using the quotient formula

$$\frac{dy}{dx} = \frac{1 \times (2x - 3) - 2(x - 1)}{(2x - 3)^2}$$

$$= \frac{2x - 3 - 2x + 2}{(2x - 3)^2}$$

$$= \frac{-1}{(2x - 3)^2}$$

Activity 10

Develop a formula for differentiating the product of **three** functions i.e.

$$y(x) = u(x)\, v(x)\, w(x)$$

Example

Find any stationary points for the curve with equation

$$y = \frac{x^3}{(1 + x)}$$

Sketch the curve.

Solution

Here $u = 3$, $v = 1 + x$, so $y = \dfrac{u}{v}$ and using the quotient formula

$$\frac{dy}{dx} = \frac{3x^2(1 + x) - x^3 \times 1}{(1 + x)^2}$$

$$= \frac{3x^2 + 2x^2}{(1 + x)^2}$$

For stationary points,

$$3x^2 + 2x^3 = 0$$

$$x^2(3 + 2x) = 0$$

$$\Rightarrow \quad x = 0, -\tfrac{3}{2}$$

To determine the nature of these stationary points, you can check the sign of $\dfrac{dy}{dx}$ each side of the points.

So, for $x = 0$

$$\frac{dy}{dx}(-0.1) = \frac{3(-0.1)^2 + 2(-0.1)^3}{(1 - 0.1)^2} = \frac{0.03 - 0.002}{(0.9)^2}$$

$$= \frac{0.028}{0.81} > 0$$

and

$$\frac{dy}{dx}(0.1) = \frac{3(0.1)^2 + 2(0.1)^3}{(1 + 0.1)^2} = \frac{0.03 + 0.002}{(1.1)^2}$$

$$= \frac{0.032}{1.21} > 0$$

So there is a point of inflection at $x = 0$ and value of function here is $y = 0$.

For $x = -1.5$,

$$\frac{dy}{dx}(-1.6) = \frac{3(-1.6)^2 + 2(-1.6)^3}{(1 - 1.6)^2} = \frac{-0.512}{0.36} < 0$$

$$\frac{dy}{dx}(-1.4) = \frac{3(-1.4)^2 + 2(-1.4)^3}{(1 - 1.4)^2} = \frac{0.392}{0.16} > 0$$

Hence there is a minimum at $x = -1.5$ of value $y = \dfrac{27}{4}$.

Also note that the function has an **asymptote** at $x = -1$, that is

$$y \to \pm\infty \text{ as } x \to -1.$$

As $x \to \pm\infty$, $y = \dfrac{x^3}{1 + x} \approx \dfrac{x^3}{x} = x^2$

All this information is marked on the diagram opposite, and the curve is sketched in dotted.

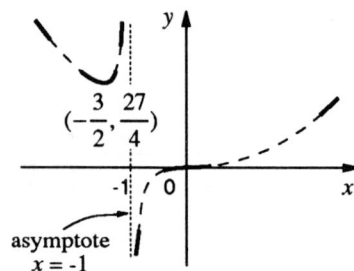

Exercise 15E

1. Differentiate the following functions with respect to x :

   (a) $y = x e^x$

   (b) $y = x^2 \ln(x)$

   (c) $y = \dfrac{2x+1}{x^2-3}$

   (d) $y = \dfrac{e^x}{1+x}$

2. Find any stationary points on the curves in Question 1.

3. Find any stationary points on the curve

   $$y = \frac{\ln(\sqrt{x+1})}{e^x}$$

4. Differentiate $\sqrt{\dfrac{x+1}{2x-1}}$

5. Differentiate $\sqrt{e^x(1+x)}$

6. Differentiate $\dfrac{\ln(x-2)}{\sqrt{x}}$

7. Sketch the curve with equation

   $$y = x^3 + \frac{3}{2}x^2 - 6x$$

   by first finding all stationary points and their nature.

8. Sketch the curve with equation

   $$y = \frac{1}{x^2} e^{-\frac{1}{x^2}}$$

9. Find any stationary points of the curve with equation

   $$y = \frac{x^2}{1-x}$$

   Hence sketch the curve, indicating any asymptotes.

15.6 Power series

This section deals with a method for writing functions like

e^x, $\ln(1+x)$ or $\sin(x)$ in the form

$$a_0 + a_1 x + a_2 x^2 + a_3 x^2 + \ldots$$

where $a_0, a_1, a_2, \ldots$, are all constants. This is called a power series and they are extremely useful, as they can be used by calculators or computers to calculate the values of these functions for specific values of x to any degree of accuracy.

Suppose that e^x can be written in the form

$$e^x \equiv a_0 + a_1 x + a_2 x^2 + a_3 x^3 + \ldots$$

Then, when $x = 0$,

$$e^0 = 1 = a_0,$$

so $a_0 = 1$.

Now differentiate the series for e^x, giving

$$e^x = a_1 + 2a_2 x + 3a_3 x^2 + 4a_4 x^3 + \ldots$$

and again put $x = 0$ to give

$$e^0 = 1 = a_1$$

So $a_1 = 1$, and differentiating again

$$e^x = 2a_2 + 3.2a_3x + 4.3a_4x^2 + \ldots$$

With $x = 0$,

$$e^0 = 1 = 2a_2 \Rightarrow a_2 = \frac{1}{2}$$

What will the values of a_3, a_4 be?

Continuing in this way,

$$a_n = \frac{1}{n!}$$

so that

$$e^x = 1 + x + \frac{x^2}{2!} + \frac{x^3}{3!} + \ldots$$

Activity 11

(a) Assume that

$$\ln(1+x) = a_0 + a_1x + a_2x^2 + \ldots \qquad (x > -1)$$

and use the method described above to find the value of the constants $a_0, a_1, a_2, \ldots$

(b) Use the approximation

$$\ln(1+x) = a_0 + a_1x + a_2x^2$$

to find (i) ln (1.001) (ii) ln (1.01) (iii) ln (1.1) (iv) ln 2

Compare your values with those found on a calculator.

As you can see from this activity, the accuracy of the approximation depends on the value of x used. So to evaluate ln 2 correct to say 4 decimal places requires further terms in the series used.

Example

How many terms are required in the series for e^x in order to estimate to 4 d.p. accuracy the value of e^2?

Solution

Using your calculator $e^{1.5} = 4.4817$ to 4 d.p.

Using the series $e^x = 1 + \dfrac{x}{1!} + \dfrac{x^2}{2!} + \dfrac{x^3}{3!} + \ldots$ gives

no.	series	$x = 1.5$
1	1	1
2	$1+x$	2.5
3	$1+x+\frac{x^2}{2!}$	3.625
4	...	4.1875
5	...	4.39844
6	...	4.46172
7	...	4.47754
8	...	4.48093
9	...	4.48157
10	...	4.48167

So ten terms are required to obtain the value correct to 4 d.p.'s.

Exercise 15F

1. Find the first five terms in the series for e^{-1}. Hence estimate the value of $e^{-0.5}$, and check the accuracy of your calculation.

2. Determine the first three non-zero terms in the power series for the function e^{x^2}.

3. By writing
$$\ln(1+x^2) = b_0 + b_1 x^2 + b_3 x^4 + \ldots$$
find the values of b_0, b_1 and b_2.

15.7 Miscellaneous Exercises

1. Find the maxima and/or minima of these functions :

   (a) $y = x^2 - 5x + 6$ (b) $y = 3x - 2x^2 + 8$

   (c) $y = x^3 + 2x^2 + x - 4$ (d) $y = 6x^2 - 2x^3 + 48x$

   (e) $y = \dfrac{e^x}{x}$

2. Differentiate these functions with respect to x

   (a) $(5 - 3x)^8$ (b) $\dfrac{1}{\left(\sqrt{x} + 1\right)}$ (c) 3^x

   (d) $\sqrt{\dfrac{1 + x}{1 - x}}$ (e) $\ln(2^x)$ (f) $e^{-x^{-2}}$

3. Differentiate e^{3x-1}. Hence write down $\displaystyle\int e^{3x-1}\,dx$

4. Find $\displaystyle\int (2x - 4)^5\,dx$.

5. Calculate $\displaystyle\int_3^6 \frac{1}{3x - 1}\,dx$.

6. Differentiate $\ln(x^2 + 1)$. Hence find $\displaystyle\int_0^1 \frac{x}{x^2 + 1}\,dx$.

16 FURTHER TRIGONOMETRY

Objectives

After studying this chapter you should

*   know all six trigonometric functions and their relationships to each other;

*   be able to solve trigonometric identities;

*   be able to solve trigonometric equations by a variety of methods.

You will need to work both in degrees and radians, and to have a working familiarity with the sine, cosine and tangent functions, their symmetries and periodic properties, and their inverse functions. In addition, you will need freely available access to graph plotting facilities.

16.0 Introduction

Until now you have worked with sines, cosines and tangents, initially as relationships between sides and angles in right angled triangles, then as functions/mappings in their own right. The clumsy notation used, for example, for $(\sin x)^2$ has been simplified to $\sin^2 x$, etc. while on the other hand $\sin^{-1} x$ has **not** been used to indicate $(\sin x)^{-1}$ i.e. $\dfrac{1}{\sin x}$ but rather as the inverse function which maps sines back into angles. (Don't forget that your calculator will yield the principle value only : $\sin^{-1} 0.5 = 30°$ according to this convention, but $\sin 150°$, for instance, also has the value 0.5.)

Three further trigonometric functions are defined as follows :

COSECANT of an angle, where $\operatorname{cosec} x = \dfrac{1}{\sin x}$,

SECANT of an angle, where $\sec x = \dfrac{1}{\cos x}$, and

COTANGENT of an angle, where $\cot x = \dfrac{1}{\tan x}$.

Hence , in terms of the ratios in a right-angled triangle,

$$\operatorname{cosec}\theta = \frac{\text{hyp}}{\text{opp}}, \quad \sec\theta = \frac{\text{hyp}}{\text{adj}}, \quad \cot\theta = \frac{\text{adj}}{\text{opp}}$$

and, for example,

$$\frac{1}{\sin^2 x} = \left(\frac{1}{\sin x}\right)^2 = \cos \mathrm{ec}^2 x, \text{ etc.}$$

16.1 Historical notes

1. The first three trig functions used were, in fact, the sine, secant and tangent.

    In a right angled triangle, the angle $\phi \ (= 90 - \theta)$ is the **complementary** angle to θ.

    Now $\cos\theta = \dfrac{\text{adj}}{\text{hyp}} = \sin\phi \ \left[\text{i.e. } \sin(90 - \theta)\right]$

    and cosine means 'the sine of the complementary angle'.

2. A circle with radius 1 is drawn and a point P taken outside this circle, then a right angled triangle can be created as shown opposite.

    The line PT is tangent to the circle, from the Latin 'tangens' for touching. The line OP is called the secant, from the Latin 'secans' for cutting.

    In trigonometric terms $PT = \tan\theta$ and $OP = \sec\theta$.

3. The derivation of the word sine is more obscure, but relates to the length of a half-chord of a circle of radius 1. Using the fact that the line through O perpendicular to the chord PQ bisects it, XP= half chord $= \sin\theta$

Some consequences that you have already met arised immediately from the basic definitions of the six trigonometric functions. They are :

(a) (i) $\dfrac{\sin\theta}{\cos\theta} = \dfrac{(\text{opp}/\text{hyp})}{(\text{adj}/\text{hyp})} = \dfrac{\text{opp}}{\text{adj}} = \tan\theta$

(ii) $\cot\theta = \dfrac{1}{\tan\theta} = \dfrac{\cos\theta}{\sin\theta}$

(b) By Pythagoras' Theorem, $(\text{opp})^2 + (\text{adj})^2 = (\text{hyp})^2$

(i) Dividing by $(\text{hyp})^2$ gives

$$\left(\frac{\text{opp}}{\text{hyp}}\right)^2 + \left(\frac{\text{adj}}{\text{hyp}}\right)^2 = 1^2$$

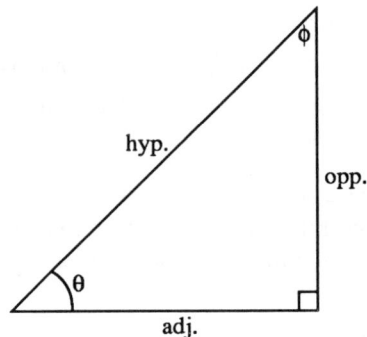

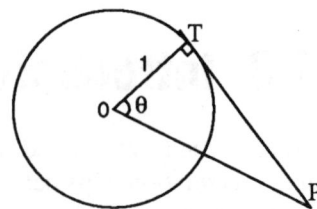

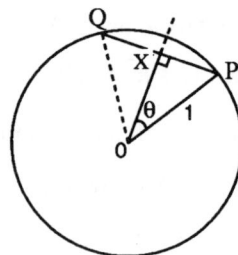

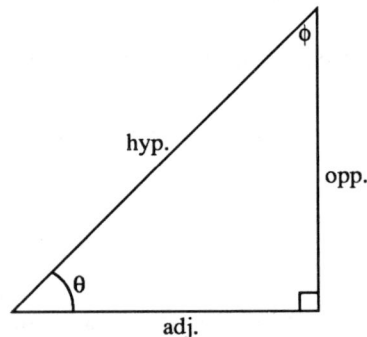

or $\boxed{\sin^2\theta + \cos^2\theta = 1}$

(ii)   Dividing by $(\text{adj})^2$ gives

$$\left(\frac{\text{opp}}{\text{adj}}\right)^2 + 1^2 = \left(\frac{\text{hyp}}{\text{adj}}\right)$$

or $\boxed{\tan^2\theta + 1 = \sec^2\theta}$

(iii)   Dividing by opp^2 gives

$$1^2 + \left(\frac{\text{adj}}{\text{opp}}\right)^2 = \left(\frac{\text{hyp}}{\text{opp}}\right)^2$$

or $\boxed{1 + \cot^2\theta = \cos ec^2\theta}$

These three results are sometimes refered to as the **Pythagorean identities**, and are true for **all** angles θ.

Activity 1

Use a graph-plotter to draw the graphs of cosec x, sec x and cot x for values of x in the range $-2\pi \le x \le 2\pi$ (remember the graph plotter will work in radians). Write down in each case the period of the functions and any symmetries of the graphs.

From the definitions of cosec, sec and cot, how could you have obtained these graphs for yourself?

16.2   Identities

An identity is an equation which is true for all values of the variable. It is sometimes disinguished by the symbol $\equiv$, rather than $=$.

Example

Establish the identity $\sin A \tan A = \sec A - \cos A$

Solution

In proving results such as this sometimes it is it helpful to follow this procedure: start with the left hand side (LHS), perform whatever manipulations are necessary, and work through a step at a time until the form of the right hand side (RHS) is obtained. In this case,

$$\text{LHS} = \sin A \ \tan A$$

$$= \sin A \ \frac{\sin A}{\cos A}$$

$$= \frac{\sin^2 A}{\cos A}$$

$$= \frac{1 - \cos^2 A}{\cos A} \quad \text{using } \sin^2 A + \cos^2 A = 1$$

$$= \frac{1}{\cos A} - \frac{\cos^2 A}{\cos A}$$

$$= \sec A - \cos A$$

$$= \text{RHS}$$

Exercise 16A

Using the basic definitions and relationships between the six trigonometric functions, prove the following identities:

1. $\sec A + \tan A = \dfrac{1 + \sin A}{\cos A}$

2. $\tan A + \cot A = \sec A \ \text{cosec} A$

3. $\sec^2 \theta + \text{cosec}^2 \theta = \sec^2 \theta \ \text{cosec}^2 \theta$

4. $\dfrac{\text{cosec}\theta - \cot \theta}{1 - \cos \theta} = \text{cosec}\theta$

4. $\dfrac{\text{cosec}\theta - \cot \theta}{1 - \cos \theta} = \text{cosec}\theta$

5. $\text{cosec } x - \sin x = \cos x \cot x$

6. $1 + \cos^4 x - \sin^4 x = 2\cos^2 x$

7. $\sec \theta + \tan \theta = \dfrac{\cos \theta}{1 - \sin \theta}$

8. $\dfrac{\sin A \tan A}{1 - \cos A} = 1 + \sec A$

16.3 The addition formulae

A proof of the formula for $\sin (A+B)$ will be given here.

Consider the diagram opposite which illustrates the geometry of the situation.

In triangle PQU,

$$\sin(A + B) = \frac{QU}{PQ}$$

$$= \frac{QS + SU}{PQ}$$

$$= \frac{QS}{PQ} + \frac{RV}{PQ} \quad (\text{since } SU = RV)$$

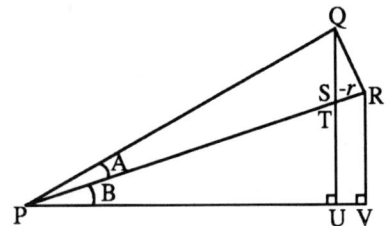

Now, notice that, since $PTU = 90 - B$, $STR = 90 - B$ also, and so

$S\hat{Q}R = B$

Then in triangle QRS

$$QS = QR \cos B$$

Also, in triangle QRS.

$$RV = PR \sin B$$

You now have,

$$\sin(A+B) = \frac{QR}{PG}\cos B + \frac{PR}{PQ}\sin B$$

But in triangle PQR, $\dfrac{QR}{PG} = \sin A$ and $\dfrac{PR}{PQ} = \cos A$ so,

$$\boxed{\sin(A+B) = \sin A \cos B + \cos A \sin B}$$

Rather than reproducing similar proofs for three more formulae, the following approach assumes this formula for $\sin(A+B)$ and uses prior knowledge of the sine and cosine functions.

Given $\sin(A+B) = \sin A \cos B + \cos A \sin B$, replacing B by $-B$ throughout gives

$$\sin(A-B) = \sin A \cos(-B) + \cos A \sin(-B) \ .$$

Now $\cos(-B) = +\cos B$ and $\sin(-B) = -\sin B$, so

$$\boxed{\sin(A-B) = \sin A \cos B - \cos A \sin B}$$

Next use the fact that $\cos\theta = \sin(90-\theta)$, so that

$$\begin{aligned}
\cos(A+B) &= \sin\big(90-(A+B)\big) \\
&= \sin\big((90-A)-B\big) \\
&= \sin(90-A)\cos B - \cos(90-A)\sin B
\end{aligned}$$

from the second result just obtained. And since

$$\cos A = \sin(90\text{-}A), \quad \sin A = \cos(90\text{-}A),$$

it follows that

$$\boxed{\cos(A+B) = \cos A \cos B - \sin A \sin B}$$

Replacing B by $-B$, as before, this gives the fourth result

$$\boxed{\cos(A-B) = \cos A \cos B + \sin A \sin B}$$

Example

Show that $\quad \sin 15° = \dfrac{\sqrt{2}}{4}\left(\sqrt{3}-1\right).$

$$\sin 15° = \sin(60°-45°)$$
$$= \sin 60° \cos 45° - \cos 60° \sin 45°$$

Now $\sin 60° = \dfrac{\sqrt{3}}{2}$, $\cos 60° = \dfrac{1}{2}$ and $\cos 45° = \dfrac{\sqrt{2}}{2}$ or $\dfrac{1}{\sqrt{2}}$

giving

$$\sin 15° = \frac{\sqrt{3}}{2} \times \frac{\sqrt{2}}{2} - \frac{1}{2} \times \frac{\sqrt{2}}{2} = \frac{\sqrt{2}}{4}\left(\sqrt{3}-1\right)$$

Exercise 16B

1. Using the values $\cos 0° = 1$, $\cos 30° = \dfrac{\sqrt{3}}{2}$, $\cos 45° = \dfrac{1}{\sqrt{2}}$, $\cos 60° = \dfrac{1}{2}$ and $\cos 90° = 0$ and related values of sine and tangent, determine, in similar form,

   (a) $\sin 75°$ (b) $\cos 15°$ (c) $\cos 105°$

   (d) $\cos 75°$ (e) $\tan 75°$ (f) $\tan 105°$

   (g) $\sin 255°$ (h) $\cos 285°$ (i) $\cot 75°$

2. Show that
   $$\cos(45°-A) - \cos(45°+A) = \sqrt{2} \sin A$$

3. Show that
   $$\sin x + \sin\left(x + \frac{2\pi}{3}\right) + \sin\left(x + \tfrac{4}{3}\pi\right) = 0$$

4. Given that $\sin(A+B) = 3\sin(A-B)$, show that $\tan A = 2 \tan B$

5. Show that
   $$\sin(x+y)\sin(x-y) = \sin^2 x - \sin^2 y.$$

16.4 Further identities

Now

$$\tan(A+B) = \frac{\sin(A+B)}{\cos(A+B)}$$
$$= \frac{\sin A \cos B + \cos A \sin B}{\cos A \cos B - \sin A \sin B}$$
$$= \frac{\left(\dfrac{\sin A \cos B}{\cos A \cos B} + \dfrac{\cos A \sin B}{\cos A \cos B}\right)}{\left(\dfrac{\cos A \cos B}{\cos A \cos B} - \dfrac{\sin A \sin B}{\cos A \cos B}\right)}$$

Here, you can divide every term of the fraction by cos A cos B, giving

$$\tan(A+B) = \frac{\tan A + \tan B}{1 - \tan A \, \tan B}$$

Replacing B by $-B$ gives

$$\tan(A-B) = \frac{\tan A - \tan B}{1 + \tan A \, \tan B}$$

Multiple Angles

If, in the formula for sin $(A+B)$, you put $B = A$, then you get

$$\sin(A + A) = \sin A \cos A + \cos A \sin A$$

or

$$\sin 2A = 2 \sin A \cos A$$

Activity 2

Use the addition formulae to find expressions for

(a)   cos $2A$ in terms of cos A and sin A;

(b)   cos $2A$ in terms of cos A only;

(c)   cos $2A$ in terms of sin A only;

(d)   tan $2A$ in terms of tan A only;

(e)   sin $3A$ in terms of powers of sin A only;

(f)   cos $3A$ in terms of powers of cos A only.

Example

Establish the identity $\dfrac{\cos 2A + \sin 2A - 1}{\cos 2A - \sin 2A + 1} \equiv \tan A$

Solution

Here, you should first look to simplify the numerator and denominator by using the identities for cos $2A$ and sin $2A$.

In the numerator, re-writing cos $2A$ as $1-2\sin^2 A$ will help cancel the -1 on the end; while cos $2A = 2\cos^2 A - 1$ will be useful in the denominator.

Thus, $\quad \text{LHS} = \dfrac{1 - 2\sin^2 A + 2\sin A \cos A - 1}{2\cos^2 A - 1 - 2\sin A \cos A + 1}$

$\qquad\qquad = \dfrac{2\sin A \cos A - 2\sin^2 A}{2\cos^2 A - 2\sin A \cos A}$

$\qquad\qquad = \dfrac{2\sin A(\cos A - \sin A)}{2\cos A(\cos A - \sin A)}$

$\qquad\qquad = \dfrac{\sin A}{\cos A}$

$\qquad\qquad = \tan A$

$\qquad\qquad = \text{RHS}$

Example

Show that $\sec^2\theta + \operatorname{cosec}^2\theta = 4\operatorname{cosec}^2 2\theta$.

Solution

To begin with, for shorthand write $s = \sin\theta$ and $c = \cos\theta$.

Then

$\qquad \text{LHS} \quad = \dfrac{1}{c^2} + \dfrac{1}{s^2}$

$\qquad\qquad\quad = \dfrac{s^2 + c^2}{s^2 c^2}$

$\qquad\qquad\quad = \dfrac{1}{s^2 c^2}$

Now notice that $2sc = \sin 2\theta$,

so $4s^2 c^2 = \sin^2 2\theta$ and $s^2 c^2 = \frac{1}{4}\sin^2 2\theta$, and

$\qquad \text{LHS} = \dfrac{1}{\frac{1}{4}\sin^2 2\theta}$

$\qquad\qquad = 4\operatorname{cosec}^2 2\theta = \text{RHS}$

Example

Prove the identity $\dfrac{\cos A + \sin A}{\cos A - \sin A} = \sec 2A + \tan 2A$

Solution

This problem is less straightforward and requires some ingenuity.
It helps to note that the

$\qquad \text{RHS} = \dfrac{1}{\cos 2A} + \dfrac{\sin 2A}{\cos 2A}$

with a common denominator of $\cos 2A$. One formula for $\cos 2A$ is

$$\cos 2A = \cos^2 A - \sin^2 A$$

$$= (\cos A - \sin A)(\cos A + \sin A)$$

by the difference of two squares.

Hence

$$\text{LHS} = \frac{\cos A + \sin A}{\cos A - \sin A}$$

$$= \frac{(\cos A + \sin A)}{(\cos A - \sin A)} \times \frac{(\cos A + \sin A)}{(\cos A + \sin A)}$$

This is done to get the required form in the denominator.

$$\text{LHS} = \frac{\cos^2 A + 2\sin A \cos A + \sin^2 A}{\cos^2 A - \sin^2 A.}$$

$$= \frac{\left(\cos^2 A + \sin^2 A\right) + 2\sin A \cos A}{\cos 2A}$$

$$= \frac{1 + \sin 2A}{\cos 2A}$$

$$= \frac{1}{\cos 2A} + \frac{\sin 2A}{\cos 2A}$$

$$= \sec 2A + \tan 2A$$

$$= \text{RHS.}$$

Exercise 16C

Prove the following identities in Questions 1 to 6.

1. $\dfrac{\sin 2A}{1 + \cos 2A} = \tan A$

2. $\tan\theta + \cot\theta = 2\csc 2\theta$

3. $\dfrac{\sin 2A + \cos 2A + 1}{\sin 2A - \cos 2A + 1} = \cot A$

4. $\cot x - \csc 2x = \cot 2x$

5. $\dfrac{\sin 3A + \sin A}{2\sin 2A} = \cos A$

6. $\dfrac{\cos 3\theta - \sin 3\theta}{1 - 2\sin 2\theta} = \cos\theta + \sin\theta$

   [You may find it useful to refer back to the results of Activity 2 for Questions 5 and 6.]

7. Use the fact that $4A = 2 \times 2A$ to show that

   $$\frac{\sin 4A}{\sin A} = 8\cos^3 A - 4\cos A.$$

8. By writing $t = \tan\theta$ show that

   $$\tan(\theta + 45°) + \tan(\theta - 45°) = \frac{1+t}{1-t} - \frac{1-t}{1+t}.$$

   Hence show that

   $$\tan(\theta + 45°) + \tan(\theta - 45°) = 2\tan 2\theta.$$

9. Using $t = \tan\theta$, write down $\tan 2\theta$ in terms of t. Hence prove the identities

   (a) $\cot\theta - \tan\theta = 2\cot 2\theta$

   (b) $\cot 2\theta + \tan\theta = \csc 2\theta$

10. Write down $\cos 4x$ in terms of $\cos 2x$, and hence in terms of $\cos x$ show that

    $$\cos 4x + 4\cos 2x = 8\cos^4 x - 3$$

11. Prove the identity

    $$\frac{\sin 4A + \cos A}{\cos 4A + \sin A} = \sec 3A + \tan 3A$$

16.5 Sum and product formulae

You may recall that

$$\sin(A+B) = \sin A \cos B + \cos A \sin B$$

$$\sin(A-B) = \sin A \cos B - \cos A \sin B$$

Adding these two equations gives

$$\sin(A+B) + \sin(A-B) = 2\sin A \cos B \qquad (1)$$

Call $C = A + B$ and $D = A - B$,

then $C + D = 2A$ and $C - D = 2B$. Hence

$$A = \frac{C+D}{2}, \ B = \frac{C-D}{2}$$

and (1) can be written as

$$\sin C + \sin D = 2\sin\left(\frac{C+D}{2}\right)\cos\left(\frac{C-D}{2}\right)$$

This is more easily remembered as

'sine plus sine = twice sine(half the sum)cos(half the difference)'

Activity 3

In a similar way to above, derive the formulae for

(a) $\sin C - \sin D$ (b) $\cos C + \cos D$ (c) $\cos C - \cos D$

By reversing these formulae, write down further formulae for

(a) $2\sin E \cos F$ (b) $2\cos E \cos F$ (c) $2\sin E \sin F$

Example

Show that $\cos 59° + \sin 59° = \sqrt{2}\cos 14°$.

Solution

Firstly, $\quad\sin 59° = \cos 31°$, since $\sin\theta = \cos(90 - \theta)$

So $\qquad$ LHS $= \cos 59° + \cos 31°$

$$= 2\cos\left(\frac{59 + 31}{2}\right)\cos\left(\frac{59 - 31}{2}\right)$$

$$= 2\cos 45° . \cos 14°$$

$$= 2 . \frac{\sqrt{2}}{2}\cos 14°$$

$$= \sqrt{2}\cos 14°$$

$$= \text{RHS}$$

Example

Prove that $\sin x + \sin 2x + \sin 3x = \sin 2x(1 + 2\cos x)$.

Solution

$$\text{LHS} \quad = \sin 2x + (\sin x + \sin 3x)$$

$$= \sin 2x + 2\sin\left(\frac{x + 3x}{2}\right)\cos\left(\frac{x - 3x}{2}\right)$$

$$= \sin 2x + 2\sin 2x \cos(-x)$$

$$= \sin 2x(1 + 2\cos x) \quad \text{since } \cos(-x) = \cos x.$$

Example

Write $\cos 4x \cos x - \sin 6x \sin 3x$ as a product of terms.

Solution

Now $\qquad \cos 4x \cos x \quad = \frac{1}{2}\{\cos(4x + x) + \cos(4x - x)\}$

$$= \frac{1}{2}\cos 5x + \frac{1}{2}\cos 3x$$

and $\qquad \sin 6x \sin 3x \quad = \frac{1}{2}\{\cos(6x - 3x) - \cos(6x + 3x)\}$

Thus, $\quad$ LHS $\quad = \frac{1}{2}\cos 5x + \frac{1}{2}\cos 3x - \frac{1}{2}\cos 3x + \frac{1}{2}\cos 9x$

$$= \frac{1}{2}(\cos 5x + \cos 9x)$$

$$= \frac{1}{2}\times 2\cos\left(\frac{5x + 9x}{2}\right)\cos\left(\frac{5x - 9x}{2}\right)$$

$$= \cos 7x \cos 2x.$$

The sum formulae are given by

$$\sin A + \sin B = 2\sin\left(\frac{A+B}{2}\right)\cos\left(\frac{A-B}{2}\right)$$

$$\sin A - \sin B = 2\cos\left(\frac{A+B}{2}\right)\sin\left(\frac{A-B}{2}\right)$$

$$\cos A + \cos B = 2\cos\left(\frac{A+B}{2}\right)\cos\left(\frac{A-B}{2}\right)$$

$$\cos A - \cos B = -2\sin\left(\frac{A+B}{2}\right)\sin\left(\frac{A-B}{2}\right)$$

and the product formulae by

$$\sin A \cos B = \tfrac{1}{2}(\sin(A+B) + \cos(A-B))$$

$$\cos A \cos B = \tfrac{1}{2}(\cos(A+B) + \cos(A-B))$$

$$\sin A \sin B = \tfrac{1}{2}(\cos(A-B) - \cos(A+B))$$

Exercise 16D

1. Write the following expressions as products:

   (a) $\cos 5x - \cos 3x$ (b) $\sin 11x - \sin 7x$

   (c) $\cos 2x + \cos 9x$ (d) $\sin 3x + \sin 13x$

   (e) $\cos\dfrac{2\pi}{15} + \cos\dfrac{14\pi}{15} + \cos\dfrac{4\pi}{15} + \cos\dfrac{8\pi}{15}$

   (f) $\sin 40° + \sin 50° + \sin 60°$

   (g) $\cos 114° + \sin 24°$

2. Evaluate in rational/surd form

   $\sin 75° + \sin 15°$

3. Write the following expressions as sums or differences:

   (a) $2\cos 7x \cos 5x$

   (b) $2\cos\left(\dfrac{1}{2}x\right)\cos\left(\dfrac{5x}{2}\right)$

   (c) $2\sin\left(\dfrac{\pi}{4} - 3\theta\right)\cos\left(\dfrac{\pi}{4} + \theta\right)$

   (d) $2\sin 165° \cos 105°$

4. Establish the following identities:

   (a) $\cos\theta - \cos 3\theta = 4\sin^2\theta\cos\theta$

   (b) $\sin 6x + \sin 4x - \sin 2x = 4\cos 3x \sin 2x \cos x$

   (c) $\dfrac{2\sin 4A + \sin 6A + \sin 2A}{2\sin 4A - \sin 6A - \sin 2A} = \cot^2 A$

   (d) $\dfrac{\sin(A+B) + \sin(A-B)}{\cos(A+B) - \cos(A-B)} = \tan A$

   (e) $\dfrac{\cos(\theta + 30°) + \cos(\theta + 60°)}{\sin(\theta + 30°) + \sin(\theta + 60°)} = \dfrac{1 - \tan\theta}{1 + \tan\theta}$

5. Write $\cos 12x + \cos 6x + \cos 4x + \cos 2x$ as a product of terms.

6. Express $\cos 3x \cos x - \cos 7x \cos 5x$ as a product of terms.

16.6 General formula

For this next activity you will find it very useful to have a graph plotting facility.   Remember, you will be working in radians.

Activity 4

Sketch the graph of a function of the form
$$y = a\sin x + b\cos x$$
(where a and b are constants) in the range $-\pi \le x \le \pi$.
From the graph, you must identify the amplitude of the function and the x- co–ordinates of

   (i)   the crossing point on the x-axis nearest to the origin, and

   (ii)   the first maximum of the function

as accurately as you can.

An example has been done for you; for $y = \sin x + \cos x$, you can see that amplitude ≈ 1.4

     crossing-point nearest to 0 at $x = \alpha = -\dfrac{\pi}{4}$

     maximum occurs at $x = \beta = \dfrac{\pi}{4}$

Try these for yourself :

(a) $y = 3\sin x + 4\cos x$ (b) $y = 12\cos x - 5\sin x$

(c) $y = 9\cos x + 12\sin x$ (d) $y = 15\sin x - 8\cos x$

(e) $y = 2\sin x + 5\cos x$ (f) $y = 3\cos x - 2\sin x$

In each case, make a note of

    R, the amplitude;

    α, the crossing - point nearest to O;

    β, the x - coordinate of the maximum.

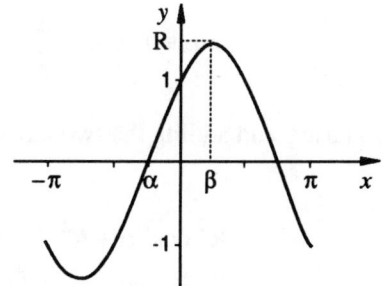

In each example above, you should have noticed that the curve is itself a sine/cosine 'wave', each of which can be obtained from the curves of either $y = \sin x$ or $y = \cos x$ by means of two simple transformations (taken in any order):

1. A **stretch** parallel to the y-axis by a factor of R, the amplitude, and

2. **A translation** parallel to the x-axis by either α or β (depending on whether you wish to start with $\sin x$ or $\cos x$ as the original function).

Consider, for example $y = \sin x + \cos x$. This can be written in the form $y = R\sin(x + \alpha)$, since

$$R\sin(x + \alpha) = R\{\sin x \cos \alpha + \cos x \sin \alpha\}$$

$$= R\cos \alpha \sin x + R\sin \alpha \cos x$$

This expression should be the same as $\sin x + \cos x$.

Thus

$$R\cos \alpha = 1 \text{ and } R\sin \alpha = 1$$

Dividing these terms gives

$$\tan \alpha = 1 \Rightarrow \alpha = \frac{\pi}{4}$$

Squaring and adding the two terms gives

$$R^2 \cos^2 \alpha + R^2 \sin^2 \alpha = 1^2 + 1^2$$

$$R^2 \left(\cos^2 \alpha + \sin^2 \alpha \right) = 2$$

Since $\cos^2 \alpha + \sin^2 \alpha = 1$,

$$R^2 = 2 \Rightarrow R = \sqrt{2} \quad \text{(negative root does not make sense)}$$

Thus

$$\sin x + \cos x = \sqrt{2} \sin\left(x + \tfrac{\pi}{4}\right).$$

Activity 5

Express the function $\sin x + \cos x$ in the form
$$\sin x + \cos x = R\cos(x - \alpha)$$

Find suitable values for R and α using the method shown above.

Another way of obtaining the result in Activity 5 is to note that

$$\sin\theta = \cos(\tfrac{\pi}{2} - \theta)$$

so that

$$\sin x + \cos x = \sqrt{2}\sin(x + \tfrac{\pi}{4})$$

$$= \sqrt{2}\cos(\tfrac{\pi}{2} - (x + \tfrac{\pi}{4}))$$

$$= \sqrt{2}\cos(\tfrac{\pi}{4} - x)$$

$$= \sqrt{2}\cos(x - \tfrac{\pi}{4})$$

since $\cos(-\theta) = \cos\theta$.

Example

Write $7\sin x - 4\cos x$ in the form $R\sin(x - \alpha)$

where $R > 0$ and $0 < \alpha < \tfrac{\pi}{2}$.

Solution

Assuming the result,

$$7\sin x \cos x = R\sin(x - \alpha)$$

$$= R\sin x \cos\alpha - R\cos x \sin\alpha$$

To satisfy the equation, you need

$$R\cos\alpha = 7$$

$$R\sin\alpha = 4$$

Squaring and adding, as before, gives

$$R = \sqrt{7^2 + 4^2} = \sqrt{65}.$$

Thus

$$\cos\alpha = \frac{7}{\sqrt{65}}, \quad \sin\alpha = \frac{4}{\sqrt{65}} \quad \left(\text{or } \tan\alpha = \frac{4}{7}\right)$$

$$\Rightarrow \alpha = 0.519 \text{ radians,}$$

so $\qquad 7\sin x - 4\cos x = \sqrt{65}\sin(x - 0.519)$

Exercise 16E

Write (in each case, $R > 0$ and $0 < \alpha < \frac{\pi}{2}$)

1. $3\sin x + 4\cos x$ in the form $R\sin(x + \alpha)$

2. $4\cos x + 3\sin x$ in the form $R\cos(x - \alpha)$

3. $15\sin x - 8\cos x$ in the form $R\sin(x - \alpha)$

4. $15\sin x - 8\cos x$ in the form $R\sin(x - \alpha)$

5. $20\sin x - 21\cos x$ in the form $R\sin(x - \alpha)$

6. $14\cos x + \sin x$ in the form $R\cos(x - \alpha)$

7. $2\cos 2x - \sin 2x$ in the form $R\cos(2x + \alpha)$

8. $3\cos\frac{1}{2}x + 5\sin\frac{1}{2}x$ in the form $R\sin\left(\frac{1}{2}x + \alpha\right)$

16.7 Equations in one function

In chapter 11 you looked at equations of the form

$$a\sin(bx + c) = d$$

for constants a, b, c, d, or similar equations involving cos or tan.

In this and the following sections you will be introduced to a variety of different types of trigonometric equation and the appropriate ways of solving them within a given range.

Here, you will be asked only to solve polynomials in one function. It is important, therefore, that you are able to determine factors of polynomials and use the quadratic formula when necessary.

Example

Solve $2\sin^2\theta + 3\sin\theta = 2$ for values of θ between $0°$ and $360°$.

Solution

Rearrange the equation as $2\sin^2\theta + 3\sin\theta - 2 = 0$, which is a quadratic in $\sin\theta$. This factorises as

$$(2\sin\theta - 1)(\sin\theta + 2) - 0$$

giving

(a) $2\sin\theta - 1 = 0 \Rightarrow \sin\theta = \frac{1}{2} \Rightarrow \theta = 30°, 150°$

or

(b) $\sin\theta + 2 = 0 \Rightarrow \sin\theta = -2$ which has no solutions.

Exercise 16F

Solve the following equations for values of x in the range given:

1. $4\sin^2 x - \sin x - 3 = 0, 0 \le x \le 2\pi$

2. $6\cos^2 x + \cos x = 1, -180° \le x \le 180°$

3. $\tan^2 x + 3\tan x - 10 = 0, 0° \le x \le 360°$

4. $\cos^2 x = 2\cos x + 1, 0° \le x \le 180°$

5. $\tan^4 x - 4\tan^2 x + 3 = 0, 0 \le x \le \pi$

6. $\frac{1}{2}\sec^2 x = \sec x + 2, 0 \le x \le \pi$

7. Use the factor theorem to factorise

$$6c^3 - 19c^2 + c + 6 = 0$$

and hence solve

$$6\cos^3 x - 19\cos^2 x + \cos x + 6 = 0$$

for $0 \le x \le \pi$

16.8 Equations in two functions reducible to one

Equations involving two (or more) trigonometric functions cannot, in general, be solved by the simple methods you have encountered up to now. However, many such equations can be tackled using some of the basic identities introduced in the first part of this chapter.

Example

Solve $5\sin\theta = 2\cos\theta$ for $0 \le \theta \le 2\pi$.

Solution

Dividing both sides by $\cos\theta$ assuming $\cos\theta \ne 0$ gives

$$5\tan\theta = 2$$
$$\Rightarrow \quad \tan\theta = 0.4$$
$$\Rightarrow \quad \theta = 0.381, \ 3.52$$

Example

Solve $2\sec^2 x + 3\tan x - 4 = 0$ for $0° \le x \le 180°$.

Solution

From earlier work, $\sec^2 x = 1 + \tan^2 x$, leading to

$$2 + 2\tan^2 x + 3\tan x - 4 = 0$$
$$\Rightarrow \quad 2\tan^2 x + 3\tan x - 2 = 0$$
$$\Rightarrow \quad (2\tan x - 1)(\tan x + 2) = 0$$

giving

(a) $2\tan x - 1 = 0 \Rightarrow \tan x = \frac{1}{2} \Rightarrow x = 26.6°$

or

(b) $\tan x + 2 = 0 \Rightarrow \tan x = -2 \Rightarrow x = 116.6°$

Example

Solve $3\sin 2\theta = 5\cos\theta$ for $0° \le \theta \le 180°$

Solution

Since $\sin 2\theta = 2\sin\theta\cos\theta$, the equation reduces to
$$6\sin\theta\cos\theta = 5\cos\theta.$$

Method 1 – divide by $\cos\theta$ to get

$$\sin\theta = \frac{5}{6}$$
$$\Rightarrow \theta = 56.4°, \ 123.6°$$

Method 2 – factorise to give

$$6\sin\theta\cos\theta - 5\cos\theta = 0$$
$$\cos\theta(6\sin\theta - 5) = 0$$

giving

(a) $\cos\theta = 0 \Rightarrow \theta = 90°$

or

(b) $\sin\theta = \frac{5}{6} \Rightarrow \theta = 56.6°, 123.6°$

You should see the error in **Method** I, which throws away the
solution for $\cos\theta = 0$. Division can only be done provided that
the quantity concerned is not zero. [You might like to check back
in the first Example to see that exactly the same division was quite
legitimate in that situation].

Exercise 16G

Solve the following equations in the required domain :

1. $2\sin^2\theta + 5\cos\theta + 1 = 0$ $-\pi \le \theta \le \pi$
2. $2\sin 2\theta = \tan\theta$ $0° \le \theta \le 180°$
3. $2\csc x = 5\cot x$ $0° \le x \le 180°$
4. $3\cos\theta = 2\cos 2\theta$ $0° \le \theta \le 360°$
5. $\sin x + \frac{1}{2}\sin 2x = 0$ $0 \le x \le 2\pi$
6. $6\cos\theta - 1 = \sec\theta$ $0° \le \theta \le 180°$

7. $\tan^2 x + 3\sec x = 0$ $0 \le x \le 2\pi$
8. $6\tan^2 A = 4\sin^2 A + 1$ $0° \le A \le 360°$
9. $3\cot^2\theta + 5\csc\theta + 1 = 0$ $0 \le \theta \le 2\pi$
10. $\csc x = \sqrt{3}\sec^2 x$ $0 \le x \le \pi$
11. $\sec^4\theta + 2 = 6\tan^2\theta$ $0° \le \theta \le 180°$
12. $\cos 2\theta\cos\theta = \sin 2\theta\sin\theta$ $-180° \le \theta \le 180°$

16.9 Linear trig equations

In this section you will be looking at equations of the form

$$a\cos x + b\sin x = c$$

for given constants a, b, and c.

Example

Solve $3\cos x + \sin x = 2$ for $0° \le x \le 360°$.

Solution

Method 1 - Note that $\cos x$ and $\sin x$ are not interchangeable, but $\cos^2 x$ and $\sin^2 x$ are (using $\cos^2 x + \sin^2 x = 1$) so a 'rearranging and squaring' approach would seem in order.

Re-arranging : $3\cos x = 2 - \sin x$

Squaring : $9\cos^2 x = 4 - 4\sin x + \sin^2 x$

$\Rightarrow \quad 9(1 - \sin^2 x) = 4 - 4\sin x + \sin^2 x$

$\Rightarrow \quad 9 - 9\sin^2 x = \sin^2 x - 4\sin x + 4$

$\Rightarrow \quad 0 = 10\sin^2 x - 4\sin x - 5$

The quadratic formula now gives $\sin x = \dfrac{4 \pm \sqrt{216}}{20}$

and $\sin x = 0.93487$ or -0.534847

giving $x = 69.2°, 110.8°$ or $212.3°, 327.7°$

Method 2 - Write $3\cos x + \sin x$ as $R\cos(x - \alpha)$

(or $R\sin(x + \alpha)$).

$$3\cos x + \sin x \equiv R\cos(x - \alpha)$$

Firstly, $R = \sqrt{3^2 + 1^2} = \sqrt{10}$

so
$$3\cos x + \sin x = \sqrt{10}\left(\cos x \frac{3}{\sqrt{10}} + \sin x \frac{1}{\sqrt{10}}\right)$$

$$\equiv \sqrt{10}(\cos x \cos a + \sin x \sin a)$$

Thus $\cos\alpha = \dfrac{3}{\sqrt{10}}$ $\left(\text{or } \sin\alpha = \dfrac{7}{\sqrt{10}} \text{ or } \tan\alpha = \dfrac{1}{3}\right) \Rightarrow \alpha = 18.43°$

The equation $3\cos x + \sin x = 2$ can now be written as

$$\sqrt{10}\cos(x - 18.43°) = 2$$

$$\Rightarrow \quad \cos(x - 18.43°) = \frac{2}{\sqrt{10}}$$

$$\Rightarrow \quad x - 18.43° = \cos^{-1}\left(\frac{2}{\sqrt{10}}\right)$$

$$\Rightarrow \quad x - 18.43° = 50.77° \text{ or } 309.23°$$

and

$$x = 50.77° + 18.43° \text{ or } 309.23° + 18.43°$$

$$x = 69.2° \text{ or } 327.7°$$

The question now arises as to why one method yields **four** answers, the other only **two**. If you check all four answers you will find that the two additional solutions in **Method 1** do not fit the equation $3\cos x + \sin x = 2$. They have arisen as extra solutions created by the squaring process. [Think of the difference between the equations $x = 2$ and $x^2 = 4$ – the second one has two solutions]. If the first approach is used then the final answers always need to be checked in order to discard the extraneous solutions.

Exercise 16H

1. By writing $7\sin x + 6\cos x$ in the form

   $R\sin(x+\alpha)$ $(R>0, 0° < \alpha < 90°)$

   solve the equation

   $7\sin x + 6\cos x = 9$

   for values of x between $0°$ and $360°$.

2. Use the 'rearranging and squaring' method to solve

   (a) $4\cos\theta + 3\sin\theta = 2$

   (b) $3\sin\theta - 2\cos\theta = 1$

   for $0° \le \theta \le 360°$.

3. Write $\sqrt{3}\cos\theta + \sin\theta$ as $R\cos(\theta - \alpha)$,

   where $R>0$ and $0 > \alpha > \pi$ and hence solve

   $\sqrt{3}\cos\theta + \sin\theta = \sqrt{2}$

   for $0 \le \theta \le 2\pi$

4. Solve

   (a) $7\cos x - 6\sin x = 4$ for $-180° \le x \le 180°$

   (b) $6\sin\theta + 8\cos\theta = 7$ for $0° \le \theta \le 180°$

   (c) $4\cos x + 2\sin x = \sqrt{5}$ for $0° \le x \le 360°$

   (d) $\sec x + 5\tan x + 12 = 0$ for $0 \le x \le 2\pi$.

16.10 More demanding equations

In this section you will need to use the identities - the Addition Formulae, the Sum and Product Formulae and Multiple Angle Identities - in order to solve the given equations.

Example

Solve $\cos 5\theta + \cos\theta = \cos 3\theta$ for $0° \le \theta \le 180°$.

Solution

Using $\cos A + \cos B = 2\cos\left(\dfrac{A+B}{2}\right)\cos\left(\dfrac{A-B}{2}\right)$,

         $\text{LHS} = 2\cos 3\theta \cos 2\theta$.

Thus $2\cos 3\theta \cos 2\theta = \cos 3\theta$

    $\Rightarrow$ $\cos 3\theta(2\cos 2\theta - 1) = 0$

Then

(a) $\cos 3\theta = 0$

    $\Rightarrow$ $3\theta = 90°, 270°, 450°$

    $\Rightarrow$ $\theta = 30°, 90°, 150°$

or

(b) $2\cos 2\theta - 1 = 0$

$\Rightarrow \qquad \cos 2\theta = \tfrac{1}{2}$

$\Rightarrow \qquad 2\theta = 60°,\ 300°,$

$\Rightarrow \qquad \theta = 30°,\ 150°$ as already found.

Solutions are $\theta = 30°\,(\text{twice}),\ 90°,\ 150°\,(\text{twice})$.

[Remember, for final solutions in range $0° \leq \theta \leq 180°$, solutions for 3θ must be in range $0° \leq \theta \leq 3 \times 180° = 540°$.]

Exercise 16I

1. Solve for $0° \leq \theta \leq 180°$.

    (a) $\cos\theta + \cos 3\theta = 0$ (b) $\sin 4\theta + \sin\theta = 0$

    (c) $\sin\theta + \sin 3\theta = \sin 2\theta$

2. Find all values of x satisfying the equation

    $$\sin x = 2\sin\left(\frac{1}{3}\pi - x\right)$$

    for $0 \leq x \leq 2\pi$.

3. By writing 3θ as $2\theta + \theta$, show that

    $$\cos 3\theta = 4\cos^3\theta - 3\cos\theta$$

    and find a similar expression for $\sin 3\theta$ in terms of powers of $\sin\theta$ only.

    Use these results to solve, for $0 \leq \theta \leq 2\pi$,

    (a) $\cos 3\theta + 2\cos\theta = 0$

    (b) $\sin 3\theta = 3\sin 2\theta$

    (c) $\cos\theta - \cos 3\theta = \tan^2\theta$

4. Show that $\tan x + \cot x = 2\operatorname{cosec}2x$

    Hence solve $\tan x + \cot x = 8\cos 2x$

    for $0 \leq x \leq \pi$.

5. Solve the equation

    $$\sin 2x + \sin 3x + \sin 5x = 0$$

    for $0° \leq x \leq 180°$

6. Find the solution x in the range $0° \leq x \leq 360°$ for which

    $$\sin 4x + \cos 3x = 0.$$

7. (a) Given $t = \tan\tfrac{1}{2}\theta$, write down $\tan\theta$ in terms of t and show that

    $$\cos\theta = \frac{1-t^2}{1+t^2}$$

    Find also a similar expression for $\sin\theta$ in terms of t.

    (b) Show that $2\sin\theta - \tan\theta = \dfrac{2t}{1-t^4}\left(1 - 3t^2\right)$

    (c) Hence solve $2\sin\theta - \tan\theta = 6\cot\tfrac{1}{2}\theta$ for values of θ in the range $0° < \theta < 360°$.

16.11 Small angle approximations

In Section 10.3 you met the small angle approximations for sin, cos and tan. These are given by

$$\sin x \approx x$$

$$\tan x \approx x$$

$$\cos x \approx 1 - \tfrac{1}{2} x^2$$

for small x, measured in radians. These approximations can be used to find approximate solution to the equations, when working in radians.

Example

Find an approximate solution, correct to 3 significant figures, to the equation

$$3\sin\theta + \cos 2\theta = 1.75$$

given that θ is small.

Solution

$\sin\theta \approx \theta$ and $\cos 2\theta \approx 1 - \tfrac{1}{2}(2\theta)^2 = 1 - 2\theta^2$

Equation becomes

$$3\theta + 1 - 2\theta^2 = 1.75$$

$$\Rightarrow \quad 0 = 2\theta^2 - 3\theta + 0.75$$

Using the quadratic formula $\theta = \dfrac{3 \pm \sqrt{3}}{4}$

and the small root is $\theta = \dfrac{3 - \sqrt{3}}{4} = 0.317$

The exact answer to 3 d.p.'s is $\theta = 0.323$ radians; so the approximation is about 98% accurate.

Exercise 16J

1. If x is sufficiently small for $4x$ to be considered small, find an approximation to the smallest positive solution of the equation $\cos 4x = 7x$.

2. Show that an approximation to the least positive root of the equation

$$\sin x + \cos x + \tan x = 1.5$$

is $\sqrt{5} - 2$.

3. Find approximately, and to three significant figures, the small root of the equation

$$\cos x = x^2 + 6\sin x.$$

4. Show that, if x is sufficiently small, then

$$\left(1 - \tan^2 2x\right)\cos 3x \approx 1 - \tfrac{17}{2}x^2.$$

Hence find an approximation to the least positive solution to the equation

$$\left(1 - \tan^2 2x\right)\cos 3x = \tfrac{1}{2}$$

giving your answer correct to 3 s.f.

5. (a) Use the small angle formulae to find, to 3 s.f., an approximation to the solution near zero of

$$4\cos x + 5\sin x = 2.$$

   (b) By writing $4\cos x + 5\sin x$ in the form $R\cos(x - \alpha)$, where $R > 0$ find the exact solution (to 3 s.f.) to the equation $4\cos x + 5\sin x = 2$.

16.12 Miscellaneous Exercises

1. Simplify the following expressions :

   (a) $\cos 37° \cos 23° - \sin 37° \sin 23°$

   (b) $\sin 28° \cos 42° + \cos 28° \cos 48°$

   (c) $\sin\left(\dfrac{\pi}{3} + x\right) - \sin\left(\dfrac{\pi}{3} - x\right)$

   (d) $\cos\left(\dfrac{\pi}{6} + x\right) + \cos\left(\dfrac{\pi}{6} - x\right)$

   (e) $\sin\left(\dfrac{2\pi}{3} - x\right) + \sin\left(x - \dfrac{\pi}{3}\right)$

2. Prove the following identities :

   (a) $\dfrac{\sin\theta + \sin 3\theta + \sin 5\theta}{\cos\theta + \cos 3\theta + \cos 5\theta} = \tan 3\theta$

   (b) $\dfrac{\sin(x - y) + \sin x + \sin(x + y)}{\cos(x - y) + \cos x + \cos(x + y)} = \tan x$

   (c) $\cos(A + B + C) + \cos(A + B - C) + \cos(A - B + C)$

   $+ \cos(-A + B + C) = 4\cos A \cos B \cos C$

3. Find all the acute angles x for which

$$\cos 9x + \cos x = \cos 5x.$$

4. Solve the following equations in the given range:

   (a) $3\cosec^4 x + 8 = 10\cosec^2 x$ $0 \le x \le 2\pi$

   (b) $3\tan\theta + \cot\theta = 5\cosec\theta$ $0 \le \theta \le 2\pi$

   (c) $2\sin^2\theta = 3\sin\theta\cos\theta + 2\cos^2\theta$ $0° \le \theta \le 360°$

   (d) $7\sin x \cos x + \cos^2 x = 2$ $0° < \theta < 180°$

17 TRIANGLES WITHOUT RIGHT ANGLES

Objectives

After studying this chapter you should

*   be able to use the sine rule;
*   be able to use the cosine rule;
*   be able to find the area of any triangle.

17.0 Introduction

In this chapter you will be introduced to the sine and cosine rules for use in any triangle. Before you start, you should be aware of the convention for referring to the sides and angles of a triangle.

In triangle ABC shown opposite, the angles are labelled as the vertices at which they occur, and are denoted by capital letters, so that

angle ABC = *B*.

Lower case letters refer to the sides of the triangle, so that

side BC = *a*,

with the convention that *a* is opposite *A* (as shown), *b* opposite *B*, and so on.

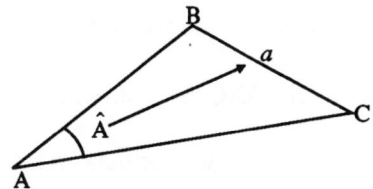

17.1 The sine rule

Activity 1   Finding a rule for sides and angles

For this activity you will need a ruler measuring in mm, an angle measurer or protractor, and a calculator.

Draw four different shaped triangles. (You should include some obtuse-angled triangles).

Label the vertices A, B and C and the opposite sides *a*, *b* and *c* corresponding to the angles.

Measure the size of angles A, B and C (an accuracy to the nearest half-degree should be possible) and the lengths of the sides *a*, *b* and *c* to the nearest mm. Then for each triangle, evaluate

$$\frac{a}{\sin A} , \frac{b}{\sin B} \text{ and } \frac{c}{\sin C}.$$

What do you notice?

A proof of the sine rule

Oddly enough, in order to work with the sine and cosine functions in a non right-angled triangle it is necessary to create a right-angle.

In the triangle ABC, a perpendicular has been drawn from A to BC, meeting BC at the point X at 90°.

Here, then, AX is the height of ABC and BC is the base.

In triangle ABX, $h = c \sin B$.

In triangle AXC, $h = b \sin C$.

By putting the two formulas for h together

$$c \sin B = b \sin C \text{ or } \frac{c}{\sin C} = \frac{b}{\sin B}.$$

If side AC had been taken as the base, the relationship

$\dfrac{c}{\sin C} = \dfrac{a}{\sin A}$ would have been obtained, and taking AB as base

would have given $\dfrac{a}{\sin A} = \dfrac{b}{\sin B}$.

Together, the set of equations obtained is

$$\boxed{\frac{a}{\sin A} = \frac{b}{\sin B} = \frac{c}{\sin C}}$$

This is called the **sine rule**, relating the sides of any triangle to the sines of its angles.

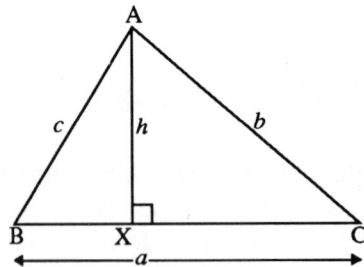

(Remember: this rule applies to **any** triangle, with or without a right angle).

Example

In triangle ABC, $A = 40°$ and $a = 17$ mm; $c = 11$ mm. Find b, B and C.

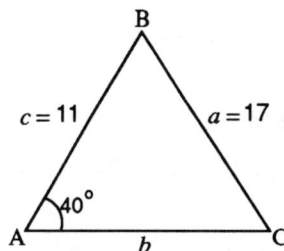

Solution

In attempting to solve a problem of this sort a sketch is necessary.

In the equation $\dfrac{a}{\sin A} = \dfrac{c}{\sin C}$,   three of the four quantities are known, or can be found. The fourth, $\sin C$, can be calculated, and hence C.

Substituting,

$$\frac{17}{\sin 40°} = \frac{11}{\sin C}.$$

Rearranging,

$$\sin C = \frac{11 \sin 40°}{17} = 0.415921 \ldots$$

$$\Rightarrow \quad C = 24.6°.$$

Knowing A and C,

$$B = 180° - A - C$$

$$= 115.4°.$$

b can now be found using

$$\frac{b}{\sin B} = \frac{a}{\sin A} \quad \left(\text{or } \frac{c}{\sin C} \right).$$

Substituting,

$$\frac{b}{\sin 115.4°} = \frac{17}{\sin 40°}.$$

Rearranging,

$$b = \frac{17 \sin 115.4°}{\sin 40°}$$

$$= 23.9 \text{ mm} \quad \text{(to 3 significant figures).}$$

Example

In triangle PQR, $P = 52°$, $R = 71°$ and $q = 9.3$ m.

Find Q, p and r.

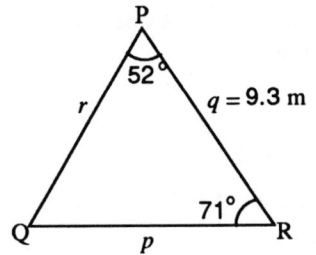

Solution

Firstly, $Q = 180° - 52° - 71° = 57°$.

Next, using $\dfrac{q}{\sin Q} = \dfrac{p}{\sin P}$,

$$\frac{9.3}{\sin 57°} = \frac{p}{\sin 52°}.$$

Rearranging,

$$p = \frac{9.3 \sin 52°}{\sin 57°} = 8.74 \text{ m}.$$

Also, $\dfrac{q}{\sin Q} = \dfrac{r}{\sin R}$,

giving $\dfrac{9.3}{\sin 57°} = \dfrac{r}{\sin 71°}.$

Rearranging,

$$r = \frac{9.3 \sin 71°}{\sin 57°} = 10.5 \text{ m} \quad \text{(to 3 significant figures)}.$$

Example

From a point P on the same level as the base of a tower, the angle of elevation of the top of the tower is 30°. From a point Q, 20 m further away than P from the tower the angle is 20°. What is the height of the tower?

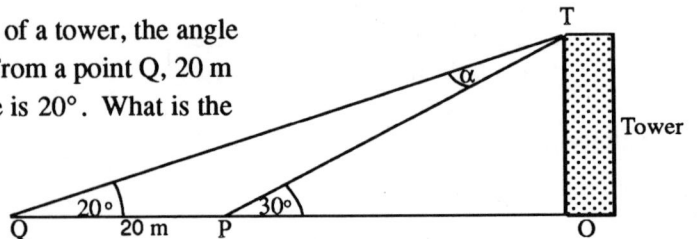

Tower

Solution

TP is first found by using the sine rule in triangle QPT;

$$\frac{TP}{\sin 20} = \frac{QP}{\sin \alpha}$$

But $\quad 20° + \alpha = 30° \quad \Rightarrow \quad \alpha = 10°$

$$TP = \frac{20 \sin 20°}{\sin 10°} = 39.39 \text{ m}.$$

Finally, from triangle TOP,

height, $\text{TO} = \text{TP} \sin 30° = 39.39 \times \sin 30° = 19.7$ m (to nearest m).

A possible difficulty

Example

Solve the triangle ABC, given $A = 33°$, $a = 20$ cm and $b = 28$ cm.

In this context 'solve' means 'find all the other sides and angles not already given'.

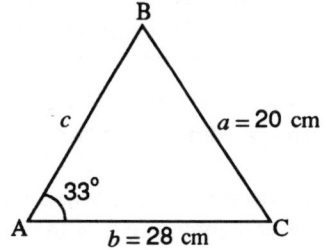

Solution

Now $\dfrac{a}{\sin A} = \dfrac{b}{\sin B}$ gives $\dfrac{20}{\sin 33°} = \dfrac{28}{\sin B}$.

Hence $\sin B = \dfrac{28 \sin 33°}{20} = 0.762\,495$,

which gives $B = 49.7°$. But B could be obtuse, and another possible solution is given by

$$B = 180° - 49.7° = 130.3°.$$

Now if $B = 49.7°$,

$$C = 180° - 33° - 49.7° = 97.3°,$$

and $\dfrac{20}{\sin 33°} = \dfrac{c}{\sin 97.3°}$

gives $c = \dfrac{20 \sin 97.3°}{\sin 33°} = 36.4$ cm (3 s.f.).

But if $B = 130.3°$,

$$C = 180° - 33° - 130.3° = 16.7°$$

and $c = \dfrac{20 \sin 16.7°}{\sin 33°} = 10.6$ cm (3 s.f.).

So there appear to be two possible solutions.

Does this make sense?

In order to visualise the reason for this ambiguity, imagine trying to draw the triangle as described:

$$A = 33°, \ a = 20, \ b = 28.$$

1. Draw the longest side first : $b = 28$.

2. Measure an angle of 33° at A - the position of B on this line is not yet known.

3. CB = 20, so B is 20 cm from C and somewhere on the line from A. Now all possible positions of a point B such that BC = 20 lie on a circle, centre at C, and radius 20. Part of this circle is drawn on the diagram.

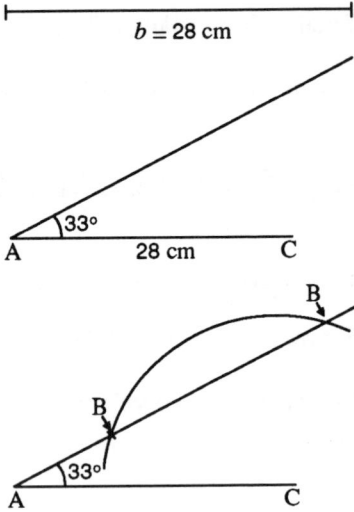

You will see that the circle and line intersect in two points corresponding to the **two** possible positions of B.

This situation arises when you are given two sides and a non-included angle (i.e. not the angle between them) because the triangle is not necessarily uniquely defined by the information given. It is called the **ambiguous case** , and you must watch out for it when using the sine rule to find angles.

Turn back to the first example in this chapter and see if you can decide why the problem did not arise there.

Exercise 17A

In the following triangles, find the sides and angles not given. Give your answers to 1 d.p. for angles and 3 s.f. for sides where appropriate.

1. In triangle LMN, $m = 32$m, $M = 16°$ and $N = 56.7°$.

2. In triangle XYZ, $X = 120°$, $x = 11$ cm and $z = 5$ cm.

3. In triangle ABC, $A = 49°$, $a = 127$ m, and $c = 100$ m.

4. In triangle PQR, $R = 27°$, $p = 9.2$ cm and $r = 8.3$ cm.

5. In triangle DEF, $E = 81°$, $F = 62°$ and $d = 4$ m.

6. In triangle UVW, $u = 4.2$ m, $w = 4$ m and $W = 43.6°$.

17.2 The cosine rule

Activity 2

Why is it that the sine rule does not enable you to solve triangles ABC and XYZ when

(a) in triangle ABC you are given :

$A = 35°$, $b = 84$ cm and $c = 67$ cm;

(b) in triangle XYZ you are given :

$x = 43$ m, $y = 60$ m and $z = 81$ m?

As with the introduction of the sine rule, it is necessary to create a right-angle in order to establish the cosine rule. Again, it is not important which side is taken as base.

The activity above shows the need for another rule in order to 'solve' triangles.

Two applications of Pythagoras' Theorem give

$$c^2 = h^2 + BX^2 \text{ in triangle ABX,}$$

and $\quad b^2 = h^2 + XC^2$ in triangle ABC.

Rearranging in terms of h^2,

$$c^2 - BX^2 = b^2 - XC^2$$

i.e. $\quad c^2 = b^2 + BX^2 - XC^2.$

Now $BX^2 - XC^2$ is the difference of two squares and can be factorised as $(BX + XC)(BX - XC)$.

Notice that $\quad BX + XC = a$, and

$$BX - XC = BX + XC - 2XC$$

$$= a - 2XC.$$

Whereas, in triangle AXC, $XC = b\cos C$.

Putting all these together gives

$$c^2 = b^2 + (BX + XC)(BX - XC)$$

$$\Rightarrow \quad c^2 = b^2 + a(a - 2b\cos C)$$

$$\Rightarrow \quad c^2 = a^2 + b^2 - 2ab\cos C.$$

Thus, given two sides of a triangle and the included angle, the **cosine rule** enables you to find the remaining side.

$$a^2 = b^2 + c^2 - 2bc \cos A$$

and $\quad b^2 = c^2 + a^2 - 2ca \cos B$

are equivalent forms of the **cosine rule**, which you could have found by choosing one of the other sides as base in the diagram.

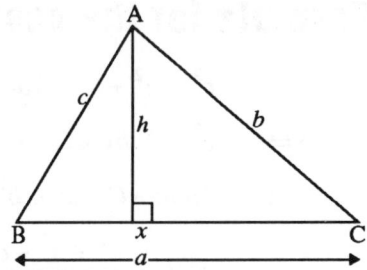

Formula for the cosine of an angle

$$c^2 = a^2 + b^2 - 2ab\cos C$$

$$\Rightarrow \quad c^2 + 2ab\cos C = a^2 + b^2$$

$$\Rightarrow \quad 2ab\cos C = a^2 + b^2 - c^2$$

$$\Rightarrow \quad \cos C = \frac{a^2 + b^2 - c^2}{2ab}$$

This arrangement (and the corresponding formulas for cos A or cos B) will enable you to find any angle of a triangle given all three sides.

What will the formulas be for cos A or cos B?

Unlike the sine rule, there is no possible ambiguity since, if C is obtuse, cos C will turn out to be negative rather than positive. (So with the cosine rule you can trust the inverse cosine function on your calculator!)

Example

Find all three angles of triangle LMN given $l = 7.2$ m, $m = 38$ m and $n = 49$ m.

Solution

To find L, use

$$\cos L = \frac{m^2 + n^2 - l^2}{2mn}$$

$$= \frac{38^2 + 49^2 - 72^2}{2 \times 38 \times 49}$$

$$= \frac{-1339}{3724} = -0.359560.$$

$$\Rightarrow \quad L = 111.1°.$$

Having found one angle, the next step could be to use either the cosine rule again or the sine rule.

$$\cos M = \frac{l^2 + n^2 - m^2}{2ln}$$

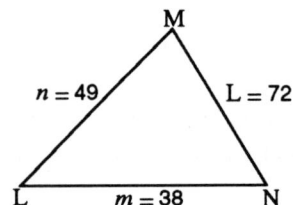

$$= \frac{78^2 + 49^2 - 39^2}{2 \times 72 \times 49} = 0.870323$$

$$\Rightarrow \quad M = 29.5°$$

$$\Rightarrow \quad N = 180° - 111.1° - 29.5°$$

$$= 39.4°.$$

Example

In triangle ABC, $b = 19$ m, $c = 8$ m and $A = 127°$.

Find a and angles B and C.

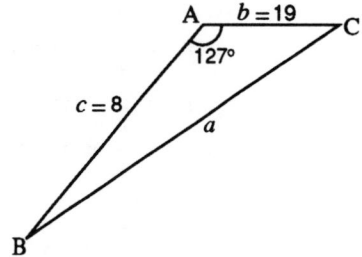

Solution

Using the cosine rule,

$$a^2 = b^2 + c^2 - 2bc \cos A$$

$$= 19^2 + 8^2 - 2 \times 19 \times 8 \cos 127°$$

$$= 425 - 304 \times (-0.601\ 815)$$

$$= 425 + 182.95176$$

$$= 607.95176$$

$$\Rightarrow \quad a = 24.7 \text{ m}$$

Next, the sine rule can be used to find B and C.

Now
$$\frac{a}{\sin A} = \frac{b}{\sin B}$$

$$\Rightarrow \quad \frac{24.7}{\sin 127°} = \frac{19}{\sin B}$$

$$\Rightarrow \quad \sin B = \frac{19 \sin 127°}{24.7} = 0.614335$$

$$\Rightarrow \quad B = 37.9°$$

and $C = 180° - 127° - 38° = 15°$ to the nearest degree.

Exercise 17B

Solve the following triangles given the relevant information :

1. In triangle ABC: $a = 18$ cm, $b = 13$ cm, $c = 8$ cm.

2. In triangle DEF: $D = 13.8°$, $e = 9.2$ m, $f = 13.4$ m.

3. In triangle LMN: $l = 33$ mm, $m = 20$ mm, $N = 71°$.

4. In triangle XYZ: $x = 4$ m, $y = 7$ m, $z = 9.5$ m.

5. In triangle PQR: $p = 9$ cm, $q = 40$ cm, $r = 41$ cm.

6. In triangle UVW: $U = 37°$, $u = 88.3$ m, $w = 97$ m.

17.3 Area of a triangle

The approach adopted in obtaining the sine rule gives an easy way
of finding a formula for the area of any triangle.

With base a, the height $h = b \sin C$ [or $c \sin B$] and the area of a
triangle is given by

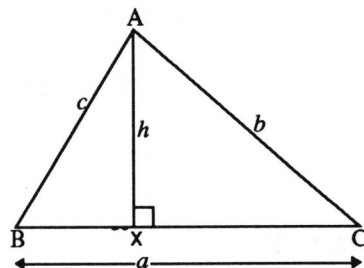

$$\text{area} = \frac{1}{2}\ \text{base} \times \text{height}$$

$$= \frac{1}{2} \times a \times b \sin C$$

$$\Rightarrow \quad \text{area} = \frac{1}{2} ab \sin C$$

$$= \frac{1}{2} ac \sin B$$

$$= \frac{1}{2} bc \sin A$$

depending upon the choice of base.

Notice again that in each case the formula requires any two sides
and the included angle.

Exercise 17C

Find the areas of the triangles in Exercise 17B,
Questions 1 to 6.

17.4 Miscellaneous Exercises

1. Find the area of triangle ABC given

    (a) $a = 42$ cm, $b = 31$ cm, $C = 58.1°$

    (b) $A = 17.6°$, $b = 127$ m, $c = 98$ m

2. A triangle has sides of length 2 cm, 3 cm and
    4 cm. Find the area of the triangle.

3. In triangle PQR, $R = 42.5°$, $p = 9$ m and
    $q = 12.2$ m. Find the area of the triangle and the
    length of the third side, r.
    Deduce the distance of R from the line PQ.

4. In triangle XYZ, $x = 17$, $y = 28$ and angle
    $X = 34°$. Find the length of the remaining side
    and the size of angles Y and Z.

5. Solve triangle PQR given

    (a) $P = 14.8°$, $R = 59.1°$, $r = 87$ m.

    (b) $P = 67°$, $p = 73$ m, $q = 42$ m.

    (c) $p = 22$ cm, $q = 89$ cm, $r = 100$ cm.

6. $a = 27$ m, $b = 32$ m and $c = 30.6$ m in triangle ABC.
    Find the smallest angle of the triangle, and its
    area.

7. In a triangle, the largest angle is twice the size of
    the smallest angle and the longest side is $1\frac{1}{2}$
    times the length of the shortest side. Find the
    angles of the triangle. Given that the length of
    the middle side is 10 cm, find the length of the
    other two sides.

ANSWERS

The answers to the questions set in the Exercises are given below. Answers to questions set in some of the Activities are also given where appropriate.

1 THE NATURE OF MATHEMATICS

Activity 3

28

Exercise 1A

1. (a) $-11°C$ (b) $-5°C$ (c) $-18°C$

2. 1016; 181.7 cm
5. (a) 4 (b) 5
6. (a) No (b) Yes

Exercise 1B

1. $r = h \approx 6.83$ cm
3. Competitor number 7
4. $a = 2$, $b = 2$

2 USING GRAPHS

Activity 1

(a) 262.5 Fr (b) £21.21 (c) 289.7 Fr
(a) 241 Fr (b) £19.21 (c) 268.7 Fr

Activity 2

(a) $86°F$ (b) $50°F$ (c) $212°F$

Activity 5

(a) $x \neq 5$ (b) $x \neq 1$ (c) $x \geq 2$
(d) all x (e) $x \neq 0$ (f) $x \neq 1$

Exercise 2A

1. (a) 6 (b) 3 (c) 2
    (d) $a^2 + 2$ (e) $(1-a)^2 + 2$
2. (a) 1 (b) -1 (c) does not exist

(d) $\dfrac{1}{a^2}$ (e) $\dfrac{1}{(1-a)}$

Exercise 2B

1. (a) odd (b) odd (c) neither
    (d) even
3. (a) odd (b) neither (c) neither
    (d) even (e) even

Miscellaneous Exercises

1. 31.5

2. (a) $x \neq 0$; $0 < f(x)$

    (b) $x > 3$; $0 < f(x)$

    (c) $x \leq 6$; $0 \leq f(x)$

    (d) $x \neq 2, -3$; $0 < f(x)$ and $f(x) \leq -6.25$

3. (a) Yes (b) No (c) Yes
    (d) Yes (e) Yes

4. (a) $-\dfrac{1}{8}$ (b) $-\dfrac{1}{7}$ (c) $-\dfrac{1}{9}$ (d) $-\dfrac{1}{16}$
5. (a) neither (b) odd (c) even

3 FUNCTIONS

Activity 1

Only $x = 1$

Exercise 3A

1. (a) $fg = x^3 - 1$, $gf = (x-1)^3$

    (b) $fg = \sqrt{x-2}$, $gf = \sqrt{x} - 2$

    (c) $fg = \dfrac{1}{x+1}$, $gf = \dfrac{1}{x} + 1$

    (d) $fg = \dfrac{1}{x^2} - 1$, $gf = \dfrac{1}{x^2 - 1}$

Answers

(e) $fg = x$, $gf = x$

(f) $fg = x$, $gf = x$

2. $fg = 1 + \dfrac{1}{x^2}$, $gf = \left(1 + \dfrac{1}{x}\right)^2$

3. $hgf = \dfrac{1}{(x-3)^2}$

Activity 2

$R = \left(\dfrac{T}{k}\right)^{\frac{2}{3}}$

Exercise 3B

1. $x - 2$
2. $\frac{1}{4}(x+1)$
3. $\frac{1}{4}(x+2)$
4. x
5. $1 - x$
6. $\dfrac{3}{x}$, $(x \neq 0)$
7. $\dfrac{1}{x} - 2$, $(x \neq 0)$
8. $5 - \dfrac{1}{(x-2)}$, $(x \neq 2)$

Activity 4

$f^{-1}(x) = \frac{1}{4}(x+3)$

Exercise 3D

1. many to one
2. one to one
3. many to one
4. many to one
5. many to one
6. one to one
7. one to one
8. many to one

Exercise 3E

1. 6000 tonnes; every 7 days; zero stock level reached

Miscellaneous Exercises

1. (a) $f^{-1}(x) = \frac{1}{3}(x+2)$

(b) $f^{-1}(x) = \frac{1}{3}(4-x)$

(d) $f^{-1}(x) = \sqrt{x+1}$, $(x \geq -1)$

(f) $f^{-1}(x) = \dfrac{1}{(x+4)}$, $(x \neq -4)$

4 GRAPH TRANSFORM

Exercise 4B

2. (a) unaltered
   (b), (c) and (d) altered

Exercise 4C

2. (a) $y = 2f(x) + 1$

(b) $y = 4 - f(x)$

(c) $y = g(x+4) + 2$

(d) $y = -2f(x) + 1$

(e) $y = f(x+1) + 41$

Miscellaneous Exercises

2. (a) $y = f(2x) + 1$

(b) $y = 1 - g(x+1)$

3. function odd

5 SOLVING PROBLEMS

Exercise 5A

1. (a) 254   (b) −15   (c) 101
   (d) 27.3   (e) 2.66   (f) 11.7

2. (a) $p = 81$, $q = 69$
   (b) $m = 23$, $n = -27$
   (c) $k = 32$, $\ell = 3$
   (d) $x = -5$, $y = -14$
   (e) $p = -0.182$, $q = 14.9$
   (f) $u = -1.55$, $v = -2.82$

3. (a) $a \leq 13$ (b) $b > 16$ (c) $c \geq 8$, $c \in \mathbf{N}$
   (d) $d > -9.43$ (b) $e > 13$ (f) $f \geq 4$, $f \in \mathbf{N}$

4. 34 tables, 66 chairs

5. 12 hardbacks, 8 paperbacks
6. 6 minutes
7. 1516 units.

Exercise 5B

1. (a) $10x+14y$ (b) a^2-ab
    (c) $18\ell^2m-15\ell m^2$ (d) $2p^3-3p^2q^2+p^2$
    (e) $2h^4k^4+3h^4k^5$

2. (a) x^2+3x+2 (b) $x^2-2x-15$
    (c) $x^2-10x+16$ (d) x^2+x-30
    (e) $x^2-3x-70$

3. (a) $2x^2-13x+21$ (b) $2x^2+5x+3$
    (c) $3x^2-4x-160$ (d) $2a^2-3ab-2b^2$
    (e) $2m^3+10m^2n^2-nm-5n^3$

4. (a) x^2+2x+1 (b) p^2-6p+9
    (c) $x^2-20x+100$ (d) $x^2+2ax+a^2$
    (e) $4x^2-20x+25$

5. (a) $2a^2-b^2$ (b) $7y^2+8y-1$
    (c) $2n+23$ (d) $m^2+2m-18$
    (e) $v^2-10uv-3u^2$

6. (a) $3(2x+5)$ (b) $u(u-3)$
    (c) $3p(p+8)$ (c) $6ab^2(2ab-1)$
    (e) $5y(x^2+3)-35$

7. (a) $(x+4)(x+2)$ (b) $(x+6)(x-5)$
    (c) $(x-2)(x-5)$ (d) $(x+4)(x+1)$
    (e) $(x-10)(x+7)$ (f) $(x-9)(x-1)$
    (g) $(x-8)(x-2)$ (h) $(x-12)(x+7)$

8. (a) $(x-4)(x+4)$ (b) $x(x-25)$
    (c) $(2x+1)(x+3)$ (d) $(2x+1)(x-4)$
    (e) $(3x+4)(x-5)$ (f) $2(x+1)(x-3)$
    (g) $(3x-10)(x+2)$ (h) $100x^2-64$

9. (a) $(5x-2)(x+1)$ (b) $(4x-3)(x+2)$
    (c) $(2x+1)(2x-3)$ (d) $3(2x-3)(x+4)$
    (e) $(3x-5)(2x+5)$ (f) $(4x-5)(3x+2)$

Activity 4

(a) $x=-9,3$ (b) $x=-m,-n$
(c) $x=\frac{1}{2},-2$

Exercise 5C

1. (a) $1,-4$ (b) $5,-10$ (c) $9,-8$
    (d) $7,11$ (e) $-2,-48$ (f) $20,30$
    (g) $\frac{3}{2},5$ (h) $\frac{7}{2},-3$ (i) $5,-3$
    (j) $\frac{7}{3},-2$

2. Length 13 m, width 10 m
3. Height $\frac{3}{2}$ m

Exercise 5D

1. (a) $3,-\frac{3}{2}$ (b) $3.27,-4.27$
    (c) $11.58,8.42$ (d) $14.58,-20.58$
2. (a) $1.70,-4.70$ (b) $6.19,0.81$
    (c) $-1.76,-6.24$ (d) $77.8,-102.8$
3. 85 m
4. 52 cm

Activity 9

(a) $p=2,\ q=7$ (b) $x=2\pm\sqrt{7}$
(c) (i) $1\pm\sqrt{2}$ (ii) $\frac{3}{2}\pm\sqrt{\frac{29}{2}}$

Exercise 5E

1. (a) $5.32,-1.32$ (b) $8.45,3.55$
    (c) $1.62,-1.62$ (d) $13.23,-23.23$
    (e) $3.57,0.13$ (f) $1.26,-2.06$
    (g) $31.7,-1.67$ (h) $38.8,-15.8$

2. (a) $x\rightarrow\infty$ (b) $-\frac{25}{4}$ (c) 190
    (d) $475,x=-15$

3. (a) 34.14 or 5.86   (b) 800 m²

Exercise 5F

1. (a) $1.22,-0.936$ (b) $0.0753,-0.781$
    (c) $3.88,-1.54$ (d) $15.6,-4.27$
    (e) $0.384,0.116$ (f) $20.8,3.06$

2. (a) none   (b) two   (c) two

(d) one   (e) two   (f) one

3. (a) 4.25   (b) 24.05 m

4. 4.82 cm

Exercise 5G

1. (a) 21   (b) 36   (c) ± 12

   (d) 5   (e) -22 (f) 9, $-\dfrac{15}{2}$

   (g) 8, -7 (h) 0, 10   (i) $\dfrac{3}{2}$, 5

   (j) 11.4, -7.43

2. 1.618

3. £1.45

4. 14p

Exercise 5H

1. (a) $x \le -6$, $x \ge 11$ (b) $\dfrac{1}{3} < x < 1$

   (c) $x \le -7$, $x \ge 9$ (d) $x < -60$, $x > -50$

2. (a) $x > 1.24$, $x < -3.24$ (b) $1.55 \le x \le 6.45$
   (c) $-18.6 < x < -5.37$ (d) $x < -9.16$, $x > 10.9$

3. (a) $-5 \le x \le 12$ (b) $-8.83 < x < -3.17$

   (c) $\dfrac{20}{3} < x < 10$ (d) $-8.61 < x < 6.78$

   (e) $0.770 \le x \le 3.90$ (f) $0.739 < x < 4.83$

Exercise 5I

1. (a) $-1 \le a \le -\dfrac{1}{5}$ (b) $4 < b < 16$

   (c) $c \le 3$, $c \ge 4\frac{1}{2}$ (d) $-27 \le d \le 6$

   (e) $-5.42 < e < 1.25$ (f) $-2.89 \le f \le 3.78$

2. (a) $x = 1$
   (b) $x = -14$, -13, -12, -11, 6, 7, 8, 9

3. (a) $1 < x < 2.27$, $5.73 < x < 7$
   (b) $x > 2.61$, $-2 < x < 0$, $x < 4.61$
   (c) $-8.37 < x < -1.13$

Miscellaneous Exercises

1. 12 short, 4 long

2. £43382.82, £21691.41

3. 10 to 40 numbers

4. (a) 7.40 am   (b) 4.47 pm

5. $0 \le Q \le 104$

6. 75 at £16 each

7. Half an hour less, 257 mph

8. 73 mph

9. 21 km/hour

10. about 4.22

11. 1.1618

12. 81 cm

13. (a) $100 - 400t$, $60 - 300t$

    (b) about $11\frac{1}{2}$ minutes

14. 60 m $\times$ 20 m

15. (a) $x = -\frac{1}{2}$ (b) $x < -9$, $-1 < x < 8$

    (c) $x = \pm 2$, ± 3 (d) $x = 4$

    (e) $x = 0$, 6, -8 (f) $x = -2$, 2

6   EXTENDING ALGEBRA

Exercise 6A

1. (a) (i) $x^2 + x - 2$ (ii) $x^2 + 5x - 3$

   (iii) $x^2 - 5x - 36$

   (b) (i) $x = 4$, 1, -2 (ii) $x = -3$, 0.54, -5.54

   (iii) $x = -4\frac{1}{2}$, -4, 9

3. (a) 1, 0.59, 3.41   (b) -1, 6, -8

   (c) $\dfrac{1}{2}$, -3, 3

4. 1 cm side

Exercise 6B

1. (b) $x = -2.34$

2. (a) -19.9, 28.4, 66.5   (b) -2.84, 5.39, -6.55

   (c) -1.86

Activity 4

(b) $(x-2)(x-1)$ (c) $x^2 - x - 2$

(c) $(qx - p)$ (e) No

Exercise 6C

2. A : 2;   B : 4;   C : 1;   D : 3

3. (a) $(x-1)$ (b) $(x-2)$ (c) $(x-1)$

   (d) $(2x+1)$ (e) $(3x-2)$

Exercise 6D

1. (a) 307   (b) 321   (c) 341   (d) 523

2. (a) x^3+5x^2+2x+6 (b) x^3-x-5

    (c) x^2+2x-4

3. (a) x^3+5x^2+4x+1 (b) x^3+x^2-6x-3

    (c) $2x^2-7$ (d) $3x^4+x^3-6x^2+4x+1$

4. (a) $x+7$ (b) $2x^2-x-1$

    (c) x^2-6x+3 (d) $x+2$

Exercise 6E

1. (a) $-1,\ -3,\ -5,\ 1$ (b) $1,\ 2,\ -2,\ -9$

2. $-3,\ -1,\ 1,\ 2$

Exercise 6F

1. (a) -40 (b) 40

2. (a) $x^2+x-10,\ 40$ (b) $x-10,\ -40$

Exercise 6G

1. (a) 7   (b) 61   (c) -488

2. $p=3$

3. $p=-3,\ q=-10$

4. $(x+2),\ (x+8)$

Exercise 6H

1. (a) $\dfrac{3}{5x}$ (b) $\dfrac{1}{(p-2)}$ (c) No

    (d) $\dfrac{(n+1)}{(n-2)}$ (e) $q+\dfrac{1}{q}$ (f) $\dfrac{4(x-3)}{(x+2)}$

    (g) $\dfrac{5(b+1)}{2ab}$ (h) $\dfrac{(2y-1)}{(y+9)}$

2. (a) $2xy$ (b) $2p^2$ (c) $x(x-1)$

    (d) $12y^2z^2$ (e) $2(x-2)(x+3)$

    (f) $(x+1)(x-2)(x-3)$

3. (a) $\dfrac{4b+5a}{ab}$ (b) $\dfrac{11}{2p}$ (c) $\dfrac{3-4m}{mn}$

    (d) $\dfrac{30x+14y}{3x^2y}$ (e) $\dfrac{2x+3}{x^2}$ (f) $\dfrac{2x-6}{x(x+2)}$

    (g) $\dfrac{16p-6}{p^2(p-1)}$ (h) $\dfrac{3r^2+8r-2}{(r-1)(r+2)}$

    (i) $\dfrac{16p^3-8q^2}{p^2q^2}$ (j) $\dfrac{13x^2+20x}{(x+2)(2x+8)}$

4. (a) $\dfrac{xp}{yq}$ (b) $\dfrac{3}{n^2}$ (c) $\dfrac{2y}{3}$

    (d) $\dfrac{a}{c^2}$ (e) $\dfrac{x-2}{x-1}$ (f) $\dfrac{(x+2)(x+1)}{(x-4)}$

    (g) $\dfrac{1}{x+6}$ (h) $\dfrac{2p}{p+4}$

Exercise 6I

1. (a) $\sqrt{12}$ (b) $\sqrt{700}$ (c) $-\sqrt{30}$

2. (a) $2\sqrt{13}$ (b) $5\sqrt{3}$ (c) $2\sqrt{30}$

    (d) $7\sqrt{5}$

3. (a) $2\sqrt[3]{2}$ (b) $3\sqrt[3]{2}$ (c) $2\sqrt[4]{3}$

4. (a) $\sqrt{2}+1$ (b) $\dfrac{\sqrt{21}+3}{4}$ (c) $\frac{2}{3}\left(\sqrt{5}+\sqrt{2}\right)$

    (d) $\dfrac{3\sqrt{2}}{2}$ (e) $\dfrac{\sqrt{14}+2}{2}$

Miscellaneous Exercises

1. (a) $(x-5)$ (b) $(x+2)$

2. (a) $-1,\ 5,\ 0.62,\ -1.62$

    (b) $3,\ 1,\ 8,\ -8$ (c) $-4,\ 2$

3. (a) $(x-7)(x^2-1)-13$

    (b) $x^4+3x^3-6x^2+9x-16+\dfrac{33}{(x+2)}$

4. (a) -15 (b) 297

5. $a=8$

6. $(x-8),\ (x+3)$

7. x^2-3x+8

8. $(x+1),\ (x-5),\ (x-6)$

9. (a) $\dfrac{(x-1)}{2}$ (b) $\dfrac{6a^2}{(2+a)}$

    (c) $\dfrac{(m-2)(m+2)}{2m}$ (d) $\frac{1}{3}(3x-1)$

    (e) x^2+x+1

10. (a) $\dfrac{12p^2+5}{10p}$ (b) $\dfrac{12-5n}{4n^2}$

   (c) $\dfrac{7x}{(x-5)(x+2)}$ (d) $\dfrac{22-3x}{(x-5)(x+5)}$

   (e) $\dfrac{5}{2b}$ (f) $\dfrac{x^2}{3x(x+5)}$

11. (a) (i) $p=2$, $q=-1$ (ii) $p=3$, $q=2$

     (iii) $p=1$, $q=-2$, $r=1$

   (b) (i) $\dfrac{1}{x}+\dfrac{3}{x-3}$ (ii) $\dfrac{5}{x+2}-\dfrac{3}{x+5}$

     (iii) $-\dfrac{3}{x}+\dfrac{5}{2(x-1)}+\dfrac{3}{2(x+3)}$

12. (a) 3 minutes   (b) $\dfrac{hc}{h+c}$

13. $\dfrac{3uv}{v+2u}$

16. (a) $x=\sqrt{\dfrac{3-\sqrt{3}}{2}}$ (b) $x=3\sqrt{2\left(4-\sqrt{5}\right)}$

   (c) $x=5\sqrt{\sqrt{15}+\sqrt{3}}$

   (d) $x=\sqrt{\dfrac{981}{221}\left(16+\sqrt{35}\right)}$

17. (a) $B=\dfrac{7}{5}$, $A=\dfrac{13}{5}$, $4x-1$

   (b) (i) $14x+7$ (ii) $83x+62$

18. (a) (i) $p-2q$ (ii) $x^2+3xy-y^2$

     (iii) $(a-b)(2a+3b)$

   (b) (i) $(x-3a)(x+a)$ (ii) $(p+12q)(p-2q)$

     (iii) $(a-b)(a-4b)(a+3b)$

19. $x(2x-3)(2x+3)\left(4x^2+9\right)$

7   STRAIGHT LINES

Exercise 7A

1. $\dfrac{20}{27}$, 0.541

2. $1.72\ \text{m s}^{-2}$

3. $0.2\ \text{m A/g}$

4. $0.7466\ \text{ms}^{-2}$

5. 1 minute 45 seconds

6. 0.106

7. $y=3x+5$; £9.5, £14, £41

Exercise 7B

1. (a) 3, −1   (b) −4, −3   (c) $\dfrac{1}{2}$, 5

   (d) $-\dfrac{6}{5}$, $-\dfrac{1}{2}$ (e) 4, 0   (f) 1, 0

   (g) −1, 0   (h) 0, 5

2. (a) −4, 9   (b) $-\dfrac{1}{2}$, 3   (c) $\dfrac{3}{2}$, −2

   (d) $\dfrac{1}{2}$, $-\dfrac{3}{2}$

3. $y=3x-14$

4. $y=-\dfrac{1}{3}x+\dfrac{22}{3}$

5. $y=2x-5$

6. $y=2x-8$

7. $y=-\dfrac{4}{3}x$

Exercise 7D

1. (a) (12, −7)   (b) (−2, 1)   (c) (−1, −1)
2. Lines are parallel.
3. (a) $x+y=300$

   (b) $23x+36y=10000$

   (c) $x=61.5$, $y=238.5$

4. (1, 2), (−2, −4), $\left(\dfrac{16}{7},\ -\dfrac{13}{7}\right)$

Exercise 7E

1. (a) 5   (b) 12   (c) 8.49   (d) 14.3

2. $\left(\dfrac{3}{2},\ 2\right)$, $\left(5,\ \dfrac{5}{2}\right)$, (1, −2), $\left(\dfrac{3}{2},\ -1\right)$

3. 5

Miscellaneous Exercises

1. 105

2. $-1.675\ \text{m s}^{-2}$

3. (a) 1 (b) $\dfrac{3}{2}$ (c) $-\dfrac{3}{2}$ (d) $-\dfrac{14}{9}$

4. (a) 5, 3 (b) -1, 1 (c) $\dfrac{1}{2}$, 0

 (d) $\dfrac{1}{3}$, 2 (e) $\dfrac{5}{2}$, $-\dfrac{11}{2}$ (f) $-\dfrac{3}{4}$, $-\dfrac{1}{4}$

5. $y = 2x - 11$

6. $y = 4x + 21$

7. $3y + 4x = 15$

8. 155 cm

9. $4y + x = 59$

10. $2y + 3x - 5 = 0$

11. $y = 5x - 20$

12. $\left(\dfrac{3}{5}, \dfrac{3}{5}\right)$

13. $16y + 10x + 71 = 0$

8. RATES OF CHANGE

Exercise 8B

1. (a) $4x$ (b) $2x + 1$ (c) $2t + 4$ (d) $2x - 1$

 (e) 12ℓ (f) $10 - \dfrac{2x}{5}$ (g) $\dfrac{2y}{9} + 3$ (h) $\dfrac{2n - 5}{2}$

 (i) $1 - 12v$ (j) $\dfrac{3}{2}x + \dfrac{1}{5}$

2. (a) -7 (b) 5 (c) $23\dfrac{5}{6}$ (d) $\dfrac{9}{4}$ (e) 71

3. (a) $16\ \text{ms}^{-1}$ (b) $-7\ \text{ms}^{-1}$

Exercise 8C

1. (a) $3x^2 + 10x + 3$ (b) $18t^2 - 20t + 2$

 (c) $10x - \dfrac{1}{x^2}$ (d) $3x^2 - 1$ (e) $\dfrac{3t + 3}{5}$

2. (a) $2(x + 2)$ (b) $3x^2 - 1$

 (c) $3s^2 + \dfrac{4}{3}s + \dfrac{1}{9}$ (d) $\dfrac{8}{3}y^2 + \dfrac{2}{3}$

 (e) $3x^2 - 5 + \dfrac{1}{x^2}$

3. (a) 9 (b) $\dfrac{13}{4}$ (c) $(2, 19)$

 (d) $(-1, 18)$ (e) $(2, 41)$, $(1, 18)$

4. (a) $\dfrac{1}{3}$ (b) $\dfrac{3}{16}$ (feet per year)

Exercise 8D

1. $(-2, 55)$ max; $(6, -201)$ min

2. (a) $\left(\dfrac{3}{2}, \dfrac{5}{2}\right)$ min

 (b) $(3, 18)$ min, $(-3, -18)$ max

 (c) $(-7, -469)$ min, $(5, 395)$ max

 (d) $(2, 12)$ min

3. (a) 50, 680

4. 47.6 mph

Exercise 8E

1. 5 m²

2. (a) 13.33 m² (b) 10.37 m²

3. 20 m

6. 1195, 2389, 1591 mm

Exercise 8F

3. 41.32 cm

Exercise 8G

1. (a) $-\dfrac{6}{x^7}$ (b) $9x^2 - \dfrac{4}{x^3}$

 (c) $20q^3 + 12q^2 + \dfrac{6}{q^4}$ (d) $8t^7 - \dfrac{18}{t^7}$

 (e) $-\dfrac{2}{x^5}$ (f) $-\dfrac{6}{5x^3}$

 (g) $-\dfrac{2}{t^2} + \dfrac{14}{t^5}$ (h) $\dfrac{x}{2} + \dfrac{9}{4x^4}$

2. (a) 2.5 (b) 11 (c) 449.5

3. (a) 4 (b) 308 (c) -1

Exercise 8H

1. (a) $y = 10x - 12$ (b) $y = x - 2$

 (c) $y = -2x - 12$

2. (a) $y = 9 - x$ (b) $8y = 128x + 527$

 (c) $7y = 24 - x$

3 (a) 4 (c) 3

Miscellaneous Exercises

1. (a) $2x + 4$ (b) $3x^2 - 8x + 17$

(c) $3x^2 + 4x + 1$ (d) $2x - \dfrac{2}{x^2}$

(e) $3x^2 - 2 - \dfrac{1}{x^2}$

2. (a) $20x^3 - \dfrac{6}{x^3}$ (b) $-\dfrac{18}{t^5}$ (c) $6y^5 + \dfrac{10}{y^3}$

(d) $-\dfrac{2}{p^2} - \dfrac{2}{p^3}$ (e) $-\dfrac{18}{25x^3} + \dfrac{8}{25}x$

3. (a) $y = 8x - 15$ (b) $y = -x - 2$

4. (b) $10x - 2$

6. 15625 cm^3

7. (a) $v = 3t^2 - 24t + 45$

(b) $t = 3$ s or 5 s   (c) $t = 4$ s

8. (a) 25

9. $9, -4$

10. (a) $-\dfrac{5}{27}$ (b) 1

11. $N = 204 - \dfrac{12}{5}v$, 42.5

12. (c) $-\dfrac{1}{2\sqrt{x}}$

14. $-\dfrac{1}{2x\sqrt{x}}$

9 POWERS

1. (a) 4 (b) 2 (c) 3

(d) $\dfrac{1}{2}$ (e) $\dfrac{2}{5}$ (f) -1

2. (a) 2.45 (b) 2.15 (c) 2.24
(d) 0.839 (e) 0.707

3. (a) $\dfrac{1}{2}$ (b) $\dfrac{1}{3}$ (c) $\dfrac{1}{4}$ (d) $-\dfrac{1}{4}$

4. (a) £409.84   (b) £219.02

5. 14.35%

6. (a) £2000   (b) £849   (c) £3167

Exercise 9B

1. (a) 8 (b) 9 (c) $100\ 000$ (d) $10\ 000$
(e) 32 (f) 4

2. (a) 1310 (b) 0.552 (c) 187 (d) 4.64
(e) 0.552 (f) 0.0421

3. (a) $\dfrac{1}{4}$ (b) $\dfrac{1}{8}$ (c) $\dfrac{1}{4}$ (d) $\dfrac{1}{625}$

4. (a) 0.316 (b) 0.405 (c) 55.9
(d) 297

5. (a) 1 (b) 2.5 (c) $-\dfrac{1}{2}$

(d) $-\dfrac{1}{3}$ (e) $\dfrac{2}{3}$ (f) $-\dfrac{3}{4}$

6. (a) $5\sqrt{p}$ (b) $\dfrac{6}{q}$ (c) $\dfrac{10}{\sqrt{z}}$

(d) $\dfrac{3}{4\sqrt{y}}$ (e) $\dfrac{1}{2}m\sqrt{m}$ (f) $\dfrac{12}{t^2\sqrt{t}}$

7. (a) 20.3% (b) 58.7%

Exercise 9C

1. (a) $1 + 2x + x^2$ (b) $1 + 3x + 3x^2 + x^3$

(c) $1 + 4x + 6x^3 + 4x^3 + x^4$

(d) $1 + 5x + 10x^2 + 10x^3 + 5x^4 + x^5$

2. (a) $a^2 + 2ax + x^2$ (b) $a^3 + 3a^4x + 3x^2a + x^3$

(c) $c^4 + 4a^2x + 6a^2x^2 + 4ax^3 + x^6$

(d) $a^5 + 5a^4x + 10a^3x^2 + 10a^2x^3 + 5ax^4 + x^5$

3. (a) $9 - 12x + 4x^2$

(b) $8 + 60p + 150p^2 + 125p^3$

(c) $625 - 250m + \dfrac{75}{2}m^2 - \dfrac{5}{2}m^6 + \dfrac{1}{16}m^4$

Exercise 9D

1. (a) 6 (b) 5 (c) 10 (d) 20 (e) 3

2. $\dbinom{3}{2}$; $\dbinom{4}{2}$

Exercise 9E

1. (a) 84 (b) 792 (c) $1\ 144\ 066$
2. $60,\ 1260,\ 210,\ 9\ 979\ 200$
3. $12\ 650$
4. 210
5. 4368
6. $13\ 860$

Exercise 9F

1. (a) $1 + 8x + 24x^2 + 32x^3 + 16x^4$

   (b) $125 - 150p + 60p^2 - 8p^3$

   (c) $46656 - 23328a + 4860a^2 - 5400a^3$
   $+ \dfrac{135}{4}a^4 - \dfrac{9}{8}a^5 + \dfrac{1}{64}a^6$

   (d) $32m^5 + 240m^4n + 720m^3n^2 + 1080m^2n^3$
   $+ 810mn^4 + 243n^5$

   (e) $16 - 16r + 6r^2 - r^3 + \dfrac{1}{16}r^4$

2. (a) $15\,063$ (b) $-16\,800$

   (c) $10\,500\,000\,p^3$ (d) $495 \times 3^4 \times 2^8 \times \ell^8$

   (e) $-1200 \times \dfrac{5^{10}}{6^7}$

3. (a) 1.05 (b) 1.22 (c) 0.932 (d) 8.12
   (e) $98\,000\,000$ (f) $1\,770\,000$

Exercise 9G

1. (a) $-\dfrac{3}{32}x^2$ (b) $-x^2$ (c) $-\dfrac{1}{2}x^2$

   (d) $-\dfrac{5}{192}x^3$ (e) $\dfrac{35}{144}x^3$

2. (a) $1 - \dfrac{x}{2} - \dfrac{x^2}{8} - \dfrac{x^3}{16}$, $|x| < 1$

   (b) $1 + 2x + 3x^2 + 4x^3$, $|x| < 1$

   (c) $1 - \dfrac{3}{2}x + \dfrac{3}{16}x^2 + \dfrac{1}{16}x^3$, $|x| < 1$

   (d) $1 - 2x + 4x^2 - 8x^3$, $|x| < \dfrac{1}{2}$

   (e) $1 + \dfrac{3}{4}x + \dfrac{3}{32}x^2 - \dfrac{1}{128}x^3$, $|x| \le 2$

   (f) $1 - \dfrac{15}{4}x + \dfrac{45}{32}x^2 + \dfrac{135}{128}x^3$, $|x| < \dfrac{1}{3}$

   (g) $1 + 3x + \dfrac{27}{2}x^2 + \dfrac{405}{6}x^3$, $|x| < \dfrac{1}{6}$

   (h) $1 + \dfrac{9}{20}x - \dfrac{27}{400}x^2 + \dfrac{569}{24000}x^3$, $|x| \le \dfrac{4}{3}$

3. (a) $\dfrac{1}{2} - \dfrac{1}{4}x + \dfrac{1}{8}x^2 - \dfrac{1}{16}x^3$, $|x| \le 2$

   (b) $2 - \dfrac{1}{4}x - \dfrac{1}{64}x - \dfrac{1}{512}x^3$, $|x| < 4$

   (c) $\dfrac{1}{2} + \dfrac{1}{16}x + \dfrac{1}{64}x^2 + \dfrac{7}{1536}x^3$, $|x| < \dfrac{8}{3}$

4. $1 - \dfrac{1}{2}x^2 - \dfrac{1}{4}x^4 - \dfrac{1}{8}x^8$, $|x| \le 1$

Exercise 9H

1. (a) 0.8333 (b) 3.015 (c) 4.080

2. $x < 100$, $k = \dfrac{11}{4800}$

Miscellaneous Exercises

1. (a) $\dfrac{1}{3}$ (b) $\dfrac{3}{2}$ (c) $\dfrac{1}{2}$ (d) $\dfrac{1}{8}$

2. (a) $\dfrac{1}{4}$ (b) $\dfrac{1}{2}$ (c) $-\dfrac{2}{3}$ (d) -1

3. (a) 1.262 (b) -0.631 (c) -5.048
   (d) 3.155

4. (a) $100\,947$ (b) $2\,240\,315$

5. (a) $1 + 5x + 10x^2 + 10x^3 + 5x^4 + x^5$

   (b) $81 + 54x + \dfrac{27}{2}x^2 + \dfrac{3}{2}x^3 + \dfrac{1}{10}x^4$

   (c) $64 - 576p + 2160p^2 - 4320p^3 + 4860p^4$
   $- 2916p^5 + 729p^6$

6. (a) $1 - x - x^2 - x^3$

   (b) $1 + \dfrac{3}{2}x - \dfrac{9}{8}x^2 + \dfrac{27}{16}x^3$

   (c) $1 - \dfrac{7}{30}x + \dfrac{21}{200}x^2 - \dfrac{91}{5400}x^3$

   (d) $1 - \dfrac{2}{3}x + \dfrac{8}{9}x^2 - \dfrac{112}{81}x^3$

7. (a) $\dfrac{1}{1000} + \dfrac{3}{200000}x + \dfrac{15}{80000000}x^3$

   (b) $\dfrac{1}{a^4} - \dfrac{4}{a^5}x + \dfrac{6}{a^6}x^2$

8. (a) 6 (b) -540 (c) 664 (d) 5760

11. (a) $(1.02)^{12} < (1.06)^4$

  (b) (i) 26.97%   (ii) 27.05%   (iii) 27.09%

10 CIRCULAR MEASURE

Activity 2

(a) $2\pi r$ (b) $r\theta$ (c) $\theta = 1$ radian

Exercise 10A

1. (a) $210°$ (b) $540°$ (c) $114.6°$ (d) $165°$
2. (a) 0.218 radians   (b) 1.265 radians
   (c) 3.665 radians   (d) 0.349 radians

Activity 4

(a) πr^2 (b) $\dfrac{\theta}{2\pi}$ (c) $\frac{1}{2}r^2\theta$

Exercise 10B

1. (a) $\dfrac{5}{3}$ radians   (b) 150000 cm^3
2. $r = 10$ cm, $\theta = 2$ radians
3. 0.4 ms^{-1}

Exercise 10C

1. (a) $3x$ (b) x^2 (c) $1 - \dfrac{x^2}{8}$ (d) $-2x$
2. (a) 0.0001   (b) 0.9999955   (c) 1
   (d) 0.01   (e) 0.999992
4. $\alpha = 0.14$

Miscellaneous Exercises

1. $\dfrac{5\pi}{12}$

2. $\dfrac{\pi}{3}$

3. 7.4 cm

4. (a) $\dfrac{\pi}{4}$ (b) 1.288 radians   (c) 3.121 radians

5. 0.04

6. $\dfrac{\pi}{2}$

7. 0.8 radians

11 MODELLING NATURAL CYCLES

Exercise 11A

1. (a) 2.6   (b) 3.2   (c) 1.88
2. (a) $-\dfrac{1}{\sqrt{2}}$ (b) $-\dfrac{1}{\sqrt{2}}$ (c) $-\dfrac{1}{\sqrt{2}}$

   (d) $\dfrac{1}{\sqrt{2}}$ (e) $-\dfrac{1}{\sqrt{2}}$ (f) 1

   (g) -1 (h) -1

Exercise 11B

1. $60°$, $300°$
2. $45°$, $225°$
3. $14.5°$, $165.5°$
4. $210°$, $330°$
5. $14.0°$, $194.0°$, $374.0°$, $554.0°$
6. -0.340 radians, $-\pi + 0.340$ radians
7. ± 1.23 radians
8. (b), (c) and (e)

Exercise 11C

1. $0, \pi, 2\pi$; 2π; 1
2. $90°$; 2π; 3
3. $0°, 180°, 360°$; 2π; 2
4. $\pm\dfrac{\pi}{2}$; π
5. No crossings; 2π; $\frac{1}{2}$
6. $\dfrac{\pi}{6}, \dfrac{\pi}{2}, \dfrac{5\pi}{6}, \dfrac{7\pi}{6}, \dfrac{3\pi}{2}, \dfrac{11\pi}{6}$; $\dfrac{2\pi}{3}$; 1
7. $0°, 360°$; 4π; 5
8. $0, \pi, 2\pi$; π
9. $\dfrac{\pi}{4}, \dfrac{3\pi}{4}$; π; 2
10. $120°, 300°$; 2π; 3

Exercise 11D

1. $134.4°$, $225.6°$
2. 1.77, 4.58 radians

3. 66.3°

4. 15°, 150°, 195°, 330°

5. $\left(\frac{\pi}{2} - 0.232\right)$, $(\pi - 0.232)$, $\left(\frac{3\pi}{2} - 0.232\right)$, $(2\pi - 0.232)$ radians

6. 0.215; $\frac{2\pi}{2} \pm 0.215$; $\frac{4\pi}{3} \pm 0.215$, $2\pi - 0.215$

7. 22.5°, 112.5°, 202.5°, 292.5°

8. 240°

9. 45°, 135°, 225°, 315°

10. 27.5°, 57.5°, 117.5°, 147.5°

11. 0, 2π

Exercise 11E

1. 0°, 180°, 360°

2. 90°, 210°, 330°

3. 41.4°, 120°, 240°, 318.6°

4. $\frac{3\pi}{4}$, $\frac{7\pi}{4}$

5. 18.4°, 198.4°

Exercise 11F

1. (a) $-\frac{\pi}{4}$　(b) $\frac{\pi}{3}$　(c) $-\frac{\pi}{2}$

2. (a) π　(b) $\frac{2\pi}{3}$　(c) $\frac{\pi}{6}$

3. (a) $-\frac{\pi}{4}$　(b) $\frac{\pi}{3}$　(c) $-\frac{\pi}{6}$

Miscellaneous Exercises

1. 5.74°, 174.26°

2. −288.4°, −108.4°, 71.6°, 251.6°

3. 3.48, 5.94 radians

4. 120°

5. 56.3°, 236.3°, 416.3°, 596.3°

7. (a) 2π, 1　(b) π　(c) 2π, 2　(d) 2π, 1
　(e) π　(f) 2π, $\frac{1}{2}$　(g) π, 4　(h) 2π, 3

(i) π　(j) π, 2　(k) $\frac{2\pi}{3}$, 1　(l) 2π

(m) $\frac{\pi}{2}$, 1　(n) π, 3

9. (a) 135°, 315°　(b) $\frac{\pi}{6}$, $\frac{5\pi}{6}$, $\frac{7\pi}{6}$, $\frac{11\pi}{6}$

(c) 129°

(d) $\frac{7\pi}{18}$, $\frac{11\pi}{18}$, $\frac{19\pi}{18}$, $\frac{23\pi}{18}$, $\frac{31\pi}{18}$, $\frac{35\pi}{36}$

10. (a) 0°, 180°, 360°　(b) 1.89, 5.03 radians

(c) 19.59°, 90°, 160.5°

(d) 120°, 240°

11. (a) 9.2°, 99.2°　(b) 0, π, 2π

(c) 0, $\frac{\pi}{2}$, $\frac{2\pi}{3}$, 2π

12 GROWTH AND DECAY

Exercise 12A

1. 148.41
2. 1.65
3. 2.12
4. 1.36
5. 7.59
6. −4.48
7. 1.1
8. 4.25
9. 1.1
10. 1.39

Exercise 12B

1. 2.32
2. 0
3. −0.415
4. 1.46
5. −2.58
6. 0.63
7. 1.68
8. 0.50
9. −4.1
10 −0.38

Exercise 12C

1. 2
2. 4
3. 3
4. −4
5. $-\frac{1}{2}$
6. $\frac{1}{2}$
7. 3
8. $\frac{1}{2}$

Miscellaneous Exercises

1. (a) 1.39   (b) −0.748   (c) 1.30   (d) 0
    (e) 0.699   (f) −2.71   (g) 0.577
    (h) 0.104   (i) 0.566   (j) 3   (k) 2
    (l) −3

2. $a = 30$, $k = 0.0995$, 10.05 hours

13 INTEGRATION

Exercise 13A

1. About 3

Exercise 13B

1. (a) 100   (b) 35 (c) 80

2. (a) $\int_0^{10}(x+5)\,dx$ (b) $\int_{20}^{30}(8-0.1t)\,dt$

    (c) $\int_{-2}^{3}(2x+15)\,dx$

3. 99
4. (a) 120   (b) 50   (c) 1800   (d) 126.16
5. 62.5 m
6. (a) 3100   (b) 6150

Exercise 13C

1. (a) x^2+7x+c (b) $10t-\frac{1}{2}t^2+c$

    (c) $2.8w+5.7w^2+c$ (d) $-14x-\frac{11x^2}{2}+c$

2. (a) 30   (b) 46.5   (c) 99   (d) 115

Exercise 13D

1. $\frac{1}{10}x^{10}+c$

2. $-\frac{1}{7x^7}+c$

3. $-\frac{1}{4x^4}+c$

4. $\frac{x^5}{5}+\frac{x^8}{8}+c$

5. $\ln x-\frac{1}{x}+c$

6. $\frac{3}{4}x^8+c$

7. $-\frac{3}{2t^2}+c$

8. $\ln w^2+c$

9. $e^p-\frac{3p^2}{2}+c$

10. $2q+\frac{1}{2q^2}+c$

11. $\frac{x^4}{8}+c$

12. $\frac{x^8}{6}+c$

13. $-\frac{3}{4y^2}+c$

14. $2x^2-\frac{2}{3}\ln x+c$

15. $\frac{1}{4}e^k-\frac{1}{2k^2}+c$

16. $\frac{3}{4}x-\frac{x^6}{12}+c$

17. $e^m+\frac{1}{6m^3}+c$

18. $\frac{5x^3}{6}-\frac{x^2}{4}+\frac{x}{2}+c$

19. $\frac{z^3}{5}-\frac{1}{5}\ln z+c$

20. $2x-\ln x+c$

Exercise 13E

1. (a) $\dfrac{250}{3}$ (b) 51.25   (c) $\dfrac{5}{6}$

   (d) $2e^3 - \dfrac{3}{e^2} + \dfrac{35}{3}$ (e) $4 + \ln 3$

   (f) 0.516 to 3 significant figures

2. (a) $41\dfrac{1}{15}$ (b) 11   (c) $\dfrac{3}{8}$

   (d) $2 + \dfrac{2}{3}\ln\dfrac{5}{6}$

3. $\dfrac{1}{6}$

4. $\dfrac{2}{11}, \dfrac{20}{11}$

Exercise 13F

1. (a) $h = 2t - 5t^2 + 1000$ (b) $t = 14.34$

2. (a) $x = 5t + \dfrac{1}{2}t^2 - \dfrac{1}{12}t^3 - 30$

   (b) $x(7.7) > 0$, $x(7.8) < 0$

3. (a) 896   (b) $4\ln 5$ (c) 200

4. (a) $C = \dfrac{1}{3}Q^3 - \dfrac{3}{2}Q^2 + 5Q + 5$

   (b) 114

Miscellaneous Exercises

1. (a) $\dfrac{5x}{2} + 2x + c$ (b) $\dfrac{t^4}{8} + \dfrac{2t^3}{9} + c$

   (c) $7\ln p - \dfrac{1}{2}p^2$ (d) $\dfrac{1}{4s^5} + c$

   (e) $\dfrac{x^3}{3} - \dfrac{4}{x} - 4x + c$

2. (a) $e^{15} + 2249$ (b) $e^3 - e^{-3} + 36$

   (c) $\dfrac{3}{10} + \ln 4$ (d) $5\dfrac{5}{6}$ (e) $-\dfrac{5}{12}$

3. $A = \dfrac{5}{12}$, $B = \dfrac{8}{3}$

4. 11

5. 3

6. (a) $\dfrac{8}{3}$ (b) $\dfrac{20}{3}$ (c) $e^2 - 5$

7. (a) $2 - \ln 3$ (b) $\dfrac{e^{-1}}{2} - \dfrac{e^{-4}}{2} + \dfrac{3}{2}\ln 4$

8. (a) (i) $0.003t^3 + 0.04\,t^2 + 0.01t$

   (ii) $36 - 0.003t^3 - 0.04t^2 - 0.01t$

   (b) (i)   0.053 litres   (ii) 1.58 litres

   (c) (i)   34.6 litres   (ii) 16.7 litres

9. (b) (i) 1.129×10^5 (ii) 8.144×10^5

10. (a) 42.41 cm³   (b) 11.7500; 3.1416

    (c) 442 cm³

11. (a) 13   (d) −14

12. (b) (i) 66   (ii) 112

13. (d) (i) 20   (ii) 25

14 SEQUENCES AND SERIES

Exercise 14A

1. (a) $3 \times 2^{n-1}$ (b) $36 \times \left(\dfrac{1}{2}\right)^{n-1}$

   (c) $2 \times (-3)^{n-1}$ (e) $90 \times \left(-\dfrac{1}{3}\right)^{n-1}$

   (e) 10^n (f) $(-1)^{n-1}6$ (g) $\dfrac{1}{4} \times \left(\dfrac{1}{3}\right)^{n-1}$

2. (a) 315   (b) 118 096   (c) 0.6641

   (d) 83.33   (e) 18 620   (f) 16.01

3. (a) 11th   (b) 21st   (c) 10th

4. (a) 9   (b) 1708

5. £10 202

Exercise 14B

1. (a) $\dfrac{320}{3}$ (b) 135   (c) 200   (d) -   (e) $\dfrac{1}{9}$

2. (a) 0.111111...   (b) $\dfrac{37}{99}$

3/ (a) $\dfrac{52}{99}$ (b) $\dfrac{358}{999}$ (c) $\dfrac{193}{990}$

4. (a) 42   (b) 0.125

5. 211.6 cm^3
6. (a) 16.963 s (b) 49.4 m

Exercise 14C

1. (a) 105 (b) 252 (c) 66 (d) 2500

2. (a) $u_n = 3n - 2$, $S_n = \dfrac{n}{2}(3n - 1)$

    (b) $u_n = 3 + 9n$, $S_n = \dfrac{n}{2}(9n + 15)$

    (c) $u_n = 65 - 5n$, $S_n = \dfrac{n}{2}(125 - 5n)$

    (d) $u_n = \dfrac{1}{2}(3n - 1)$, $S_n = \dfrac{n}{4}(1 + 3n)$

3. 736 cm

Exercise 14D

1. (a) 536 (b) 45 (c) 1162
    (d) 277.2 (e) 385 (f) 409.125
2. -8, 621
3. (a) 10 (b) 19
4. 5.5
5. £76 300

Exercise 14E

1. (a) $5^2 + 6^2 + 7^2 + \ldots + 15^2$

    (b) $1 + 3 + 5 + \ldots + 19$

    (c) $1 + 2 + 3 + \ldots + n$

    (d) $\dfrac{1}{3} + \dfrac{2}{4} + \dfrac{3}{5} + \ldots + \dfrac{8}{10}$

    (e) $4^2 + 5^2 + 6^2 + \ldots + 98^2$

2. (a) $\displaystyle\sum_{r=1}^{25} \dfrac{1}{r}$ (b) $\displaystyle\sum_{r=0}^{40} (10 + r)$ (c) $\displaystyle\sum_{r=1}^{n} r^3$

    (d) $\displaystyle\sum_{r=0}^{12} 3^r$ (e) $\displaystyle\sum_{r=1}^{n} (5r + 1)$ (f) $\displaystyle\sum_{r=0}^{16} 14 + 3r$

    (g) $\displaystyle\sum_{r=0}^{n} 5 \times 10^r$ (h) $\displaystyle\sum_{r=1}^{20} \dfrac{r}{(r+1)(r+2)}$

3. (a) $\displaystyle\sum_{r=10}^{99} r$ (b) $\displaystyle\sum_{r=1}^{60} (2r - 1)$,

    (c) $\displaystyle\sum_{r=10}^{20} r^2$ (d) $\displaystyle\sum_{r=0}^{14} (7r + 1)$

4. (a) $\displaystyle\sum_{r=0}^{18} (19 - r)$ (b) $\displaystyle\sum_{r=1}^{40} \dfrac{1}{r}$ (c) $\displaystyle\sum_{r=0}^{6} (r - 3)^2$

Exercise 14F

1. (a) 10 (b) 100 (c) 731 (d) 8.8×10^{16}

    (e) $\dfrac{7}{3}$ (f) 3

2. (a) $\displaystyle\sum_{r=1}^{6} \left(-\dfrac{1}{r}\right)^{r+1}$ (b) $\displaystyle\sum_{r=1}^{12} (-1)^r r^2$

    (c) $\displaystyle\sum_{r=0}^{50} 12 \, (-0.2)^r$

3. (a) 84 (d) 32 (e) 44 (f) 36

Exercise 14G

1. (a) $\displaystyle\sum_{r=1}^{20} r = 210$ (b) $\displaystyle\sum_{r=1}^{10} r^2 = 385$

    (c) $2\displaystyle\sum_{r=1}^{15} r^2 = 2480$ (d) $2\displaystyle\sum_{r=1}^{50} r = 2550$

    (e) $\displaystyle\sum_{r=1}^{13} (2r - 1) = 169$ (f) $\displaystyle\sum_{r=1}^{10} r^3 = 3025$

2. 165
3. (a) 4515 (b) 13959 (c) 861
4. 6461

Exercise 14H

1. (a) $(2n + 1) - 9 = 2(n - 4)$

    (b) $n(n + 1) - 12 = (n + 4)(n - 3)$

    (c) $9n^2 - 29n - 10$

    (d) $(n + 1)(2n + 1) - 15(n + 1) + 24$; $(n - 5)(n - 1)$

    (e) $3n^2 - 27n + 26$

3. (a) 315 (b) 816 (c) $10\,385$
    (d) $1\,668\,550$ (e) 2460

4. (a) $n(n + 1)^2$ (b) $\dfrac{n}{3}(n^2 - 3n - 9)$

    (c) $\dfrac{n}{3}(n + 4)(n - 1)$

(d) $\frac{1}{4} n (n+1)(n+5)(n-4)$

(e) $\frac{1}{4} n (n+1)(n+2)(n+3)$

Exercise 14I

1. (a) $\frac{7}{2}$ (b) 1 (c) 0 (d) $-\frac{5}{2}$ (e) 3

   (f) 1 (g) $\frac{12}{5}$ (h) -6

Miscellaneous Exercises

1. (a) 102.4 mm (b) 15 (c) 42

2. (a) $4 \times (1.5)^{n-1}$ (b) $256-6n$

   (c) $10 \times (0.2)^{n-1}$ (d) $0.32n - 0.15$

3. (a) 810 (b) 1062 880 (c) -330 (d) 960
4. 21
5. (a) 249 (b) 22

6. (a) $\displaystyle\sum_{r=1}^{62} (2r-1)$ (b) $\displaystyle\sum_{r=1}^{25} \frac{r}{r^2+2}$

7. (a) divergent (b) $\frac{60}{7}$ (c) -4650

9. (a) $\frac{n}{6}(n+1)(2n+19)$

   (b) $\frac{n}{4}(n+1)(n+4)(n-3)$

10. (a) 2 (b) $\frac{1}{16}$ (c) 6 (d) 2

11. (a) $u_r = 3r^2 - 3r + 1$

12. (a) $\frac{1}{4} n (n+1)(n^2+5n+6)$

15 USING CALCULUS

Exercise 15A

1. (a) $6x$ (b) $12x^2$ (c) 2 (d) 0 (e) $\frac{2}{x^3}$

   (f) $24x - 24$ (g) 0 (h) e^x (i) $-\frac{1}{x^2}$

2. (a) $24x$ (b) $-\frac{72}{x^5}$ (c) e^x (d) $\frac{2}{x^3}$ (e) 0

Exercise 15B

1. (a) $x = -\frac{3}{4}$, minimum

   (b) $x = 2$, minimum; $x = -2$, maximum

   (c) $x = \frac{-2+\sqrt{19}}{3}$, minimum

   $x = \frac{-2-\sqrt{19}}{3}$, maximum

   (d) $x = 1$, point of inflection
   (e) $x = \ln 4$, minimum
   (f) none

2. (a) maximum at $x = \frac{2+a}{3}$

   minimum at $x = a$

   (b) minimum at $x = \frac{2+a}{3}$

   maximum at $x = a$

Exercise 15C

1. (a) $16(2x-5)^7$ (b) $10\left(x^2+x^3\right)^{10}\left(2x+3x^2\right)$

   (c) $-\frac{1}{(x-2)^2}$ (d) $-\frac{6}{(3x+1)^3}$

   (e) $\frac{1}{2}\frac{1}{\sqrt{x+1}}$ (f) $6\left(e^x-x\right)^5\left(e^x-1\right)$

   (g) $\frac{3}{(3x+4)}$ (h) $\frac{1}{2\sqrt{x}} e^{\sqrt{x}}$

2. (a) $x = 0$ (b) $x = -4$ (c) $x = 1$
   (d) $x = \pm 1$

3. $-0.964°\text{C}/\min$; $-0.290°\text{C}/\min$
4. 0.48 millions / year

5. (a) $2e^{2x}$; $\frac{e^{2x}}{2} + c$

   (b) $-e^{-x}$; $-e^{-x} + c$

6. $2xe^{x^2}$; $\frac{1}{2}e^{x^2} + c$

7. $\ln(x+2) + c$

Exercise 15D

1. (a) $\dfrac{(2x+1)^5}{10}+c$ (b) $-\dfrac{1}{(x-5)}+c$

(c) $2\sqrt{x+1}+c$ (d) $\dfrac{3}{8}(4x-1)^{\frac{3}{2}}+c$

2. (a) $\ln 2$ (b) $\frac{1}{2}\ln 3$

3. (a) $\frac{1}{3}\ln(3x+1)+c$ (b) $\frac{1}{2}\ln\left(x^2+1\right)+c$

(c) $\ln\left(1+e^x\right)+c$

4. $\dfrac{1}{2}\left(1-\dfrac{1}{e}\right)$

Exercise 15E

1. (a) $e^x(x+1)$ (b) $x+2x\ln x$

(c) $\dfrac{-2\left(x^2+x+3\right)^2}{\left(x^2-3\right)^2}$ (d) $\dfrac{xe^x}{(1+x)^2}$

2. (a) $x=-1$ (b) $x=0$, $x=0.607$
(c) $x=0$

3. Root of $(1+x)\ln(1+x)=0$

4. $\dfrac{-3}{2(2x-1)^2}\sqrt{\dfrac{2x-1}{x+1}}$

5. $\dfrac{e^x(2+x)}{2\sqrt{e^x(x+1)}}$

6. $\dfrac{1}{\sqrt{x}}\left(\dfrac{1}{(x-2)}-\dfrac{\ln(x-2)}{2x}\right)$

7. $x=1$, minimum
$x=-2$, maximum

9. $\dfrac{x(2-x)}{(1-x)^2}$

Exercise 15F

1. $1-x+\dfrac{x^2}{2}-\dfrac{x^3}{6}+\dfrac{x^4}{24}$;   0.60677;   0.60653

2. $1+x^2+\dfrac{x^4}{2}$

3. $b_0=0$, $b_1=1$, $b_2=-\dfrac{1}{2}$

Miscellaneous Exercises

1. (a) $x=\dfrac{5}{2}$, minimum

(b) $x=\dfrac{3}{4}$, maximum

(c) $x=-1$, maximum; $x=-\dfrac{1}{3}$, minimum

(d) $x=\dfrac{3+\sqrt{105}}{4}$, maximum

$x=\dfrac{3-\sqrt{105}}{4}$, minimum

(e) $x=1$ minimum

2. (a) $-24(5-3x)^7$ (b) $\dfrac{-1}{2\sqrt{x}\left(\sqrt{x}+1\right)^2}$

(c) $(\ln 3)\,3^x$ (d) $\dfrac{1}{2}\sqrt{\dfrac{1-x}{1+x}}\times\dfrac{1}{(1-x)^2}$

(e) $\ln 2$ (f) $\dfrac{2e^{-x^{-2}}}{x^3}$

3. $3e^{3x-1}$; $\dfrac{1}{3}e^{3x-1}+c$

4. $\dfrac{(2x-4)^6}{12}+c$

5. $\dfrac{1}{3}\ln\left(\dfrac{17}{8}\right)$

6. $\dfrac{2x}{x^2+1}$; $\dfrac{1}{2}\ln 2$

16 FURTHER TRIGONOMETRY

Exercise 16B

1. (a) $\dfrac{1}{2\sqrt{2}}\left(\sqrt{3}+1\right)$ (b) $\dfrac{1}{2\sqrt{2}}\left(\sqrt{3}+1\right)$

(c) $\dfrac{1}{2\sqrt{2}}\left(1-\sqrt{3}\right)$ (d) $\dfrac{1}{2\sqrt{2}}\left(\sqrt{3}-1\right)$

(e) $\left(2+\sqrt{3}\right)$ (f) $-\left(2+\sqrt{3}\right)$

(g) $-\dfrac{1}{2\sqrt{2}}\left(\sqrt{3}+1\right)$ (h) $\dfrac{1}{2\sqrt{2}}\left(\sqrt{3}-1\right)$

(i) $2-\sqrt{3}$

Exercise 16D

1. (a) $-2\sin 4x \sin x$ (b) $2\cos 9x \sin 2x$

   (c) $2\cos\dfrac{11}{2}x\,\cos\dfrac{7}{2}x$ (d) $2\sin 8x\,\cos 5x$

   (e) $\cos\dfrac{\pi}{5}\cos\dfrac{\pi}{3}\cos\dfrac{\pi}{5}$ (f) $2\cos 90\cos 24 = 0$

2. $\dfrac{\sqrt{3}}{\sqrt{2}}$

3. (a) $\cos 12x + \cos 2x$ (b) $\cos 3x + \cos 2x$

   (c) $\cos 2\theta + \cos 4\theta$

   (d) $\sin 270° + \cos 60° = -0.5$

5. $4\cos 3x \cos 4x \cos 5x$

6. $2\sin 4x \sin 8x$

Exercise 16E

1. $5\sin(x+0.927)$

2. $5\cos(x-0.644)$

3. $17\sin(x-0.490)$

4. $\sqrt{40}\cos(x+0.322)$

5. $29\sin(x-0.810)$

6. $\sqrt{197}\cos(x-0.071)$

7. $\sqrt{3}\cos(2x+0.464)$

8. $\sqrt{34}\sin\left(\tfrac{1}{2}x+0.540\right)$

Exercise 16F

1. $\dfrac{\pi}{2}$, 5.435 radians

2. $\pm 70.5°$, $\pm 120°$

3. $63.4°$, $101.3°$, $243.4°$, $281.3°$

4. $114.5°$

5. $\dfrac{\pi}{4}$, 1.25

6. 1.257, 2.513 radians

7. 0.841, 2.094 radians

Exercise 16G

1. $\pm\dfrac{2\pi}{3}$

2. $60°$, $120°$

3. $66.4°$

4. $115.2°$, $244.8°$

5. 0, π, 2π

6. $60°$, $109.5°$

7. 1.878, 4.405 radians

8. $41.7°$, $138.3°$, $221.7°$, $319.3°$

9. $150°$, $210°$

10. 0.474, 2.667 radians

11. $45°$, $60°$, $120°$, $135°$

12. $\pm 60°$, $\pm 90°$, $\pm 150°$

Exercise 16H

1. $36.87°$, $61.93°$

2. (a) $103.29°$, $330.45°$

   (b) $49.79°$, $197.59°$

3. $2\cos\left(\theta-\dfrac{\pi}{3}\right)$; $\theta=\dfrac{7\pi}{12}$

4. (a) $23.69°$, $104.89°$ (b) $82.5°$

   (c) $123.43°$, $3.43°$ (d) 2.04 radians

Exercise 16I

1. (a) $45°$, $135°$, $90°$

   (b) $0°$, $72°$, $144°$, $180°$

   (c) $0°$, $60°$, $90°$, $180°$

2. 0.714, 3.835 radians

3. $\sin 3\theta = 3\sin\theta - 4\sin^3\theta$

   (a) $60°$, $90°$, $120°$, $270°$, $330°$

   (b) $0°$, $98.7°$, $180°$, $261.3°$, $360°$

   (c) $0°$, $50.9°$, $180°$, $309.1°$, $360°$

4. $\dfrac{\pi}{24}$, $\dfrac{5\pi}{24}$, $\dfrac{13\pi}{24}$, $\dfrac{17\pi}{24}$

5. $0°$, $90°$, $180°$

6. $45°$, $105°$, $135°$, $165°$, $225°$, $285°$, $315°$, $345°$

7. (a) $\sin\theta = \dfrac{2t}{1+t^2}$ (b) $120°$, $240°$

Exercise 16J

1. 0.125 radians
3. 0.16 radians
4. 0.242 radians
5. (a) -0.351 radians   (b) -0.357 radians

Miscellaneous Exercises

1. (a) $\cos 60 = 0.5$ (b) $\sin 70$ (c) $\sin x$

   (d) $\sqrt{3}\cos x$ (e) $\sin x$

3. $15°$, $18°$, $54°$, $75°$, $90°$

4. (a) $\dfrac{\pi}{6}$, $\dfrac{5\pi}{6}$, $\dfrac{7\pi}{6}$, $\dfrac{11\pi}{6}$; $\dfrac{\pi}{4}$, $\dfrac{3\pi}{4}$, $\dfrac{5\pi}{4}$, $\dfrac{7\pi}{4}$

   (b) $\dfrac{\pi}{3}$, $\dfrac{5\pi}{3}$

   (c) $63.45°$, $153.45°$, $243.45°$, $333.45°$

   (d) $8.49°$, $73.39°$

17 TRIANGLES WITHOUT RIGHT ANGLES

Exercise 17A

1. $n = 97.0$ m, $L = 107°$, $\ell = 111$

2. $y = 7.62$ cm, $Z = 23.2°$, $V = 36.8°$

3. $C = 36.4°$, $B = 94.6°$, $b = 168$ m

4. $P = 30.2°$, $Q = 123°$, $q = 15.4$ cm

5. $D = 37°$, $e = 6.56$ m, $f = 5.87$ m

6. $U = 46.4°$ $V = 90°$, $v = 5.8$ m

Exercise 17B

1. $A = 116°$, $B = 40.5°$, $C = 23.6°$

2. $d = 4.98$, $E = 26.2°$, $F = 140°$

3. $n = 32.5$ mm, $L = 73.5°$, $M = 35.5°$

4. $X = 22.1°$, $Y = 41.1°$, $Z = 116.8°$

5. $P = 77.3°$, $Q = 12.7°$, $R = 90°$

6. $v = 143$ m, $V = 102°$, $W = 41.4°$

Exercise 17C

1. 46.8 cm²
2. 14.7 m²
3. 312.0 mm²
4. 12.5 m²
5. 180 cm²
6. 4195 m²

Miscellaneous Exercises

1. (a) 553 cm²   (b) 1882 cm²
2. 2.905 cm²
3. 37.1 m², 8.42 m, 9.00 m
4. $z = 29.8$, $Y = 67.1°$, $Z = 78.9°$
5. (i) $Q = 106°$, $p = 35.9$ m, $q = 97.4$ m

   (ii) $Q = 32.0°$, $R = 81.0°$, $r = 78.4$ m

   (iii) $P = 11.6°$, $Q = 54.4°$, $R = 114°$

6. $51.0°$,   381 m²
7. $41.4°$, $82.8°$, $55.8°$;   8 cm, 12 cm

INDEX

Heinemann Educational
a division of Heinemann Educational Books Ltd.
Halley Court, Jordan Hill, Oxford OX2 8EJ

OXFORD   LONDON   EDINBURGH
MADRID   ATHENS   BOLOGNA   PARIS
MELBOURNE   SYDNEY   AUCKLAND   SINGAPORE
TOKYO   IBADAN   NAIROBI   HARARE
GABORONE   PORTSMOUTH NH (USA)

ISBN 0 435 51550 0
First Published 1992
© CIMT, 1992

Typeset by ISCA Press, CIMT, University of Exeter
Printed and bound by The Bath Press, Avon